Numerical Methods and Applications

Edited by
Guri I. Marchuk

Numerical Methods and Applications

CRC Press
Boca Raton   Ann Arbor   London   Tokyo

Library of Congress Cataloging in Publication Data

Numerical methods and applications /Guri I. Marchuk, Editor.
    p.   cm.
    Includes bibliographical references and index.
    ISBN 0-8493-8947-X
    1. Numerical analysis.   I. Marchuk, G.I.,
  1925–   .
  QA297.N8619   1994
  519.4–dc20

                                                93-27093
                                                    CIP

©1994 by CRC Press, Inc.

No claim to original U.S. Government works
International Standard Book Number 0-8493-8947-X
Library of Congress Card Number 93-27093
Printed in the United States of America 1 2 3 4 5 6 7 8 9 0
Printed on acid-free paper

The Editor's Preface

This book contains works by acknowledged Russian Academy of Sciences mathematicians, working at the Institute of Numerical Mathematics, Moscow; Institute of Applied Mathematics, Moscow; Computer Center, Novosibirsk; Kurchatov Institute, and Moscow State University. The papers in the book deal with contemporary numerical mathematics, which nowadays is based on many ideas from differential equations theory and functional analysis, and on many approaches developed in various fields of classical numerical analysis: linear algebra, numerical integration, and difference approximation theory.

Contemporary numerical mathematics studies problems which were inherited from previous development stages of science and are arising in new sciences. Development of efficacious algorithms for numerical solution of stiff systems of ordinary differential equations is typically one of the classical problems of numerical mathematics. Stiff systems arise in radio technology, chemical kinetics, reactor and radiation physics, hydrodynamics, and so on. Methods of solving these equations have been thoroughly covered in the literature (Yu.V. Rakitski, H. Stetter, E.L. Wachspress, V.I. Lebedev, R.P. Fedorenko et al.), concentrating basically on implicit methods. Contemporary tendencies in science, however, led a number of researchers to investigate the efficacy of explicit difference schemes. Results of investigation in this area are represented in the article by V.I. Lebedev. The author examines explicit schemes with variable step sizes in time; investigates their efficacy; shows that they allow stable integration of nonstationary problems with a much larger mean step size than in schemes where the step size is constant; and points at obvious possibilities to parallelize the algorithms. R.P. Fedorenko discusses other effective methods of solving stiff systems, suggesting special methods to solve the so-called regular stiff systems and principles of the B-theory of stiff system integration.

Stiff elliptic problems arise in the theory of composite materials, representing solid media with great numbers of nonhomogeneities. In order to solve such problems, special algorithms based on the averaging of composite materials' characteristics are needed (Sanchez-Palencia, N.S. Bakhvalov, J.-L. Lions, G. Papanicoloau et al.). These methods are well established in practical computations of nuclear reactors (L.N. Usachev, V.V. Smelov, V.Ya. Pupko et al.). A range of applications is growing ever wider in shipbuilding, space technology, and so forth. Computation of efficient characteristics of composite materials requires solving certain periodic problems, as a rule, for differential equations of the elliptic type, with a broad spread of characteristics. N.S. Bakhvalov and A.V. Knyazev present new contemporary efficient algorithms to solve such stiff problems.

Contemporary numerical mathematics emphasizes investigations into methods for the solution of nonlinear mathematical models which adequately describe real physical processes. Approximation of nonlinear models by finite differences or by finite-element methods leads, as a rule, to nonlinear grid systems, requiring effective solutions. This problem is discussed by E.G. Dyakonov, who is continuing his research in this area on the efficacy of iterative processes for approximate solutions of nonlinear grid systems, and on the analysis of these methods under a general condition of reversibility of employed linear operators. The results can be applied to several classes of nonlinear systems related to grid systems which appear while solving the Navier-Stokes equations. Another competing class of efficacious algorithms for solution of the Navier-Stokes problem is presented in the work by G.M. Kobelkov.

Along with the above works and results, the book contains an article with original results related to the analysis of errors of the finite-element method in hyperbolic problems with initial data and right-hand parts not sufficiently smooth (A.A. Zlotnik). The author obtains extremely accurate error estimates for this class of problems and methods.

Publishing the articles selected for this book in English makes it possible for the authors to present their ideas and statements of problems to a broad readership. The book will hopefully interest university students, researchers, engineers, and specialists in mathematical numerical methods, mathematical physics, and their applications.

Academician Guri Marchuk

The Editor

Guri I. Marchuk, academician, is a director of the Institute of Numerical Mathematics, Russian Academy of Sciences.

From 1962 to 1980 he worked in the Siberian Branch of the USSR Academy of Sciences, first as a director of the Computer Center and then as Chairman of this branch and Vice President of the USSR Academy of Sciences. From 1980 to 1986 he was a Deputy Prime Minister of the USSR and Chairman of the State Committee of Science and Technology. From 1986 to 1991 he was President of the USSR Academy of Sciences.

Academician Marchuk is a prominent scientist in numerical and applied mathematics; winner of the Fridman, Keldysh, and Carpinski prizes; member of the Academy of Sciences of Bulgaria, Czechoslovakia, Europe, Finland, France, Germany, India, Poland, and Rumania; Honorary Professor of Calcutta, Houston, Karlov, Tel-Aviv, Toulouse, and Oregon Universities, and Budapest and Dresden Polytechnic Universities; a member of the editorial boards of six foreign (France, Germany, Italy, Sweden, the United States) and several Russian journals; and Editor-in-Chief of the Russian Journal of Numerical Analysis and Mathematical Modelling published by the Institute of Numerical Mathematics RAS in the Netherlands.

Academician Marchuk is the author of a series of monographs on numerical mathematics, numerical simulation of nuclear reactors, numerical techniques for problems of the atmosphere and ocean dynamics, immunology, medicine, and environment protection. For notable progress in scientific and organizational activities he has received prestigious state rewards.

Contents

1. E.G. Dyakonov

Iterative Methods Based on Linearization for
Nonlinear Elliptic Grid Systems   1

  1. Preconditioning and Symmetrization of Linear Systems .   2

  2. Two-Stage Iterative Methods for Nonlinear Systems . .   11

  3. Choice of Model Grid Operators: Optimal and
Nearly Optimal Preconditioning   19

  4. Examples of Applications of Two-Stage Iterative
Methods for Nonlinear Elliptic Problems   34

  References   40

2. V.I. Lebedev

How to Solve Stiff Systems of Differential Equations
by Explicit Methods   45

  1. Problem Statement and Assumptions   46

  2. Formulation of Four Methods Involving Explicit
Difference Schemes with Time-Variable Steps   48

  3. Implicit Schemes for Linear Equations   50

  4. Investigation of Stability and Approximation of
Explicit Difference Schemes   51

  5. Investigation of Stability and Approximation of
Implicit Difference Schemes   53

  6. Some Estimates for Polynomials   56

  7. Determination of Parameters of Difference Equations . .   66

  8. One Realization of Stable Algorithms   72

  9. On Passing Boundary Layers, Estimating Quantities
Cou and τ and the DUMKA Code   73

  10. Cases of Stiff System Computations   75

  References   79

3. G.M. Kobelkov

On Numerical Methods of Solving the Navier-Stokes
Equations in "Velocity-Pressure" Variables   81

  1. Numerical Methods for Solution of the Stationary
Stokes Problem   82

  2. Methods for Solving Nonstationary Stokes Problems . .   97

  3. Numerical Methods for Nonlinear Equations 106
  References   114

4. R.P. Fedorenko

  Stiff Systems of Ordinary Differential Equations 117
  1. Linear Stiff Systems   118
  2. A-Stability of Difference Schemes 123
  3. Singularly-Perturbated Systems 126
  4. Numerical Integration of a Singularly-Perturbated System 128
  5. On Regular Stiff Systems   131
  6. Some Applications of the Theory 138
  7. Slow Processes in a Nuclear Reactor 144
  8. On the Accuracy of Euler's Implicit Scheme 146
  References   153

5. A.A. Zlotnik

  Convergence Rate Estimates of Finite-Element
  Methods for Second-Order Hyperbolic Equations . . 155
  1. Initial-Boundary Value Problem for Second-Order
     Multidimensional Hyperbolic Equations 156
  2. Three-Level Finite-Element Method with Weight 165
  3. Second-Order Accuracy A Priori Error Estimates 174
  4. Fractional-Order Error Estimates in
     Nonsmooth Data Classes 179
  5. The $W_{2,h}^1$ and C_h Error Estimates for the
     One-Dimensional Case 189
  6. Second-Order Hyperbolic Equations of General Form . . 196
  7. Finite-Element Method with the Splitting Operator . . . 204
  8. Two-Level Finite-Element Method 210
  9. Abstract Second-Order Hyperbolic Equations:
     Applications to Dynamic Problems of Mechanics . . 216
  References   217

6. N.S. Bakhvalov and A.V. Knyazev

  Fictitious Domain Methods and Computation of
  Homogenized Properties of Composites with a Periodic
  Structure of Essentially Different Components 221
  1. Homogenization 223
  2. Fictitious Gradients Method: Perforated Composites . . 228
  3. Composites with Inclusions of a Soft Material 239
  4. Composites with Inclusions of Soft Materials
     and with Cavities 252
  5. On a Function Extension on a Torus 259
  References   265

Index 267

Iterative Methods Based on Linearization for Nonlinear Elliptic Grid Systems

E.G. D'yakonov

Solving large elliptic grid systems resulting from discretization of elliptic boundary value problems on the basis, for example, of finite-element or difference methods, is one of the most important problems of computational mathematics. Iterative methods are an indispensable tool in dealing with this problem, and a very large number of investigations have been devoted to the questions of improving their theoretical and practical efficiency for finding appropriate approximations to the desired solutions. We only mention publications connected (to some degree) with construction of asymptotically optimal or nearly optimal preconditioners and some of their generalizations for nonsymmetric or nonlinear operators (see references 1–45). The usefulness of such iterative methods for effective implementation on modern vector and parallel computers is widely recognized. In the case of a complicated nonlinear operator, L, even the problem of evaluating a residual, $r^n \equiv L(u^n) - f$, with a given iterate, u^n, and the right-hand term, f, needs considerable computational work. As usual, it will be characterized by the number of required arithmetical operations. Therefore, for such problems it is especially reasonable not only to apply an iterative method with the rate of convergence independent of the grid (with a parameter h) but also to try to decrease the number of iterations required to obtain the desired accuracy.

Sometimes a considerable effect can be achieved using various continuation procedures, in particular the multigrid acceleration procedure (multigrid predictor–corrector method) where the approximation obtained on a coarser grid serves as an initial iterate for the solution of the system on a finer grid. Sometimes it is reasonable to apply iterative methods based on linearization of the given operator and on the approximate solution, using some effective inner iterations of the obtained linearized system. Of course, the selection of such inner iterations, and the realization that only a very restricted number of them is really performed, is of fundamental importance. This was demonstrated in the theory of two-stage iterative methods in the case of positive operators (see references 19, 28, 46, and 47). On the other hand, it is well known that in the case of continuously differentiable operators very promising linearizations, leading even to quadratic convergence, can be constructed due to the Newton–Raphson (Newton–Kantorovich) method if the solutions of the linearized systems are found exactly.[7,48,49]

Precisely such methods, but with approximate solutions of linearized systems, are investigated in this paper under a general and natural condition of uniform boundedness of inverse operators to ones obtained through the linearization. The same general questions of the theory of such methods are considered in Section 2. Because these results are closely connected with some results of other authors' papers, Section 1 serves as a brief summary of them. The most important statements are given with proofs to enable the reader to start with an analysis of a simpler but very important case of linear systems and be prepared for a more complicated analysis of nonlinear systems in Section 2.

Both these sections deal with linear and nonlinear systems regarded as abstract operator equations in a Euclidean space H, under the assumption that in some sense good preconditioners (model operators), B, are available. In Section 3 some new results connected with construction of such model operators are described, and a brief review of other existing approaches in the case of some elliptic boundary value problems of the second order is presented. Some concrete examples of applications of these methods for elliptic systems of the second and fourth order are given in Section 4.

1.   PRECONDITIONING AND SYMMETRIZATION OF LINEAR SYSTEMS

1.1 NOTATION

Throughout the paper we let $H = H_h$ denote a Euclidean space of grid functions $u = u_h$, with the simplest inner product. $\mathcal{L}(H)$ is the space of linear operators mapping the Euclidean space H into itself. I is the identity operator; for $L \in \mathcal{L}(H)$ we use

$$\mathcal{L}^+(H) \equiv \{B : B \in \mathcal{L}(H), \qquad B = B^* > 0\},$$

that is, the set of symmetric (self-adjoint) and positive (positive definite) operators. If $B \in \mathcal{L}^+(H)$, then $H(B)$ denotes a Euclidean space differing from H only in the form of the inner product, defined by the equality $(u, v)_B \equiv (Bu, v)$. The norm in the Euclidean space $H(B)$ takes the form $\|u\|_B \equiv (u, u)_B^{1/2}$. If a linear operator, L, is regarded as a mapping of the Euclidean space $H(B_1)$ into the Euclidean space $H(B_2)$, with $B_r \in \mathcal{L}^+(H)$, $r = 1, 2$, then

$$\|L\|_{H(B_1) \mapsto H(B_2)} \equiv \max_{v \neq 0} \frac{\|Lv\|_{B_2}}{\|v\|_{B_1}} .$$

The particular case of this norm with $B_1 = B_2 = B$ is denoted by $\|L\|_B$, and the simplest case with $B = I$ corresponds to $\|L\|$. We also denote by $\lambda(L)$ and by $\mathrm{sp}L \equiv \{\lambda(L)\}$ any eigenvalue and the spectrum of L, respectively.

1.2 SPECTRALLY EQUIVALENT OPERATORS

Our main concern here is to find solutions of operator equations

$$Lu = f, \tag{1.1}$$

with $u \in H$, $L \in \mathcal{L}(H)$, $f \in H$, and $H \equiv H_h$ obtained from some grid approximations for a given elliptic boundary value problem. Once a basis for H has been chosen, operator Equation (1.1) leads immediately to a system of linear algebraic equations. Therefore, we need not distinguish Equation (1.1) from the corresponding linear system if the basis is a standard one, as in the Euclidean space $\mathbf{R}^N$. If the linear system (1.1) is such that

$$L \in \mathcal{L}^+(H), \qquad \delta_0 B \leq L \leq \delta_1 B, \ \delta_0 > 0, \tag{1.2}$$

where $B \in \mathcal{L}^+(H)$ is a preconditioner (model operator), then the modified Richardson method

$$B(u^{n+1} - u^n) = -\tau_n(Lu^n - f) \tag{1.3}$$

is the classical Richardson method, $u^{n+1} - u^n = -\tau_n B^{-1}(Lu^n - f)$, for the preconditioned system $B^{-1}Lu = B^{-1}f$ with the operator $B^{-1}L \in \mathcal{L}^+(H(D))$, where the most useful choices of the Euclidean space $H(D)$ are associated with either $D = B$ or $D = L$ (see references 17–19). Furthermore, $\mathrm{sp}(B^{-1}L) \subset [\delta_0, \delta_1]$, and the convergence of the modified Richardson method, under a proper choice of iterative parameters τ_n, depends only on the condition number $\kappa(B^{-1}L) \equiv \delta_1/\delta_0$ of the operator $B^{-1}L$. The same is true for modifications of other classical iterative methods like the well-known conjugate gradient method which is very often used in practice (see references 7, 14, 32, 44, 46, and 50).

Nowadays, the importance of choosing the grid operators $B \equiv B_h$ such that the condition number $\kappa(B^{-1}L)$ either does not depend on the grid parameter, h, or depends very weakly, such as $O(\|\ln h\|^k)$, $k \geq 0$, is widely recognized. In these cases it is said that operators L and B are spectrally equivalent, or nearly spectrally equivalent. And, of course, B can be called a good preconditioner only when the solution of every system

$$Bw = g \qquad (1.4)$$

with a given g is easy to obtain. More precisely, we assume that there exists a computational algorithm leading from a given vector, g, to the solution, w, of Equation (1.4) at the cost of $W(N)$ arithmetical operations, with estimates of the type

$$W(N) = O(N), \qquad W(N) = O\big(N(\ln N)^k\big), \qquad k \geq 0, \qquad (1.5)$$

where $N \equiv N_h$ is the number of unknowns in system (1.4).

The operators B which are spectrally equivalent to L and require the computational work $W(N) = O(N)$ for finding solutions of system (1.4) can be regarded as asymptotically optimal preconditioners; other combinations of the previously mentioned dependencies on h and N define the nearly optimal preconditioners. It should be mentioned that the most important case of grid system (1.1) corresponds to finite element approximations of an operator equation with a bounded, symmetric, and positive definite operator, $\mathbf{L}$, in a Hilbert space, $G \equiv G(\Omega)$, where Ω is a bounded region in $\mathbf{R}^d$. If $\hat{\psi}_1, \ldots, \hat{\psi}_N$ is a basis of the chosen subspace $\hat{G} \equiv \hat{G}_N$ in the Hilbert space G, expansions

$$\hat{u} = u_1\hat{\psi}_1 + \cdots + u_N\hat{\psi}_N, \qquad \forall \hat{u} \in \hat{G} \qquad (1.6)$$

define a one-to-one correspondence,

$$\hat{u} \leftrightarrow \equiv [u_1, \ldots, u_N]^T \in \mathbf{R}^N \equiv H, \qquad (1.7)$$

of the Euclidean spaces $\hat{G}$ and H. If, in addition, we introduce the Gram matrices

$$J \equiv J_h \equiv \big[(\hat{\psi}_j, \hat{\psi}_i)_G\big], \qquad L \equiv L_h \equiv \big[(\mathbf{L}\hat{\psi}_j, \hat{\psi}_i)\big], \qquad (1.8)$$

we can maintain isometry of the Euclidean spaces $\hat{G}$ and $H(J)$,

$$\|\hat{u}\|_G = \|u\|_J, \qquad (1.9)$$

and the spectral equivalence of operators L and J. Therefore, any grid operator B which is spectrally equivalent to L defines a norm in the Euclidean space $H(B)$ which is equivalent uniformly with respect to h to the norm in the Euclidean space $\hat{G}$ (e.g., see references 20 and 28). In this case we write $\|u\|_B \sim \|\hat{u}\|_G$. This simple fact is nevertheless of fundamental importance and emphasizes the role of the $H(B)$-geometry as consistent with the geometry of the original Hilbert space, G. We return to constructing optimal preconditioners in Section 3.

1.3 SYMMETRIZATION OF THE SYSTEMS

Consider now the general case of system (1.1) under the only assumption that there exists an inverse operator L^{-1}. It is well known that the problem can be reduced to a symmetrized one, $L^*Lu = L^*f$, with the operator $L^*L \in \mathcal{L}^+(H)$. This Gaussian symmetrization can be combined with preconditioning in several ways and gives rise,[18,28] for example in the case of the Richardson method, to modifications of the form

$$B^2 \left(u^{n+1} - u^n \right) = -\tau_n \, L^* \left(Lu^n - f \right), \qquad (1.10)$$

$$u^{n+1} - u^n = -\tau_n \, L^* B^{-2} \left(Lu^n - f \right), \qquad (1.11)$$

$$B(u^{n+1} - u^n) = -\tau_n \, L^* B^{-1} \left(Lu^n - f \right). \qquad (1.12)$$

Here,
$$C_1 \equiv B^{-2} L^* L \in \mathcal{L}^+ \left(H(B^2) \right),$$
$$C_2 \equiv L^* B^{-2} L \in \mathcal{L}^+(H),$$
$$C_3 \equiv B^{-1} L^* B^{-1} L \in \mathcal{L}^+ \left(H(B) \right).$$

Notice that the operators $C_k (k = 1, 2, 3)$ are obtained due to the Gauss symmetrization of the operator L being regarded as a mapping of the Euclidean space V_k into F_k, where $V_1 = H(B^2)$, $F_1 = H$; $V_2 = H$, $F_2 = H(B^{-2})$; $V_3 = H(B)$, $F_3 = H(B^{-1})$. The spectra of C_1, C_2, and of C_3 belong to the interval $[\delta_0, \delta_1]$ if the inequalities

$$\|L\|_{H(B^2) \mapsto H} \le \delta_1^{1/2}, \qquad \|L^{-1}\|_{H \mapsto H(B^2)} \le \delta_0^{-1/2}, \qquad (1.13)$$

$$\|L^*\|_{H \mapsto H(B^2)} \le \delta_1^{1/2}, \qquad \|(L^*)^{-1}\|_{H(B^2) \mapsto H} \le \delta_0^{-1/2}, \quad (1.14)$$

$$\|L\|_{H(B) \mapsto H(B^{-1})} \le \delta_1^{1/2}, \qquad \|L^{-1}\|_{H(B^{-1}) \mapsto H(B)} \le \delta_0^{-1/2} \quad (1.15)$$

are valid, respectively.[18,28] These inequalities with constants δ_k independent of h were analyzed by the author in references 18 and 28; in reference 17 the inequalities of Equation (1.15) were written in the form

$$\delta_0 \|v\|_B^2 \le \|Lv\|_{B^{-1}}^2 \le \delta_1 \|v\|_B^2.$$

For B to satisfy the consistency property previously mentioned, it is reasonable to consider Equation (1.15) as a consequence of correctness of the original differential problem in the Hilbert space G. Therefore, iterations of Equation (1.12) should be regarded as the most suitable for general problems (see also references 11, 32, and 51).

1.4 MODIFIED RICHARDSON METHODS IN A GENERAL SETTING

Using two operators, B_1 and B_2 from $\mathcal{L}^+(H)$, one can rewrite the iterative methods considered previously in a unified form,

$$u^{n+1} - u^n = -\tau_n B_2^{-1} L^* B_1^{-1}\big(L(u^n) - f\big), \qquad n = 0, 1, \ldots, k-1,$$

$$(1.16)$$

which is equivalent to the application of the Richardson method to the transformed system

$$Cu = B_2^{-1} L^* B_1^{-1}(Lu^n - f) \qquad (1.17)$$

with the operator

$$C \equiv B_2^{-1} L^* B_1^{-1} L \in \mathcal{L}^+\big(H(B_2)\big). \qquad (1.18)$$

The pairs B_1 and B_2 chosen to use in Equation (1.17), namely $B_1 = I$, $B_2 = B^2$; $B_1 = B^2, B_2 = I$; $B_1 = B_2 = B$, lead to Equations (1.10), (1.11), and (1.12), respectively. It is possible to include formally the method of Equation (1.3), with $L \in \mathcal{L}^+(H)$, into Equation (1.16), by choosing either $B_2 = B$, $B_1 = L$ or $B_2 = L$, $B_1 = B$.

Lemma 1.1. Let the operators $B_r \in \mathcal{L}^+(H)$, $r = 1, 2$ and an invertible linear operator, L, be such that

$$\|L\|_{H(B_2) \mapsto H(B_1^{-1})} \leq \delta_1^{1/2}; \qquad \|L^{-1}\|_{H(B_1^{-1}) \mapsto H(B_2)} \leq \delta_0^{-1/2}. \quad (1.19)$$

Then the spectrum of operator C from Equations (1.17) and (1.18) can be localized as

$$\mathrm{sp}\,C \subset [\delta_0, \delta_1], \qquad \delta_o > 0. \qquad (1.20)$$

Proof. Inequalities (1.19) are equivalent to

$$\delta_0 \|v\|_{B_2}^2 \leq \|Lv\|_{B_1^{-1}}^2 \leq \delta_1 \|v\|_{B_2}^2, \qquad \delta_0 > 0, \qquad (1.21)$$

$$\delta_0 B_2 \leq L^* B_1^{-1} L \leq \delta_1 B_2, \qquad \delta_0 > 0, \qquad (1.22)$$

which lead to the desired conclusion, Equation (1.20).

Lemma 1.2. Let $B \in \mathcal{L}^+(H)$. Then inequalities (1.19), with $B_2 = B^2$, $B_1 = I$, are equivalent to Equation (1.13), and inequalities (1.19), with $B_1 = B^2$, $B_2 = I$, are equivalent to Equation (1.14). In the case of $L \in \mathcal{L}^+(H)$, inequalities (1.19), with either $B_1 = B$, $B_2 = L$ or $B_1 = L$, $B_2 = L$, are equivalent to Equation (1.2).

Proof. Under the choice $B_2 = B^2$, $B_1 = I$, inequalities (1.21) are equivalent to $\delta_0 \|Bv\|^2 \leq \|Lv\|^2 \leq \delta_1 \|Bv\|^2$, $\delta_0 > 0$, and lead directly to Equation (1.13). The choice $B_1 = B^2$, $B_2 = I$ in Equation (1.21) leads to the operator inequalities $\delta_0 I \leq L^* B^{-2} L \leq \delta_1 I$. They imply that $\mathrm{sp} L^* B^{-2} L \subset [\delta_0, \delta_1]$. Making use of the well-known fact from linear algebra that for any A and A' the spectra of $A'A$ and AA' coincide if they do not contain point $\lambda = 0$, we can write $\mathrm{sp} L^* B^{-2} L = \mathrm{sp} B^{-2} L L^* \subset [\delta_0, \delta_1]$. Therefore, $\delta_0 B^2 \leq L L^* \leq \delta_1 B^2$, and

$$\delta_0 \|Bv\|^2 \leq \|L^* v\|^2 \leq \delta_1 \|Bv\|^2, \qquad \delta_0 > 0.$$

The latter inequalities lead to Equation (1.14). If $L = L^* > 0$, after the change $v = L^{-1}w$ the choice $B_1 = B$, $B_2 = L$ in Equation (1.21) implies that

$$\delta_0(L^{-1}w, w) \leq (B^{-1}w, w) \leq \delta_1(L^{-1}w, w)$$

and $\qquad \delta_0 L^{-1} \leq B \leq \delta_1 L^{-1}, \qquad \delta_0 > 0.$

Therefore, Equations (1.2) and (1.19) are equivalent. The case of $B_1 = L$, $B_2 = B$ is even simpler.

For iterations of Equation (1.16) we use the standard error, $z^n \equiv u^n - u$, and the residual, $r^n \equiv Lu^n - f = Lz^n$. By $R_k \equiv Q_k(C)$ (see Equation [1.18]) with

$$Q_k(\lambda) \equiv (1 - \tau_0 \lambda) \cdots (1 - \tau_{k-1} \lambda)$$

we denote the reducing operator leading to the representation $z^k = R_k z^0$. It is easy to see that $r^k = R_{r,k} r^0$ with $R_{r,k} = Q_k(D)$, where $D \equiv L B_2^{-1} L^* B_1^{-1}$ and $D \in \mathcal{L}^+(H(B_1^{-1}))$.

Lemma 1.3. Assume that the operators $B_r \in \mathcal{L}^+(H)$, $r = 1, 2$, and L be an invertible one. Then $\mathrm{sp} C = \mathrm{sp} D$.

Proof. All eigenvalues of the operators C and D are positive. Besides, $C = E_1 E_2$ and $D = E_2 E_1$, where $E_1 = B_2^{-1} L^* B_1$ and $E_2 = L$. Because the spectra of $E_1 E_2$ and of $E_2 E_1$ coincide, the same is true for those of C and D.

Theorem 1.1. Let the conditions of Lemma 1.1 be satisfied. We perform a cycle of k iterations (Equation [1.16]) with iterative parameters such that

$$\{\tau_n^{-1}\} = \{t_i\}, \qquad t_i \equiv \varphi\left(\cos\frac{\pi(2i+1)}{2k}\right),$$

$$i = 0, \ldots, k-1, \qquad (1.23)$$

where the function $\varphi(t) \equiv 2^{-1}[\delta_1 + \delta_0 + (\delta_1 - \delta_0)t]$ maps the interval $[-1, 1]$ onto the given one, $[\delta_0, \delta_1]$. Then the reducing operators, R_k and $R_{\tau,k}$, are symmetric as elements of $\mathcal{L}(H(B_2))$ and $\mathcal{L}(H(B_1^{-1}))$, respectively, and

$$\|R_k\|_{B_2} = \|R_{\tau,k}\|_{B_1^{-1}} \leq q_k, \qquad (1.24)$$

$$\|z^k\|_{B-2} \leq q_k\|z^0\|_{B_2}, \quad \|r^k\|_{B_1^{-1}} \leq q_k\|r^0\|_{B_1^{-1}}, \qquad (1.25)$$

where $\quad q_k \equiv \left[T_k\left(\dfrac{\delta_1 + \delta_0}{\delta_1 - \delta_0}\right)\right]^{-1} = \dfrac{2\rho^k}{1 + \rho^{2k}} < 1, \qquad (1.26)$

$$\rho \equiv \frac{\delta^{1/2} + 1}{\delta^{1/2} - 1}, \quad \delta \equiv \frac{\delta_1}{\delta_0} \; ; \qquad (1.27)$$

$$T_k(t) \equiv \frac{\left[t + (t^2 - 1)^{1/2}\right]^k + \left[t - (t^2 - 1)^{1/2}\right]^k}{2},$$

$$|t| \geq 1, \qquad (1.28)$$

is a standard Chebyshev polynomial of degree k.

Proof. We need to prove only Equation (1.24), because Equation (1.25) follows directly from it; other statements of the theorem follow directly from the theory of the classical Richardson method (for example, see references 7 and 44). Because the points $\cos\left(\pi(2i+1)/2k\right)$, with $i = 0, \ldots, k-1$, are roots of the same Chebyshev polynomial, $T_k(t) \equiv \cos(k \arccos t)$, written in this form for $|t| \leq 1$, the choice of Equation (1.23) of the iterative parameters leads to the polynomial

$$Q_k(\lambda) = \frac{T_k\left(\dfrac{2\lambda - \delta_1 - \delta_0}{\delta_1 - \delta_0}\right)}{T_k\left(-\dfrac{\delta_1 + \delta_0}{\delta_1 - \delta_0}\right)} = q_k T_k\left(\dfrac{\delta_1 + \delta_0 - 2\lambda}{\delta_1 - \delta_0}\right).$$

Therefore, $R_k = Q_k(C)$ and $R_{\tau,k} = Q_k(D)$, being polynomials of operators C and D, respectively, which are symmetric mappings of the Euclidean spaces $H(B_2)$ and $H(B_1^{-1})$ into themselves, respectively, must also be symmetric operators. Thus, norms of R_k and $R_{r,k}$ are estimated by

$$\max_{\lambda \in [\delta_0, \delta_1]} \|Q_k(\lambda)\| \le q_k,$$

and the theorem has been proved.

Several years ago (see references 3–5, 25) it was noticed that in some inner iterations it is reasonable to use not the classical polynomials, $T_k(t)$, but polynomials $T_k^+(t)$, leading to error reducing operators, R_k, such that $R_k \ge 0$.

Theorem 1.2. In Theorem 1.1 let the set $\{t_i\}$ of Equation (1.23) be replaced by a set $\{t_i^+\}$, where $t_i^+ \equiv \varphi^+(\alpha_i)$ with $\alpha_i \equiv \cos(\pi(2i+1)/k)$ and $i = 0, 1, \ldots, k-1$ (for even, $k = 2m$), $\|i-m\| = 0, 1, \ldots, m$ (for odd, $k = 2m+1$), and

$$\varphi^+(t) \equiv 2^{-1}[\delta_1 + \delta_0 - (\delta_1 - \delta_0)t]. \tag{1.29}$$

Then estimates (1.24) and (1.25) remain valid, with q_k replaced by $q_k^+ \equiv 2q_k(1 + q_k)^{-1} < 1$. The reducing operators, $R_k \equiv R_k^+$ and $R_{r,k} \equiv R_{r,k^+}$, are symmetric and nonnegative as mappings of the Euclidean spaces $H(B_2)$ and $H(B_1^{-1})$ into themselves, respectively, and

$$0 \le Z_k^+ \le q_k^+ I, \qquad 0 \le Z_{k,r}^+ \le q_k^+ I. \tag{1.30}$$

Proof. Consider a set $\omega \equiv \{\alpha_i \equiv (\pi(2i+1)/k)\} \subset [0, 2\pi]$. If $k = 2m$, this set does not contain π; all its points are displaced on $(0, 2\pi)$ symmetrically with respect to π. Thus, the set $\{\alpha_i\}$ is a twice-repeated set of the points $\alpha_0, \ldots, \alpha_m$, which are the roots of the polynomial $T_m(t) = \cos(m \arccos t)$, $|t| < 1$. Also, each of these α_i is a two-multiple root of the polynomial $T_k(t) + 1 \equiv T_k^+(t)$. Therefore, in this case we deal with the nonnegative polynomial, $Q_k(\lambda) \equiv Q_k^+(\lambda)$, on $[\delta_0, \delta_1]$, which can be written in the form

$$Q_k(\lambda) = \frac{\left(T_m\left(\dfrac{\delta_1 + \delta_0 - 2\lambda}{\delta_1 - \delta_0}\right)\right)^2}{\left(T_m\left(\dfrac{\delta+1}{\delta-1}\right)\right)^2} = \frac{T_k\left(\dfrac{\delta_1 + \delta_0 - 2\lambda}{\delta_1 - \delta_0}\right) + 1}{T_k\left(\dfrac{\delta+1}{\delta-1}\right) + 1}$$

$$= q_k^+ \left(T_k\left(\frac{\delta_1 + \delta_0 - 2\lambda}{\delta_1 - \delta_0}\right) + 1\right).$$

Hence, $$\max_{\lambda \in [\delta_0, \delta_1]} |Q_k(\lambda)| = \frac{2}{T_{2m}\left(\dfrac{\delta+1}{\delta-1}\right) + 1}$$

$$= \frac{2q_k}{1 + q_k} = q_k^+.$$

Similarly, if $k = 2m+1$, the set of α_i with $i = 0, \ldots, 2m$ consists of the twice-repeated points $\cos\big(\pi(2i + 1)/2m + 1\big)$ with $i = m-1, \ldots, 0$ and the single point $\alpha_m = -1$. These points are the roots of the polynomial $T_k(t) + 1 = T_k^+(t)$. Because $\varphi^+(-1) = \pi$, it is easy to see that $Q_k(\lambda) = (1 - \lambda/\delta_1)[r_m(\lambda)]^2$ and is nonnegative on $[\delta_0, \delta_1]$. Also, it implies that the representation

$$Q_k(\lambda) = \frac{T_k\left(\dfrac{\delta_1 + \delta_0 - 2\lambda}{\delta_1 - \delta_0}\right) + 1}{T_k\left(\dfrac{\delta+1}{\delta-1}\right) + 1}$$

$$= q_k^+\left(T_k\left(\dfrac{\delta_1 + \delta_0 - 2\lambda}{\delta_1 - \delta_0}\right) + 1\right)$$

is valid for all k, and the same reasoning as in Theorem 1.1 yields the desired estimates.

Recall that from the computational point of view, especially when either k or δ is comparatively large and the influence of the rounding errors is significant, we should pay special attention to the order in which the elements of the sets $\{t_i\}$ and $\{t_i^+\}$ are being used as the inverses of the iterative parameters.[7,44,51] It is also possible to obtain the final iterate, u^k, in another way, using a recurrence relation of type[7,44]

$$u^{n+1} - u^n = \omega_n\omega_{n-1}\left[u^n - u^{n-1}\right] - \frac{2(1 + \omega_n\omega_{n-1})}{\delta_1 + \delta_0}\left[Cu^n - g\right],$$

$$n = 1, \ldots, k - 1.$$

Finally, recall here a practical aspect of applying modified Richardson methods in cases when approximate and sometimes even wrong constants δ_0 and δ_1 are used.[18,28,52] The optimal values of these constants are denoted by δ_0^* and δ_1^*; for example, in the case of Equation (1.2) they correspond to the minimal and maximal eigenvalues of the operator $B^{-1}L$. The positive constants δ_0, δ_1, corresponding to the iterative parameters from Equation (1.23), we denote now by $\delta_0^{(m)}, \delta_1^{(m)}$. Our wish is to reexamine and possibly to improve these constants after performing a cycle of k iterations, taking into account estimate (1.25) for the residual. More precisely, if parameters $\tau_0, \ldots, \tau_{k-1}$ from Equation (1.23), with given $\delta_k \equiv \delta_k^{(m)}$ $k = 0, 1$, are used in iterations (1.16), we define $t^{(m)} \equiv \delta_1^{(m)} + \delta_0^{(m)} / \delta_1^{(m)} - \delta_0^{(m)}$. If the cycle of iterations yields the residual, $r_?^k$ we obtain the readily observed reducing factor $\rho_k^{(m)} \equiv \|r^k\|_D / \|r^0\|_D$, with $D \equiv B_1^{-1}$. Compute $\alpha_k^{(m)} \equiv \rho_k^{(m)} / q_k = \rho_k^{(m)} T_k(t^{(m)})$; this gives a ratio of the obtained reducing factor and of the theoretical one, provided that the estimates used were valid. If $\alpha_k^{(m)} \leq 1$, there is no reason to reject the constants; and we can perform a new cycle of either k iterations or even of $2k$ iterations. If the contrary case, with $\alpha_k^{(m)} > 1$, takes place and we are sure that this is not a result of rounding errors, then certainly we should reexamine the constants at hand. We do this with the help of the inverse function, $T_k^{(-1)}(t)$, defined for $t \geq 1$ (for such t, the polynomial $T_k(t)$ is monotonically increasing). $\lambda'_m > 1$ with $T_k(\lambda'_m) = t$ can be found explicitly, namely, $\lambda'_m = a^2 + 1/2a$, where $a \equiv \left(t + [t^2 - 1]^{1/2} \right)^{1/k}$. $a^k \equiv x$ is a solution of the equation $x + x^{-1} = 2t$.

Theorem 1.3. If $\rho_k^{(m)} \geq 1$, then the inequality

$$\delta_1^* \geq 2^{-1} \left[\delta_1^{(m)} + \delta_0^{(m)} + \lambda'_m \left(\delta_1^{(m)} - \delta_0^{(m)} \right) \right]$$

is valid; if $\rho_k^{(m)} < 1$, $\alpha_k^{(m)} > 1$, and $\delta_1^* \leq \delta_1^{(m)}$, then

$$\delta_0^* < 2^{-1} \left[\delta_1^{(m)} + \delta_0^{(m)} - \lambda'_m \left(\delta_1^{(m)} + \delta_0^{(m)} \right) \right].$$

The proof of this Theorem can be found in reference 28.

2. TWO-STAGE ITERATIVE METHODS FOR NONLINEAR SYSTEMS

2.1 INNER ITERATIONS

Consider now, in the Euclidean space H a nonlinear system

$$L(u) = f, \tag{2.1}$$

assuming only that it has a unique solution, u, in a subset S of H; usually S will be a closed ball in a certain Euclidean space, $H(B_2)$). Sometimes it can be obtained as a result of an a priori estimate of the solution; for some rather weak nonlinear problems, there is no need for this in any localization of the solution, and we can even take $S = H$. In the general case, we can assume only that the center of the ball (unknown as a rule) coincides with the desired solution, u, that is,

$$S \equiv S_{B_2}(u, r) \equiv \{v : \|v - u\|_{B_2} \leq r\}, \qquad r > 0. \qquad (2.2)$$

We linearize $L(u) - L(v)$ by $A_v(u - v)$ with a linear operator, A_v, defined by $v \in S$. For a continuously differentiable operator, L, it is possible to use the standard choice, $A_v = L'_v,$[7,48,49] with L'_v being the Frechet or the Gateaux derivative evaluated at point v of nonlinear operator L. For our purposes it will even do to take L'_v as the Jacobian matrix. We are interested in iterative methods such as

$$A_{u^m}(u^{m+1} - u^m) = f - L(u^m), \qquad m = 0, 1, \ldots \qquad (2.3)$$

Note that if $A_{u^m} = L'_{u^m}$, then Equation (2.3) is the classical Newton–Raphson method, and the analysis of it is well known.[7,48,49] To perform one iteration of Equation (2.3) we must solve the system

$$Av = g \qquad (2.4)$$

with $A = A_{u^m}$, $v = u^{m+1} - u^m$, and $g = f - L(u^m)$ (see Equation [1.1]).

What we want to investigate is a usage of the modified Richardson method,

$$B_2(v^{n+1} - v^n) = -\tau_n A^* B_1^{-1}(Av^n - g), \qquad n = 0, \ldots, k-1 \qquad (2.5)$$

(see Equation [1.16]), as an inner iterative method for obtaining an approximation to v from Equation (2.4) and, of course, to estimate the effect of such inner iterations on the convergence of iterative method (2.3), based on linearization of the given nonlinear operator.

Lemma 2.1. Let the operators A, B_1, and B_2 satisfy conditions (1.19). Let iterative parameters $\tau_0, \ldots, \tau_{k-1}$ in Equation (2.5) be defined by δ_0 and δ_1 from Theorem 1.1. Let the initial approximation v^0 in Equation (2.5) be taken as $v^0 = 0$. Then the vector v^k, obtained as the result of k iterations of Equation (2.5), satisfies the relation

$$A(I - R_k)^{-1} v^k = g, \qquad (2.6)$$

with the reducing error operator, R_k, from Theorem 1.1.

Proof. Use again the basic property of R_k: $v^k - v = R_k(v^0 - v) = -R_k v$. Because $\|R_k\|_{B_2} \leq q_1 < 1$, we can find that $v = (I - R_k)^{-1} v^n$. Substituting this expression into Equation (2.4), we obtain Equation (2.6).

If we introduce u^{n+1} not from Equation (2.3) but define it by the relation $u^{n+1} = u^n + v^k$, then the inner iterations result in the change of Equation (2.3) by

$$A_{u^m}(I - R_k)^{-1}(u^{m+1} - u^m) = f - L(u^m). \tag{2.7}$$

If we want to apply more general methods, with the number of inner iterations depending on m, then we must consider two-stage iterations,

$$A_{u^m}\left(I - R^{(m)}\right)^{-1}(u^{m+1} - u^m) = f - L(u^m) \tag{2.8}$$

$$\|R^{(m)}\|_{H(B_2) \mapsto H(B_1)} \le q^{(m)} < 1, \tag{2.9}$$

where the operator $R^{(m)}$ is R_k, with k replaced by the desired number, k_m. Recall that two-stage iterative methods with inner ADI-iterations for difference elliptic systems were introduced by Douglas, Wachspress, Gunn, and D'yakonov as early as 1961–1964.[17,28,46,47] Some other inner iterations with factorized grid operators for similar systems were analyzed by Dupont and D'yakonov in 1968 and 1969.[19] Similar iterative methods for finite element systems were analyzed, probably for the first time, by D'yakonov in 1971.[20] Two-stage preconditioners,

$$B = A(I - R_k)^{-1} \tag{2.10}$$

with $A = A > 0$, actually have been applied by many authors (see references 3–5, 10, 16, 33–39, 44 and 45), in particular for some variants of multigrid and domain decomposition methods.

2.2 CONVERGENCE OF TWO-STAGE ITERATIVE METHODS

Investigate the convergence of the two-stage iterative method (Equation [2.7]), assuming that $\|u^n - u\|_{B_2} \le r$, with u being the solution of system (2.1) under consideration.

Lemma 2.2. Let the operator L and its linearization, A_v, considered on S from Equation (2.2), be such that the inequalities of Equation (1.19) are satisfied with $A \equiv A_v$ for any v from S. For

$$g_v \equiv L(u) - L(v) + A_v(v - u), \qquad v \in S, \tag{2.11}$$

let the estimate

$$\|g_v\|_{B_1^{-1}} \le \kappa \|v - u\|_{B_2}^{1+\alpha}, \qquad 0 < \alpha \le 1 \tag{2.12}$$

be valid. Let u^m from Equation (2.8) belong to S, let condition

(2.9) be satisfied, and let u^{n+1} be found from Equation (2.8). Then, for $z^{m+1} \equiv u^{m+1} - u^n$, the estimate

$$\|z^{m+1}\|_{B_2} \le \delta_0^{-1/2} \left[\kappa \|z^m\|_{B_2}^{1+\alpha} + \xi^{(m)} \right] \tag{2.13}$$

holds, with

$$\xi^{(m)} \equiv q^{(m)} [1 - q^{(m)}]^{-1} \delta_1^{1/2} \|u^{m+1} - u^m\|_{B_2}. \tag{2.14}$$

Proof. From Equations (2.1), (2.9), and (2.11) it follows that

$$A_{u^m} z^{m+1} = g_v + g^{(m)},$$

with $\qquad g^{(m)} \equiv A_{u^m} \left[(I - R^{(m)})^{-1} - I \right] (u^{m+1} - u^m).$

Thus, it is not difficult to verify that

$$\|z^{m+1}\|_{B_2} \le \delta_0^{-1/2} \left[\|g_v\|_{B_1^{-1}} + \|g^{(m)}\|_{B_1^{-1}} \right]. \tag{2.15}$$

For $g^{(m)}$ we can write

$$\|g^{(m)}\|_{B_1^{-1}} \le \|A_{u^m}\|_{H(B_2) \mapsto H(B_1^{-1})} \times$$

$$\|(I - R^{(m)})^{-1} - I\|_{H(B_2)} \|u^{m+1} - u^m\|_{B_2}.$$

Therefore, $\|g^{(m)}\|_{B_1^{-1}} \le \xi^{(m)}$. This inequality, together with Equations (2.14), (2.15), and (2.12), leads to Equation (2.13).

Theorem 2.1. Let the conditions of Lemma 2.2 with respect to the operators L and A_v be satisfied and iterative methods (2.8) and (2.9) be applied, with $u^0 \in S$ and all $q^{(m)}$ from Equation (2.9) sufficiently small in the sense that

$$q^{(m)} (1 - q^{(m)})^{-1} \delta_1^{1/2} \delta_0^{-1/2} \le q < 1$$

$$(1 - q)^{-1} \left[\delta_0^{-1/2} \kappa r^\alpha + q \right] \le \rho < 1$$

for some $\rho < 1$. Then iterative methods (2.8) and (2.9) are convergent with the estimate of the rate of convergence,

$$\|z^s\|_{B_2} \le \rho^s \|z^0\|_{B_2} \le \rho^s \delta_0^{-1/2} \|r^0\|_{B_1^{-1}}.$$

Proof. Assuming $\|z^s\|_{B_2} \le r$ and taking into account Equations (2.13) and (2.14), we obtain

$$\|z^{m+1}\|_{B_2} \le \delta_0^{-1/2} \kappa r^\alpha \|z^m\|_{B_2} + q \left[\|z^{m+1}\|_{B_2} + \|z^m\|_{B_2} \right].$$

This, together with Equation (2.16), gives

$$\|z^{m+1}\|_{B_2} \le \rho \|z^m\|_{B_2}.$$

Therefore, all u^m in Equation (2.8) belong to S, and the estimate (2.17) is valid.

2.3   NEWTON'S LINEARIZATION

Consider the classical case $A_v = L'_v$ under the assumption

$$\|A_v - A_w\|_{H(B_2) \mapsto H(B_1^{-1})} \leq l \|v - w\|_{B_2} \qquad \forall v \in S, \forall w \in S. \quad (2.18)$$

Then it is known that Equation (2.12) holds with $\alpha = 1$, $\kappa = 2^{-1} l$.[48,49] Consequently, condition (2.16) is satisfied for sufficiently small r and $q^{(m)}$. Moreover, Equation (2.16) can be taken for the arbitrary positive value, $\rho < 1$. If, on the other hand, we additionally assume that $q^{(m)} = 0$, then Equation (2.13) leads to the standard recurrence relation from the theory of the Newton–Raphson iterations, that is,

$$\left\| z^{m+1} \right\|_{B_2} \leq \delta_0^{-1/2} \, 2^{-1} \, l \left\| z^m \right\|_{B_2}^2 .$$

Note that $\xi^{(m)}$ from Equation (2.14) is easily computed and

$$\|u^m - u\|_{B_2} \leq \delta_0^{-1/2} \left\| L(u^m) - f \right\|_{B_2}^{-1} .$$

The constants δ_0 and δ_1 can be obtained due to the adaptation procedure mentioned in Lemma 1.4 if the invertibility of L'_v and L'_u takes place. The constant, l, from Equation (2.18) is more difficult to estimate. In practical cases it is sometimes possible to get an approximation to l by analyzing values of $\|L(u^p) - f\|_{B_1^{-1}}$, with $p = m$ and $p = m + 1$.

In the preceding analysis we considered inner iterations associated with the modified Richardson method from Theorem 1.1. It is almost obvious that from the theoretical point of view such selection of iterative parameters is much better than the one used in Theorem 1.2 leading to the reducing operator, R_k^+. Nevertheless, in some practical problems the latter choice of iterative parameters might be of help, especially if we want to avoid dealing with systems with an almost degenerate matrix.

Lemma 2.3.   Let $A = A^* > 0$ and the operator R be a symmetric and nonnegative operator being regarded as an element of $\mathcal{L}(H(A))$, with $0 \leq R \leq q^+ I$ where $q^+ < 1$. Then the operator, $B \equiv A(I - R)_?^{-1}$ is a symmetric and positive operator, being regarded as a mapping of the Euclidean space, H, into itself, and $A \leq B \leq 1/1 - q^+ A$.

Proof. It is obvious that

$$B^{-1} = (I - R)A^{-1} = A^{-1} - RA^{-1}.$$

Take arbitrary $v \in H, w \in H$ and notice that

$$(RA^{-1}v, w) = (RA^{-1}v, A^{-1}w)_A$$

$$= (A^{-1}v, RA^{-1}w)_A = (v, RA^{-1}w).$$

This means that $(RA^{-1})^* = RA_?^{-1}$ On the other hand, for $w = v$

we have $(RA^{-1}v, v) = (RA^{-1}v, A^{-1}v)_A \geq 0$, and

$$(RA^{-1}v, v) = (RA^{-1}v, A^{-1}v)_A \leq q^+ (A^{-1}v, A^{-1}v)_A$$
$$= q^+ (A^{-1}v, v).$$

Thus, $0 \leq RA^{-1} \leq q^+ A^{-1}$ and $(1 - q^+)A^{-1} \leq B^{-1} \leq A^{-1}$, which directly yields the desired inequalities.[25]

This lemma enables one, in the case of positive operators A_{u^m} with the smallest eigenvalue almost equal to zero, to use iterations (2.8) with operators $B^{(m)} \equiv A_{u^m}(I - R^{(m)})^{-1} \geq A_{u^m}$. It might be useful in tracing solution branches in the vicinity of bifurcation points when the Jacobian matrices become singular. Note that this topic is very important in many applications[53,54] but is beyond the scope of this paper.

2.4   MODIFIED SIMPLE ITERATIONS

The two-stage iterative method of Equations (2.8) and (2.9) usually works well for sufficiently good approximations, u^0, to the solution. Sometimes to make such approximations it helps to use the simpler iterative method

$$B_2(u^{n+1} - u^n) = -\tau (L'_{u^n})^* B_1^{-1} [L(u^n) - f]. \qquad (2.19)$$

To investigate this we need

Lemma 2.4.   Let A be a linear operator and $\|A\|_{H(B_2) \mapsto H(B_1^{-1})} \leq K$. Then

$$\left\| B_2^{-1/2} A^* B_1^{-1/2} \right\| \leq K.$$

Proof. The assumed conditions on A imply that for any u, $\|B_1^{-1/2} Au\| \leq K \|B_2^{-1/2} u\|$. After substituting $u = B^{-1/2}v$ we find that for each v, $\|Qv\| \leq K \|v\|$, with $Q \equiv B_1^{-1/2} AB_2^{-1/2}$ and thus $\|Q\| \leq K$. The adjoint operator Q^* to Q is $B_2^{1/2 A^* B_1^{1/2}}$, and it is known that $\|Q\| = \|Q^*\|$ for any Q.

Theorem 2.2.   For all v from the ball S (see Equation [2.2]) let conditions (2.18) and (1.19), with $A \equiv A_v \equiv L'_v$, be satisfied and $l < \delta_0/\delta_1$, $0 < \tau < 2/(\delta_1 + \kappa)$, with $\kappa \equiv l\delta_1^{1/2}$. Let $u^0 \in S$. Then iterative method (2.19) is convergent and

$$\|z^{n+1}\|_{B_2} \leq \rho_1(\tau) \|z^n\|_{B_2}, \qquad (2.20)$$

where $\rho_1(\tau) \equiv \max\{|1 - \tau\delta_0|, |1 - \tau\delta_1|\} + \tau\kappa < 1$.

Proof. From Equation (2.19) it follows that

$$z^{n+1} = \mathcal{P}(u^n) - \mathcal{P}(u),$$
$$\mathcal{P}(v) \equiv v - \tau B_2^{-1}(L'_{u^n})^* B_1^{-1} L(v).$$

The nonlinear operator $\mathcal{P}$ is continuously differentiable in S. Its derivative at the point w is $\mathcal{P}'_w$, and

$$\mathcal{P}_w v = v - \tau B_2^{-1}(L'_{u^n})^* B_1^{-1} L'_w v.$$

Having used the classical functional analysis inequality for

$$\|\mathcal{P}(u^n) - \mathcal{P}(u)\|_{B_2},$$

we write
$$\|z^{n+1}\|_{B_2} \le \|\mathcal{P}'_w\|_{B_2} \|z^n\|_{B_2}$$

with $w = u + \theta z^n$; θ is a number from $[0, 1]$. Write $\mathcal{P}'_v \equiv R = R_1 + R_2$, with

$$R_1 = I - \tau B_2^{-1}(L'_{u^n})^* B_1^{-1} L'_{u^n},$$

$$R_2 \equiv I - \tau B_2^{-1}(L'_{u^n})^* B_1^{-1}(L'_{u^n} - L'_w).$$

Then $\quad \|R_1\|_{B_2} \le \max\{|1 - \tau\delta_0|, |1 - \tau\delta_1|\},$

$$\|R_2 v\|_{B_2} \le \tau \left\| B_2^{-1/2}(L'u^n)^* B_1^{-1/2} \right\| \times$$

$$\left\| B_1^{-1/2}(L'_{u^n} - L'_{u^n - (1-\theta)z^n}) v \right\|_{B_2}$$

(we have replaced w by $u^n - (1 - \theta)z^n$). Having used Lemma 2.4 and the conditions of Theorem 2.2, we deduce that

$$\|R_2 v\|_{B_2} \le \tau \delta_1^{1/2} l \|z^n\|_{B_2} \|v\|_{B_2}.$$

The estimates for $\|R_i\|_{B_2}$, $i = 1, 2$, yield Equation (2.20). The conditions on l and τ imply $\rho_1(\tau) < 1$.

It is easy to verify that
$$\min \rho_1(\epsilon) = \rho_0 \equiv 1 - \frac{2(\delta_0 - \kappa)}{\delta_1 + \delta_0},$$

which is obtained when $\tau = \tau_0 = 2/(\delta_1 + \delta_0)$. (This theorem, with $B_1 = I$, $B_2 = B^2$, was proved in reference 18; with $B_1 = B_2 = B$ it was proved in reference 28.) This yields a way to prove the existence and uniqueness of the solution of Equation (2.1) in a ball,

$$S' = S_{B_2}(w, r) \equiv \{v : \|v - w\|_{B_2} \le r\},$$

if the residual at a given point, w, is sufficiently small.

Theorem 2.3. Let the conditions of Theorem 2.2 be satisfied for all $v \in S'$ and

$$\|L(w) - f\|_{B_1}^{-1} \leq r\delta_1^{1/2}\tau_0^{-1}(1 - \rho_0).$$

Then the system (2.1) has a unique solution in the ball S'.

Proof. For all $v \in S'$ define a mapping,

$$R(v) \equiv v - \tau_0 B_2^{-1}(L_w')^* B_1^{-1}[L(v) - f].$$

Then $\|R(v) - w\|_{B_2} \leq r_1 + r_2$,

with $r_1 \equiv \|v - w - \tau_0 B_2^{-1}(L_w')^* B_1^{-1}[L(v) - L(w)]\|_{B_2}$,

and $r_2 \equiv \tau_0 \|B_2^{-1/2}(L_w')^* B_1^{-1}[L(w) - f]\|_{B_2}$

$$\leq \tau_0 \|B_2^{-1/2}(L_w')^* B_1^{-1/2}\| \|L(w) - f\|_{B_1^{-1}}.$$

Using estimates from the proof of Theorem 2.2, we get

$$r_1 \leq \rho_0 r, \quad r_2 \leq \tau_0 \delta_1^{1/2} \|L(w) - f\|_{B_1^{-1}}.$$

Hence, $r_1 + r_2 \leq r$. This means that we are considering a mapping, R, of the ball S' into itself. Notice that

$$\|R(v_2) - R(v_1)\|_{B_2} \leq \rho_0 \|v_2 - v_1\|_{B_2},$$

with $\rho_0 < 1$ for all v_2 and v_1 from this ball. Therefore, R is a contracting operator, and the statement of Theorem 2.3 is a particular case of the classical contraction principle.

Note that to solve Equation (2.1), in $H(B_2)$ we can minimize the functional

$$\Phi(v) \equiv \|L(v) - f\|_{B_1^{-1}}^2$$

(see references 18, 32, and 50) on the basis of the very popular conjugate gradient method. But estimates of convergence of this method are no better than those obtained from the significantly simpler method analyzed in Theorem 2.2.

If we consider system (2.1) rewritten in the form $Au + \mathcal{P}(u) = f$, where A is a linear operator and the nonlinear part, $\mathcal{P}$, is small in a sense, then sometimes we can use the iterative method

$$B_2(u^{n+1} - u^n) = -\tau A^* B_1^{-1}[Au^n + \mathcal{P}(u^n) - f]. \qquad (2.21)$$

If for any z with $\|z\|_{B_2} \leq r$, it is possible to maintain that $\|\mathcal{P}(u + z) - \mathcal{P}(u)\|_{B_1^{-1}} \leq l\|z\|_{B_2}$ (see Equation [2.18]), then the convergence of Equation (2.21) is analyzed in much the same way as in the proof of Theorem 2.2.

3. CHOICE OF MODEL GRID OPERATORS: OPTIMAL AND NEARLY OPTIMAL PRECONDITIONING

3.1 BASIS APPROACHES TO CONSTRUCTING EFFECTIVE PRECONDITIONERS

Section 1 stressed that finding model operators $B = B^* > 0$ such that on the one hand the systems $Bv = g$ (see Equation [1.4]) are easily solvable and on the other hand lead to inequalities of type (1.2) and (1.13)–(1.15), with constants $\delta_1 \geq \delta_0 > 0$ either independent on h or at least weakly dependent on this parameter, is of fundamental importance. Their construction for a given class of grid operators $L_h \equiv L$ is not a simple matter. It has attracted the attention of many mathematicians since the early 1960s, and many significant notions and ideas have become better and better understood.

Of course, some elliptic boundary value problems, such as typical ones for the convection-diffusion equation with a dominating convection term, need additional basic ideas in order to understand the nature of operators with a dominating antisymmetric part and to be able to use nonsymmetric model operators. But this will be an open problem for at least several years. The most difficult part of the problem is how to construct nonsymmetric model operators leading to the easily solvable systems,

$$Bv = g, \qquad B^*v = g,$$

because in the case of the preconditioned system, $B^{-1}Lu = B^{-1}f$, it seems natural to symmetrize it in the Euclidean space H and to work with the symmetrized system

$$Au \equiv L^*(B^{-1})^* B^{-1} Lu = L^*(B^{-1})^* B^{-1}f. \qquad (3.1)$$

If we notice that $(B^{-1})^* B^{-1} = (BB^*)^{-1}$ and introduce a symmetric and positive operator, $D \equiv (BB^*)^{1/2}$, then we can rewrite system (3.1) in the form $Au \equiv L^*D^{-2}Lu = L^*D^{-2}f$. Therefore, inequalities (1.14), with D instead of B, will ensure that $\delta_0 I \leq L^*D^{-2}L \leq \delta_1 I$. Thus, for example, the rate of convergence of the modified Richardson method,

$$u^{n+1} - u^n = -\tau_n L^*(B^*)^{-1} [B^{-1}(Lu^n - f)],$$

will be completely defined by these estimates.

What concerns the grid problems being approximations of a well-posed operator equation in a Hilbert space such as $G \subset W_2^1(\Omega)$ (approximations of second-order elliptic boundary value problems associated with a given bounded region in the Euclidean space, $\mathbf{R}^d$),[28,55] now we can choose from among a variety of fruitful approaches to construct optimal and nearly optimal preconditioners. Some widely applied concepts and classes of methods are:

1) selecting the principal and at the same time sufficiently simple part, $\Lambda \equiv \Lambda_\Omega$, of the given grid operator $L_h \equiv L_{h,\Omega}$ such that solving a system $\Lambda v = g$ with an arbitrary given g by some fast direct methods based on the separation of variables and the fast discreet Fourier transform is possible;

2) instead of direct methods, some effective inner iterations can be found leading to the two-stage preconditioners, B, of the type in Equation (2.10);

3) instead of Λ_Ω a simpler model operator, Λ_Q, defined on a grid topologically equivalent to the original one but for a significantly simpler model region, Q, can be used;

4) a partition of the given closed region, $\bar{\Omega}$, or a model closed region, $\bar{Q}$, can be used for construction of B (domain decomposition preconditioners and methods of the Schwartz type);

5) on the basis of a partition of the given grid and splitting of the original finite-element subspace (for $\bar{\Omega}$ or $\bar{Q}$), algebraic multigrid constructions of model operators B can be obtained;

6) if a rather simple closed region, $\bar{\Pi}$, can be obtained from $\bar{Q}$ by adding some $\bar{F}$ (a fictitious region, $\bar{\Pi} = \bar{Q} \cup \bar{F}$), then B_Q can be constructed via B_Π (iterative methods of the fictitious grid region and capacitance methods).

It is noteworthy that many of these methods are associated with the use, in one form or another, of the block elimination of unknowns. To review this, consider a linear system, $\Lambda u = f$, with $u \in H \equiv \mathbf{R}^N$, $f \in H$, $\Lambda \in \mathbf{R}^{N \times N}$. Assume that it can be rewritten in block form, that is,

$$\Lambda u \equiv \begin{bmatrix} \Lambda_{1,1} & \Lambda_{1,2} \\ \Lambda_{2,1} & \Lambda_{2,2} \end{bmatrix} \begin{bmatrix} u_1 \\ u_2 \end{bmatrix} = \begin{bmatrix} f_1 \\ f_2 \end{bmatrix}, \tag{3.2}$$

where $u \equiv [u_1, u_2]^T \in H \equiv H_1 \times H_2$, $u_r \in H_r$, $\dim H_r \equiv N_r$, $r = 1, 2$; $N_1 + N_2 = N$, and the block $\Lambda_{1,1} \in \mathbf{R}^{N \times N}$ is an invertible matrix such that systems $\Lambda_{1,1} v = g$ may be regarded as easily solvable. Then,

$$\Lambda_{1,1}\, u_1 = f_1 - \Lambda_{1,2}\, u_2, \tag{3.3}$$

and elimination of u_1 leads to the system

$$S_2(\Lambda)u_2 \equiv S_2 u_2 \equiv (\Lambda_{2,2} - \Lambda_{2,1}\, \Lambda_{1,1}^{-1}\, \Lambda_{1,2})u_2 = g_2, \tag{3.4}$$

$$g_2 \equiv f_2 - \Lambda_{2,1}\, \Lambda_{1,1}^{-1} f_1. \tag{3.5}$$

The original system and systems (3.4) and (3.3), with g_2 from Equation (3.5), which also may be written in the block-triangular form,

$$\Lambda_t u \equiv \begin{bmatrix} \Lambda_{1,1} & \Lambda_{1,2} \\ 0 & S_2 \end{bmatrix} \begin{bmatrix} u_1 \\ u_2 \end{bmatrix} = \begin{bmatrix} f_1 \\ g_2 \end{bmatrix},$$

are equivalent. Thus, the described method (sometimes referred to as the tearing method, or the bordering method) can be reduced to

1) solve the system $\Lambda_{1,1}v_1 = f_1$, and after that evaluate the vector $g_2 = f_2 - \Lambda_{2,1}v_1$;

2) solve matrix equation $\Lambda_{1,1}X = \Lambda_{1,2}$. This requires solution of N_2 systems with the same matrix $\Lambda_{1,1}$ and the right-hand sides coinciding with the $(N_1 + k)$th column of Λ, $k = 1, \ldots, N_2$ and is perfectly suited for implementation on vector and parallel computers;

3) find the matrix $S_2 = \Lambda_{2,2} - \Lambda_{2,1}X$;

4) solve the system $S_2 u_2 = g_2$;

5) evaluate the vector $g_1 = f_1 - \Lambda_{1,2}u_2$;

6) solve the system $\Lambda_{1,1}u_1 = g_1$.

The required computational work may be characterized as $W \approx (N_2 + 2)W_1 + W_2$, where W_1 and W_2 are upper bounds for the computational work required to solve the systems with matrices $\Lambda_{1,1}$ and S_2, respectively.

Emphasize right away that the simplest algorithm described is well suited only for blocks with a relatively small number N_2. In many contemporary variants of this dealing with large N_2, stages 2–4 are replaced by a separate procedure leading directly to $u_2 = S_2^{-1}g_2$ without finding the matrix S_2. This is achieved when S_2 is chosen in the form of Equation (2.10), and therefore solving a system $S_2 u_2 = g_2$ amounts to performing a chosen number of inner iterations. The same applies to the systems with $\Lambda_{1,1}$. Especially attractive are blocks such that the solutions of the systems with them can be reduced to independent solutions of some smaller subsystems.

Notice that the procedure may be connected with a block-triangular factorization of the matrix

$$\Lambda = \begin{bmatrix} \Lambda_{1,1} & 0 \\ \Lambda_{2,1} & \Lambda_{2,2} \end{bmatrix} \begin{bmatrix} I_1 & \Lambda_{1,1}^{-1}\Lambda_{1,2} \\ 0 & I_2 \end{bmatrix}.$$

The matrix $S_2(\Lambda) \equiv S_2 \equiv \Lambda/\Lambda_{1,1}$ is often referred to as the Shur complement to the block $\Lambda_{1,1}$, or simply as the Shur matrix.

Operators (matrices) B of the same block structure

$$B \equiv \begin{bmatrix} B_{1,1} & \Lambda_{1,2} \\ \Lambda_{2,1} & B_{2,2} \end{bmatrix}, \tag{3.6}$$

with $B_{1f,2} \equiv \Lambda_{1,2}$ and leading to relatively easily solvable systems with matrices $B_{1,1}$ and $S_2(\Lambda) \equiv D_{2,2}$, were used as preconditioners in references 14 and 63 and as optimal preconditioners in references 2, 17, 22, and 24. They may be referred to as model block-triangular factorized operators, or cooperative operators. The latter notion emphasizes the connection with game theory: joint efforts of both players are required to obtain the desired solution; each of them makes decisions associated with Euclidean space either H_1 or H_2. Such operators, besides the previous factorization, can be written also in the form

$$B = \begin{bmatrix} B_{1,1} & 0 \\ 0 & D_{2,2} \end{bmatrix} \begin{bmatrix} I_1 & 0 \\ D_{2,2}^{-1}\Lambda_{2,1} & I_2 \end{bmatrix} \begin{bmatrix} I_1 & B_{1,1}^{-1}\Lambda_{1,2} \\ 0 & I_2 \end{bmatrix}$$

with $D_{2,2} \equiv B_{2,2} - \Lambda_{2,1}B_{1,1}^{-1}\Lambda_{1,2}$ (this was used in reference 22 for construction of some optimal preconditioners). Similar model saddle-point operators are considered in references 27, 28, and 57.

A simple theorem may be of help in investigating properties of such operators.

Theorem 3.1. Let the operators $\Lambda \in \mathcal{L}^+(H)$ and $B = B^*$ be defined by Equations (3.2) and (3.6) and

$$(\Lambda v, v) \geq s^2(\Lambda_{1,1}v_1, v_1), \qquad s^2 > 0 \tag{3.7}$$

for all $v \in H$. Let there exist positive constants, $q < 1$, and

$$\kappa_0 \leq 1 \leq \kappa_1$$

such that

$$(1 - q)B_{1,1} \leq \Lambda_{1,1} \leq B_{1,1},$$

$$\kappa_0 S_2(B) \leq S_2(\Lambda) \leq \kappa_1 S_2(B). \tag{3.8}$$

Then $\quad \dfrac{\kappa_0}{\xi_1} B \leq \Lambda \leq \kappa_1 B, \qquad \xi_1 \equiv 1 + \dfrac{q}{s^2(1-q)}. \tag{3.9}$

(This theorem can be found in references 29 and 30 and has some points in common with similar statements from references 3, 12, 33, and 34.)

Now we reconsider the classes of iterative methods enumerated previously and give some specifications. Very frequently in the role of a model operator, B, for grid operators associated with approximations of an elliptic second-order operator, we can use an operator spectrally equivalent to the operator $J \equiv \Lambda_\Omega$ associated with approximations of the operator $-\Delta$, where Δ refers to the Laplacian being regarded as a mapping of the Hilbert space $G \equiv W_2^1(\Omega; \Gamma_0) \subset W_2^1(\Omega)$.[28,55] For several rather simple situations, when $\bar{\Omega}$ is a rectangle or a d-dimensional parallelepiped and the grid at hand has a similar simple structure, some very fast direct methods are used to solve the systems with this operator (see references 14, 18, 19, 23, 28, and 44 and references therein). In particular, marching methods and methods of cyclic reduction based on recurrent use of the block elimination of unknowns should be mentioned. Sometimes, in order to apply these direct methods, further simplifications of the operator are necessary; this is the case when the region is taken in the d-dimensional Euclidean space $\mathbf{R}^d$, with $d \geq 3$.[28] In the case of strongly elliptic systems, such as widely known elasticity systems, it is possible to use block-diagonal operators with diagonal blocks equal to Λ_Ω.[17,18,20,22,28]

For more general situations, we can apply not the direct methods but some effective inner iterations (such as alternating direction iterations (ADI) in the so-called commutative case, or some more general iterations) leading to two-stage, nearly optimal preconditioners B, of the type such as Equation (2.10) (see references 17–24; 28, 44, 46, 47).

Instead of Λ_Ω, very often a simpler model operator, Λ_Q, defined on a grid topologically equivalent to the original one but for a significantly simpler model region, Q, can be used. This approach, for difference methods and nearly orthogonal curvilinear grids on two-dimensional manifolds, has been used with success from the beginning of the 1960s for problems mentioned previously, and even for some elliptic boundary value problems of the fourth order, including many practically very important problems of the theory of plates and shells.[18,28] In the case of the projective-grid methods on composite grids with local refinements, we will describe some generalizations of results[23,28] a little later.

So-called domain decomposition, or cutting methods, when a partition of $\bar{\Omega}$ or of a model closed region, $\bar{Q}$, can be used for construction of B, are currently very popular (see references 12, 15, 16, 28, 34, 39, 43, and 58 and references therein). These actually are based on the use of operators of the block structure, Equation (3.6), with $B_{1,1} = \Lambda_{1,1}$. The Euclidean space H_2 is associated with the grid nodes belonging to the cutting lines (surfaces), and the Euclidean space H_1 is associated with the remaining nodes. A great deal of attention has been paid to the construction of operators B_2 nearly spectrally equivalent to an operator such as $S_2(\Lambda)$. The corresponding theory dealing with grid approximations of the Sobolev–Slobodeckii spaces, $W_2^{1/2}(C)$, where C refers to the manifold comprising all cuttings lines (surfaces), has been developed.[12,15,16,43,58,59]

We discuss some very interesting and relatively unknown results of Siganevich[59] leading to the estimates

$$\delta_0 B_2 \leq S_2(\Lambda) \leq \delta_1 B_2, \qquad \delta \equiv \frac{\delta_1}{\delta_0} = O\big(|\ln h|^3\big).$$

There are indications that the given asymptotic estimates can be improved. What is of importance is that solving a system with the previous model operator, B_2, can be reduced to solving similar and independent systems on separate parts of the manifold, C. Because each such part is a $(d-1)$-dimensional parallelepiped, the cosine Fourier transform can be effectively applied. We also discuss important results dealing with the operator, B, having the block $B_{1,1}$ spectrally equivalent to block $\Lambda_{1,1}$ (see reference 33 and also Theorem 3.1).

During the past decade the understanding of the importance of spectrally equivalent operators in the theory of multigrid methods has considerably widened. An appropriate recurrent splitting of the original finite-element subspace and usage of effective inner iterations, such as those considered in Theorems 1.1 and 1.2, can produce model operators B from Equation (1.2), with $L \equiv \Lambda$ not only spectrally equivalent to Λ but having the constants δ_k, $k = 0, 1$, in Equation (1.2) very close to 1.[3-5,9,29,30,34,35] We will present a key theorem estimating the angles between the addends in the splittings previously mentioned for the general case of d-dimensional problems. This technique works well not only for operators such as those associated with approximations of the Laplacian but also for approximations of some elliptic second-order operators with strongly varying coefficients.

Finally, speaking about fictitious grid region methods,[1,39] we stress that they are actually based on the block structure of the operator Λ_{Π}, with $\bar{\Pi} = \bar{Q} \cup \bar{F}$) (see Equation [3.2]). The Euclidean space H_1 is associated with the nodes not belonging to $\bar{\Omega}$, and in the role of the model operator, B, serves the Shur matrix, $S_2(\Lambda_{\Pi})$.[28,45,59]

3.2   SPECTRALLY-EQUIVALENT OPERATORS ON COMPOSITE GRIDS WITH LOCAL REFINEMENTS

Now we describe some results connected with optimal preconditioning of finite-element (projective-grid) approximations of some second-order elliptic boundary value problems associated with triangulation of the given planar region, Ω, and with piecewise linear basis functions.

Under a triangulation of the region Ω, more accurately of its closure, $\bar{\Omega}$, we understand a partition, $T_h(\bar{\Omega})$, of $\bar{\Omega}$ into a finite set of triangles, $T_k \in T_h(\bar{\Omega})$, such that two different triangles do not have common inner points and neither side of one triangle can be a part of a side of another. Different triangles may have one common vertex or one common side; in other words, a triangular grid is being used. Of course, a boundary is assumed to be composed of a number of straight line segments.

At the present time more general triangulations, so-called composite triangulations, are frequently used[12,16,35,37,42,59] in order to refine the original grid or the triangulation in subregions, in which the solution of the problem changes very rapidly, and thereby improve the accuracy of the method. Any such grid or triangulation can be obtained as a result of a refinement process consisting of p levels. We start with an initial triangulation, $T^{(0)}(\bar{\Omega}) \equiv T^0$. Then we take a closed subset, $\bar{\Omega}^{(1)} \subset \bar{\Omega}$, consisting of a number of triangles from $T^{(0)}$, and an integer $t_1 > 1$, which will be called a refinement ratio. Next, each triangle from $T^{(0)}$ and belonging to $\bar{\Omega}_1$ is partitioned into a set of t_1^2 equal smaller triangles. This is accomplished by subdividing each side of the given triangle into t_1 equal parts and drawing through the points of division straight lines parallel to other sides of the triangle. The collection of cells of the old triangulation not belonging to $\bar{\Omega}^{(1)}$, and of all new smaller cells belonging to $\bar{\Omega}^{(1)}$, define the composite triangulation, $T^{(1)}$, of the first level. If $p > 1$, then a similar procedure is carried out with respect to $\bar{\Omega}^{(2)} \subset \bar{\Omega}^{(1)}$ and the refinement ratio $t_2 > 1$ and so on. Subsets $\bar{\Omega}^{(l)} \subset \bar{\Omega}^{(l-1)}$, with $l = 2, \ldots, p$, usually can be taken such that $\bar{\Omega}^{(l)}$ does not have common points with the closure of $\bar{\Omega}^{(l-2)} \setminus \bar{\Omega}^{(l-1)}$. Under this assumption, the lines separating $\bar{\Omega}^{(l)}$ and $\bar{\Omega}^{(l-1)} \setminus \bar{\Omega}^{(l)}$ consist of the sides of triangles from $T^{(l-1)}$ subdivided into t_l equal parts. These cutting lines define decomposition as

$$\bar{\Omega} = F_1 \cup F_2, \ldots, F_p, \tag{3.10}$$

where F_l is the closure of $\bar{\Omega}^{(l-1)} \setminus \bar{\Omega}^{(l)}$ for $l = 1, \ldots, p$, $\quad \Omega^{(0)} \equiv \Omega$. Note that these closed sets may have common points only belonging to their boundaries. Thus after all these consequent p local refinements of the initial triangulation, we come to composite triangulations with local refinement, $T^{(p)} \equiv T_{c,h}(\bar{\Omega})$, consisting of standard triangulations of subsets $F_1, \ldots, F_p$, which is a particular case of more general composite grids with local refinement.

We say triangulations $T_{c,h}(\bar{\Omega})$ and $T_{c,h}(\bar{Q})$ are topologically equivalent triangulations if there exists a one-to-one piecewise affine mapping, $z = \Pi^{-1}x$, continuous with respect to both directions of $\bar{\Omega}$ onto $\bar{Q}$, and such that each elementary triangle, T_Ω, from $T_{c,h}(\bar{\Omega})$ is mapped onto a corresponding triangle, T_Q, from $T_{c,h}(\bar{Q})$. Figures 1 and 2 provide the simplest illustrations of topologically equivalent triangulations; there exists a great variety of them.

Suppose now that, given a composite triangulation $T_{c,h}(\bar{\Omega})$, we wish to construct a subspace, $\hat{G}_h$, of the Hilbert space $G \equiv W_2^1(\Omega; \Gamma_0)$, with Γ_0 covered by a collection of some sides of elementary triangles. Recall that we have

$$(u, v)_G \equiv (u, v)_{1,\Omega} \equiv (u_{x_1}, v_{x_1})_{0,\Omega} + (u_{x_2}, v_{x_2})_{0,\Omega} = (g, v)_{0,\Omega}, \tag{3.11}$$

$$(u, v)_{0,\Omega} \equiv \int_\Omega uv d\Omega \equiv (u, v)_{L_2(\Omega)}, \tag{3.12}$$

$$|u|_{0,\Omega} \equiv (u, u)_{0,\Omega}^{1/2}, \qquad |u|_{1,\Omega}^{1/2} \equiv (u, u)_{1,\Omega}^{1/2}. \tag{3.13}$$

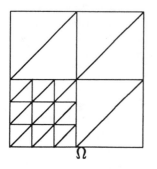

FIGURE 1. Triangulation of a model region.

Notice that for $\Gamma_0 = \emptyset$, $(u, v)_{1,\Omega}$ is only a semiinner product, which satisfies all axioms of inner product with the exception of the axiom $\|u\| = 0 \Rightarrow u = 0$. The space $\hat{G}_h$ previously mentioned may be described as the space of functions $\hat{u}$, continuous on $\bar{\Omega}$, vanishing on Γ_0, and linear on each triangle $T_\Omega \in T_{c,h}(\bar{\Omega})$.

Consider now the set of vertices of triangles from $T_{c,h}(\bar{\Omega})$ belonging to the closed region $\bar{\Omega}$ and not belonging to Γ_0. In the case of a certain standard triangulation (not composite), all these points have equal rights. They are called grid nodes and are represented in the basis of $\hat{G}_h$ by their own basis functions, $\hat{\psi}_P$. In the case of composite triangulations, the previous points must be subdivided into two subsets. The first set, denoted by Ω_h, consists of points (nodes) with the properties previously mentioned. The second one consists of points (seminodes) at which values of grid functions have to be specified by interpolation procedures involving some values of $\hat{u}$ at nodes from Ω_h.

With each node P from Ω_h we associate a standard basis function, $\hat{\psi}_P(x)$, defined by certain conditions: $\hat{\psi}_P(x) = 1$ if $x = P$, $\hat{\psi}_P(x) = 0$ if x coincides with any node from Ω_h different from P, and the restriction of $\hat{\psi}_P(x)$ to an arbitrary triangle from $T_{c,h}(\bar{\Omega})$ is a linear function. The support of each $\hat{\psi}_P$ consists of triangles T_Ω with the common vertex P.

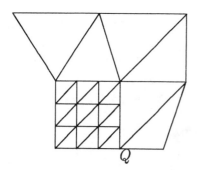

FIGURE 2. A topologically equivalent triangulation to that represented in Figure 1.

Consider the two topologically equivalent triangulations, $T_{c,h}(\bar{\Omega})$ of the original region Ω, and $T_{c,h}(\bar{Q})$ of a chosen model region Q, for which

$$(u, v)_{1,Q} \equiv \int_Q \left(\frac{\partial u}{\partial z_1} \frac{\partial v}{\partial z_1} + \frac{\partial u}{\partial z_2} \frac{\partial v}{\partial z_2} \right) dz_1 dz_2. \qquad (3.14)$$

The previous one-to-one correspondence, $x = \Pi z$, between the closed regions defines an isomorphism between finite element spaces, $\hat{G}_{Q,h}$ and $\hat{G}_{\Omega,h}$, of the elements

$$\hat{u}^{\Omega}(x) = \sum_{i=1}^N u_i \hat{\psi}_{\Omega,i}(x), \quad \hat{u}^Q(z) = \sum_{i=1}^N u_i \hat{\psi}_{Q,i}(z), \qquad (3.15)$$

where numbers u_i correspond to equal values of functions at the equivalent nodes and $\hat{\psi}_{\Omega,i}, \hat{\psi}_{Q,i}$ are the piecewise linear basis functions defined previously.[28,59] We define grid operators (matrices)

$$\Lambda_\Omega \equiv \Lambda_{\Omega,h} \equiv \left[\sum_{r=1}^2 \left(1, \frac{\partial \hat{\psi}_{\Omega,j}}{\partial x_r} \frac{\partial \hat{\psi}_{\Omega,i}}{\partial x_r} \right)_{0,\Omega} \right], \qquad (3.16)$$

$$\Lambda_Q \equiv \Lambda_{Q,h} \equiv \left[\sum_{r=1}^2 \left(1, \frac{\partial \hat{\psi}_{Q,j}}{\partial z_r} \frac{\partial \hat{\psi}_{Q,i}}{\partial z_r} \right)_{0,Q} \right]. \qquad (3.17)$$

Of course, both operators are symmetric and positive.

For the previous mapping, Π, of $\bar{Q}$ onto $\bar{\Omega}$, let the image of each triangle $T' \in T_{c,h}(\bar{Q})$ be a corresponding triangle, $T \in T_{c,h}(\bar{\Omega})$, that is, $\Pi\{T'\} = T$, $\Pi^{-1}\{T\} = T'$. We assume that all triangles T' have a right angle and two equal sides. Then, for an arbitrary triangle $T \equiv P_1 P_0 P_2$ we denote by P_0 the vertex corresponding to the right angle of the triangle T' and define

$$\mu(T) \equiv \frac{S_1 + S_2}{2 S_{1,2}}, \qquad (3.18)$$

where $S_1 \equiv |P_0 P_1|^2$, $S_2 \equiv |P_0 P_2|^2$, and $S_{1,2} \equiv |[P_0 P_1, P_0 P_2]|$. S_r is an area of the square with a side equal to $P_0 P_r$, $r = 1, 2$, and $S_{1,2}$ is an area of a parallelogram with sides $P_0 P_1$, $P_0 P_2$. It is obvious that $\mu(T) \geq 1$ and $\mu(T) = 1$ corresponds only to half of a square. We also will be making use of

$$\mu\big(T_{c,h}(\bar{\Omega})\big) \equiv \max \mu(T), \qquad T \in T_{c,h}(\bar{\Omega}). \qquad (3.19)$$

Theorem 3.2. For each $h \in \{h\}$, let topologically equivalent triangulations $T_{c,h}(\bar{\Omega})$ and $T_h(\bar{Q})$ define grid operators, Λ_Ω and Λ_Q, by Equations (3.16) and (3.17). For all $h \in \{h\}$, let

$$\sup \mu\big(T_{c,h}(\bar{\Omega})\big) \leq \mu < \infty.$$

Then Λ_Ω and Λ_Q are spectrally equivalent operators, and

$$\delta_{0,\Omega}\,\Lambda_Q \leq \Lambda_\Omega \leq \delta_{1,\Omega}\,\Lambda_Q, \tag{3.20}$$

where $\qquad \delta_{1,\Omega} \equiv \mu + (\mu^2 - 1)^{1/2} \equiv \delta_{0,\Omega}^{-1}. \tag{3.21}$

Proof. Expansions (3.15) lead to the representations

$$(\Lambda_\Omega u, u) = \left(1, \left|\nabla u\right|^2\right)_{0,\Omega}, \quad (\Lambda_Q u, u) = \left(1, \left|\nabla v\right|^2\right)_{0,Q}, \tag{3.22}$$

where $\qquad \left|\nabla u\right|^2 \equiv \sum_{r=1}^{2}\left[\dfrac{\partial \hat{u}^\Omega}{\partial x_r}\right]^2, \tag{3.23}$

$$\left|\nabla v\right|^2 \equiv \sum_{r=1}^{2}\left[\dfrac{\partial \hat{u}^Q}{\partial z_r}\right]^2. \tag{3.24}$$

Integrals over Ω, Q in Equation (3.22) can be regarded here as sums of integrals over all possible triangles T, T'. Thus, to get the desired inequalities it suffices to obtain them simply for $|\hat{u}^\Omega|_{1,T}^2$ and $|\hat{u}^Q|_{1,T'}^2$ (see Equations [3.22]–[3.24]). The desired local inequalities with the previously defined constants (see Equations [3.19]–[3.21]) were proved in reference 23.

Notice that given estimates remain true for $\Gamma_0 = \emptyset$ when $(u,v)_{1,\Omega}$ is a semiinner product and the operators Λ_Ω and Λ_Q become nonnegative ones. Generalizations of the theorem are almost obvious with respect to more general operators associated with equalities

$$(\Lambda_\Omega u, u) = \sum_{T \in T_{c,h}(\bar{\Omega})} a(T) \int_T \left|\nabla \hat{u}^\Omega\right|^2 dx_1 dx_2, (\Lambda_Q u, u)$$

$$= \sum_{T' \in T_{c,h}(\bar{Q})} a(T') \int_{T'} \left|\nabla \hat{u}^Q\right|^2 dz_1 dz_2$$

with constants $a(T) = a(T') \geq 0$ on the corresponding triangles T' and $T = \Pi\{T'\}$.

3.3   ESTIMATES OF ANGLES BETWEEN SOME FINITE-ELEMENT SUBSPACES

Suppose that the space G is either a Hilbert space or a Euclidean space with a given inner product. Consider its subspace $\hat{G}$ such that $\hat{G} \equiv \hat{G}_1 \oplus \hat{G}_2$ with a basis $\hat{\psi}_1, \ldots, \hat{\psi}_{N_1+N_2}$ and $\hat{G}_1 \equiv \text{lin}\{\hat{\psi}_1, \ldots, \hat{\psi}_{N_1}\}$,

$\hat{G}_2 \equiv \text{lin}\{\hat{\psi}_{N_1+1}, \ldots, \hat{\psi}_{N_1+N_2}\}$. Denote by $\Lambda, \Lambda_{1,1}, \Lambda_{2,2}$ the Gram matrices defined by three bases previously mentioned. Then, for matrix Λ and its block-diagonal part D, we write

$$\Lambda \equiv \begin{bmatrix} \Lambda_{1,1} & \Lambda_{1,2} \\ \Lambda_{2,1} & \Lambda_{2,2} \end{bmatrix}; \qquad D \equiv \begin{bmatrix} \Lambda_{1,1} & 0 \\ 0 & \Lambda_{2,2} \end{bmatrix}. \tag{3.25}$$

The notion of the angle between subspaces $\hat{G}_1$ and $\hat{G}_2$ is very helpful for studying such matrices and the corresponding Shur complements. Recall that the angle $\alpha(\hat{u}_1; \hat{u}_2)$ between nonzero elements $\hat{u}_1$ and $\hat{u}_2$ is defined by the equality

$$\cos \alpha(\hat{u}_1; \hat{u}_2) \equiv \frac{|(\hat{u}_1, \hat{u}_2)|}{\|\hat{u}_1\| \, \|\hat{u}_2\|} \, ,$$

and the angle α between the subspaces is the minimal angle between their elements, that is, for all elements $\hat{u}_r \in \hat{G}_r$ with $r = 1, 2$, we have

$$|(\hat{u}_1, \hat{u}_2)| \leq \cos \alpha \, \|\hat{u}_1\| \, \|\hat{u}_2\|. \tag{3.26}$$

These notions can be directly generalized for spaces with semiinner products: it is useful in estimating the angles between some finite-element subspaces through local analysis on a cell of the grid. Lemma 3.1 is a simple result used in references 24 and 28 and later in references 2 and 37.

Lemma 3.1. Let α be the angle between the subspaces $\hat{G}_1$ and $\hat{G}_2$ (with an inner or semiinner product), the bases of which define matrices Λ and D from Equation (3.25). Then

$$(1 - \cos \alpha)D \leq \Lambda \leq (1 + \cos \alpha)D. \tag{3.27}$$

Now consider the closure, $\bar{Q}$, of a model given domain Q in the Euclidean space $\mathbf{R}^d$, and assume that $\bar{Q}$ consists of a finite number of d-dimensional simplexes, $T_{0,k}$. The collection of all of them defines a triangulation, $T^{(0)}(\bar{Q})$. Recall that in order to use necessary geometrical notions and illustrations it is reasonable, following Weyl, to think about $\mathbf{R}^d$ as a point-vector space (affine-vector space) with standard axioms connecting the notions of points and vectors. Then, if we identify the zero vector with the point O serving as the origin of a Descartes coordinate system with the standard orthonormal basis consisting of vectors $e_1, \ldots, e_d$, then we have a one-to-one correspondence between points M with coordinates $[x_1, \ldots, x_d] \equiv x$ and vectors $\overrightarrow{OM} \equiv x_1 e_1 + \ldots + x_d e_d$, being elements of the Euclidean space . For j given points $P_0, P_1, \ldots, P_j$ with $j \leq d$, and linearly independent vectors $\overrightarrow{P_0 P_1}, \ldots, \overrightarrow{P_0 P_j}$, the j-dimensional simplex $[P_0 P_1 \ldots P_j]$ is defined as the set of points P corresponding to vectors $\overrightarrow{OM}$ from the convex hull of vectors $\overrightarrow{OP_0}, \ldots, \overrightarrow{OP_j}$, that is,

$$T \equiv P_0 P_1 \ldots P_j \equiv \left\{ M : \overrightarrow{OM} = \sum_{i=0}^{j} \alpha_i \overrightarrow{OP_i}, \alpha_i \geq 0, \right.$$

$$\left. i = 1, \ldots, j, \sum_{i=0}^{j} \alpha_i = 1 \right\}.$$

Each point P_i is a vertex of the simplex; for $1 \leq k \leq j-1$, k-dimensional faces of the simplex are k-dimensional simplexes with $k+1$ different vertices, $P_{i_1}, \ldots, P_{i_k}$, from the given set of P_i. One-dimensional faces are called edges. We will be dealing with special d-dimensional simplexes, which may be referred to as regular parts of some cubes in $\mathbf{R}^d$. More precisely, let us consider a cube

$$Q_a \equiv \{x : 0 \leq x_i \leq a, \qquad i = 1, \ldots, d, a > 0\}$$

with 2^d vertices $[\alpha_1, \ldots, \alpha_d]$, where each number α_i is either 0 or a. Given the direction $[1, 1, \ldots, 1]$ of one of its diagonals, we take

$$P_0 \equiv (0, 0 \ldots, 0), \quad P_1 \equiv P_0 + a e_{j_1},$$

$$P_2 \equiv P_1 + a e_{j_2}, \ldots, P_d \equiv (a, a, \ldots, a),$$

where all $j_1, \ldots, j_d$ are different integers and each $j_r \in [1, d]$. Then these points $P_0, P_1, \ldots, , P_d$ define a d-dimensional simplex, $T \equiv [P_0 P_1 \ldots P_d]$. The cube Q_a contains $d!$ such different congruent simplexes; one of them can match the other by an isometric mapping. Any such simplexes and similar ones corresponding to other $2^{d-1} - 1$ directions of its diagonals may be called a regular part of the cube Q_a. Figure 3 represents $\bar{Q}$ consisting of such simplexes (triangles); Figure 4 schematically represents such a three-dimensional simplex (tetrahedron), which must be partitioned into 2^3 smaller congruent simplexes with edges reduced by half.

Similar general partitioning of a d-dimensional simplex into 2^d or even t^d congruent subsimplexes can be proved to exist, but we simply assume this and consider the $(l+1)$th level of triangulation, $T^{(l+1)}(\bar{Q})$, as obtained from the triangulation $T^{(l)}(\bar{Q})$ as a result of the previous global refinement procedure.

Let $Q_h^{(l)}$ be a set of vertices, $P_i^{(l)}$, of simplexes $T_{l,k}$ from $T^{(l)}(\bar{Q})$ such that each of them corresponds to the standard continuous piecewise linear basis function, $\hat{\psi}_i^{(l)}(x) : \hat{\psi}_i^{(l)}(P_i^{(l)}) = 1$, $\hat{\psi}_i^{(l)}(P_j^{(l)}) = 0$ for $i \neq j$, and $\hat{\psi}_i^{(x)}$ is linear on each simplex $T_{l,k}$ from the triangulation $T^{(l)}(\bar{Q})$. Note that in approximating the space $W_2^1(Q; \Gamma_0)$ we usually identify the set of grid nodes, $Q_h^{(l)}$, with the set of all vertices (of all triangles $T_{l,k}$)

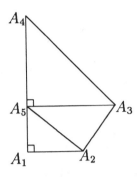

FIGURE 3. A representation of $\bar{Q}$ consisting of simplexes (triangles) which correspond to regular parts of cubes (squares).

not belonging to Γ_0. We consider this space as a Hilbert space G with the inner product

$$(u, v)_{1,Q} \equiv \int_Q \left(\frac{\partial u}{\partial x_1} \frac{\partial v}{\partial x_1} + \frac{\partial u}{\partial x_2} \frac{\partial v}{\partial x_2} \right) dx_1 dx_2.$$

Define the finite-element subspaces of G,

$$\hat{G}^{(l)} \equiv \left\{ \hat{u} : \hat{u} = \sum_{P_i^{(l)} \in Q_h^{(l)}} u_i \hat{\psi}_i^{(l)}(x) \right\}, \tag{3.28}$$

with $l = 0, \ldots, p$. Along with the basis $\hat{\psi}_i^{(l+1)}(x)$ for $\hat{G}^{(l+1)}$ consider a new basis, $\bar{\psi}_i^{(l+1)}(x)$, with

$$\bar{\psi}_i^{(l+1)} \equiv \hat{\psi}_i^{(l+1)} \quad \text{for } P_i^{(l+1)} \in Q_h^{(l+1)} \setminus Q_h^{(l)}$$

and $\qquad \bar{\psi}_i^{(l+1)} \equiv \hat{\psi}_i^{(l)} \quad \text{for } P_i^{(l+1)} \in Q_h^{(l)},$

assuming that the number of nodes from $Q_h^{(l+1)} \setminus Q_h^{(l)}$ is less than those

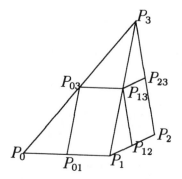

FIGURE 4. A schematic representation of a three-dimensional simplex (or tetrahedron) which must be partitioned into 2^3 smaller congruent simplexes with edges reduced by half.

from $Q_h^{(l)}$. The indicated choice of the new basis, often referred to as an hierarchical basis, leads to splittings

$$G^{(l+1)} = G_1^{(l+1)} \oplus G_2^{(l+1)}, \qquad l = 0, \ldots, p-1, \quad (3.29)$$

where

$$\hat{G}_2^{(l+1)} = \hat{G}^{(l)} \qquad (3.30)$$

$$\hat{G}_1^{(l+1)} \equiv \left\{ \hat{u} : \ \hat{u} \in \hat{G}^{(l+1)}, \qquad \hat{u}(P_i^{(l)}) = 0 \text{ for all } P_i^{(l)} \in Q_h^{(l)} \right\}. \quad (3.31)$$

Notice that the Gram matrices for these two bases of the Euclidean space $\hat{G}^{(l+1)}$ take the form

$$J^{(l+1)} \equiv \begin{bmatrix} J_{1,1}^{(l+1)} & J_{1,2}^{(l+1)} \\ J_{2,1}^{(l+1)} & J_{2,2}^{(l+1)} \end{bmatrix}; \quad \bar{J}^{(l+1)} \equiv \begin{bmatrix} \bar{J}_{1,1}^{(l+1)} & \bar{J}_{1,2}^{(l+1)} \\ \bar{J}_{2,1}^{(l+1)} & \bar{J}_{2,2}^{(l+1)} \end{bmatrix}. \quad (3.32)$$

Theorem 3.3. Let all simplexes from the triangulations $T^{(l)}(\bar{Q})$ be regular parts of some cubes in $\mathbf{R}^d$ and the finite element subspaces, $\hat{G}_2^{(l)}$ and $\hat{G}_1^{(l)}$, in the splittings of Equation (3.29) be defined by Equations (3.30) and (3.31). Then the angle α_{l+1} between these Euclidean spaces is such that

$$\cos \alpha_{l+1} \leq \left[1 - 2^{1-d} \right]^{1/2} \equiv \gamma < 1. \quad (3.33)$$

Proof. Denote by $\hat{u}_1$ and $\hat{u}_2$ arbitrary elements of $\hat{G}_1^{(l)}$ and $\hat{G}_2^{(l)}$, respectively. Then

$$(\hat{u}_1, \hat{u}_2) = \sum_{T_{l,k} \in T^{(l)}(\bar{Q})} (\hat{u}_1, \hat{u}_2)_{1, T_{l,k}}, \quad (3.34)$$

with $\quad (\hat{u}, v)_{1, T_{l,k}} \equiv \int_{T_{l,k}} \sum_{r=1}^{d} \frac{\partial u}{\partial x_r} \frac{\partial v}{\partial x_r} \, dx_1 \cdots dx_d$

being a semiinner product of u and v. It suffices to prove that

$$|(\hat{u}_1, \hat{u}_2)_{1, T_{l,k}}| \leq \gamma |\hat{u}_1|_{1, T_{l,k}} |\hat{u}_2|_{1, T_{l,k}} \quad (3.35)$$

(see Equations [3.25] and [3.34]). To do this use the invariance of $(u, v)_{1,T}$ with respect to a change of Descartes coordinates, and choose them in such a way that the edges $P_0 P_1, P_1 P_2, \ldots,$ $P_{d-1} P_d$ of simplex $T_{l,k} = [P_0 P_1 \cdots P_d]$ are parallel to elements of the new basis associated with coordinates $y_1, \ldots, y_d$. Then

$$
(\hat{u}_1, \hat{u}_2)_{1,T_{l,k}}
$$
$$
\equiv \int_{T_{l,k}} \left(\frac{\partial \hat{u}_1}{\partial y_1} \frac{\partial \hat{u}_2}{\partial y_1} + \cdots + \frac{\partial \hat{u}_1}{\partial y_d} \frac{\partial \hat{u}_2}{\partial y_d} \right) dy_1 \cdots dy_d.
$$

Notice again that for proving Equation (3.35) it is sufficient to show that

$$
|X| \equiv \left| \int_{T_{l,k}} \frac{\partial \hat{u}_1}{\partial y_r} \frac{\partial \hat{u}_2}{\partial y_r} dy_1 \cdots dy_d \, y_d \right|
$$

$$
\leq \gamma \left| \frac{\partial \hat{u}_1}{\partial y_r} \right|_{0,T_{l,k}} \left| \frac{\partial \hat{u}_2}{\partial y_r} \right|_{0,T_{l,k}} \tag{3.36}
$$

where
$$
X = \sum_{T_{l+1}} \int_{T_{l+1} \subset T_{l,k}} \frac{\partial \hat{u}_1}{\partial y_r} \frac{\partial \hat{u}_2}{\partial y_r} dy_1 \cdots dy_d \tag{3.37}
$$

and T_{l+1} refers to any subsimplex from the partition of $T_{l,k}$. Since $\hat{u}_2$ is a linear function on the simplex $T_{l,k}$, its derivative, $\partial \hat{u}_2 / \partial y_r$, is a constant, D_r. Similarly, $\partial \hat{u}_1 / \partial y_r$ is a constant, $d_{r,l+1}$, on the subsimplex T_{l+1}. If simplex $T_{l,k}$ is a regular part of a cube, Q_{2h}, then its volume is $|T_{l,k}| = 2h)^d / d!$, and $|T_{l+1}| = h^d / d!$. $|\partial \hat{u}_2 / \partial y_r|^2_{0,T_{l,k}} = (2h)^d / d! D_r^2$, and

$$
\left| \frac{\partial \hat{u}_1}{\partial y_r} \right|^2_{0,T_{l+1}} = \frac{h^d}{d!} d_{r,l}^2, \quad \left| \frac{\partial \hat{u}_1}{\partial y_r} \right|^2_{0,T_{l,k}} = \sum_{T_{l+1} \subset T_{l,k}} \frac{h^d}{d!} d_{r,l}^2.
$$

Now notice that in the sum $X = |T_{l+1}| \sum_{T_{l+1} \subset T_{l,k}} D_r d_{r,l+1}$ (see Equation [3.37]) at least two addends will be opposite numbers. Such addends correspond to subsimplexes having some edges which are halves of the edge parallel to the y_r-axis of the simplex $T_{l,k}$. (On Figure 4 such an edge for $r = 1$ is $[P_0 P_1]$.) This means that

$$
|X| \leq \frac{h^d}{d!} \left[(2^d - 2) D_r^2 \right]^{1/2} \left[\sum_{T_{l+1} \subset T_{l,k}} d_{r,l+1}^2 \right]^{1/2}
$$

$$
\leq \left(\frac{2^d - 2}{2^d} \right)^{1/2} \left| \frac{\partial \hat{u}_2}{\partial y_r} \right|_{0,T_{l,k}} \left| \frac{\partial \hat{u}_1}{\partial y_r} \right|_{0,T_{l,k}}
$$

which yields Equation (3.36) and consequently Equation (3.33).

We note that the given estimate holds for more general inner products. For example, we might replace $(u, v)_{1,Q}$ by

$$(u, v) = \sum_{T_{l,k} \in T^{(l)}(\bar{Q})} a(T_{l,k})(u, v)_{1, T_{l,k}}, \tag{3.38}$$

with arbitrary nonnegative constants $a(T_{l,k})$. Moreover, if all simplexes were regular parts of cubes with edges parallel to the original coordinate axes, then the given estimate remains true for

$$(u, v)_G \equiv \sum_{r=1}^{d} a_r(T_{l,k}) \left(\frac{\partial u}{\partial x_r}, \frac{\partial v}{\partial x_r} \right)_{0,Q} \tag{3.39}$$

with arbitrary nonnegative constants $a_r(T_{l,k})$. The proven theorem, together with Theorem 3.1 and results stated in references 4, 30, and 31, leads to the construction of asymptotically optimal preconditioners for grid systems associated with projective-grid approximations of d-dimensional elliptic boundary value problems mentioned previously. Of special importance is the fact that constants δ_0 and δ_1 from Equation (1.2) for such model operators, B, even in the case of A associated with strongly discontinuous coefficients from Equations (3.38) and (3.39), can be made independent of these coefficients and very close to 1. Note also that in the case of the composite grids with local refinements with a bounded number of refinements, the desired preconditioner can be constructed on the basis of Theorem 3.1.

4. EXAMPLES OF APPLICATIONS OF TWO-STAGE ITERATIVE METHODS FOR NONLINEAR ELLIPTIC PROBLEMS

4.1 QUASILINEAR SECOND-ORDER ELLIPTIC EQUATIONS AND SYSTEMS

For elements u and v of the Hilbert space $G \equiv W_2^1(\Omega; \Gamma_0)$ with

$$\|u\|^2 \equiv \left(1, |\nabla u|^2 \right)_{0,\Omega} \tag{4.1}$$

we define a quasibilinear form

$$b(u, v) \equiv \sum_{r=0}^{d} \left(a_r(x, Du), D_r v \right)_{0,\Omega} \tag{4.2}$$

where Ω is a given bounded region in the Euclidean space $\mathbf{R}^d (d = 2, d = 3)$ with a sufficiently good boundary, $\partial\Omega \equiv \Gamma$. Γ_0 is a part of Γ having a positive $(d-1)$-dimensional measure. $D_j v \equiv \partial v / \partial x_j$ if $j = 1, \ldots, d$. $D_0 v \equiv v$, and $Du \equiv [D_0 u, D_1 u, \ldots, D_d u]$. Functions

$$a_r(x, \xi) \equiv a_r(x, \xi_0, \xi_1, \ldots, \xi_d), \tag{4.3}$$

defined for all $x \in \bar{\Omega}$ and all $\xi \in \mathbf{R}^{d+1}$ are bounded piecewise continuous with respect to all x and have bounded and continuous derivatives of type $\partial a_r / \partial \xi_j$, $\partial^2 a_r / \partial \xi_i \partial \xi_j$, and $a_r(x, 0) = 0$. Also, we assume that there exist constants $\nu_1 \geq \nu_0 > 0$, $c_0 \geq 0$ such that, for all u, v and z from G,

$$|b(u; z) - b(v, z)| \leq \nu_1 \|u - v\| \|z\|, \tag{4.4}$$

$$b(u + z; z) - b(u; z) \geq \nu_0 \|z\|^2 - c_0 |z|_{0,\Omega}^2. \tag{4.5}$$

Given b and a linearly bounded functional, l_G, we consider the original elliptic boundary value problem written in the standard weak form: find $u \in G$ such that

$$b(u; v) = l_G(v), \qquad \forall v \in G. \tag{4.6}$$

Using a triangulation of $\bar{\Omega}$ (which may be a composite one with local refinements) and the finite-element subspace, $\hat{G}_h$, associated with the piecewise linear basis functions (see Section 2), we approximate Equation (4.6) and look for $\hat{u} \in \hat{G}_h$ such that

$$b(\hat{u}; \hat{v}) = l_G(\hat{v}), \qquad \forall \hat{v} \in \hat{G}_h. \tag{4.7}$$

Using expansions (1.6) and isomorphism (1.7), we rewrite Equation (4.7) in the algebraic form

$$L_h(u_h) = f_h, \tag{4.8}$$

with $u_h \equiv u$ (see Equation [1.7]) being an element of the Euclidean space H,

$$f_h \equiv f \equiv [l_G(\hat{\psi}_1), \ldots, l_G(\hat{\psi}_N)]^T \in H,$$

$$L_h(u_h) \equiv \left[(L_h(u))_1, \ldots, (L_h(u))_N \right]^T \in H,$$

and $\qquad (L_h(u))_i \equiv b(\hat{u}; \hat{\psi}_i), \qquad i = 1, \ldots, N.$

Thus, $\qquad (L_h(u), v) = b(\hat{u}; \hat{v}), \qquad \forall \hat{v}, \forall \hat{u}. \tag{4.9}$

Under the conditions of Equations (4.1)–(4.5) and additionally that $c_0 \geq 0$, we have a standard situation with correct operator Equations (4.6)–(4.8), since the classical results of the theory of strongly monotone nonlinear operators[28,50,61] are applicable. In investigating two-stage iterative methods for Equation (4.8) we will do without such assumptions, but invertibility of all Jacobian matrices, $L'_{h,w}$, defined by the relation

$$(L'_{h,w}u, v) = \sum_{r=0}^{d}\sum_{j=0}^{d}\left(\frac{\partial a_r(x, D\hat{w})}{\partial \xi_j}\, D_j\hat{u}, D_r\hat{v}\right)_{0,\Omega}, \qquad (4.10)$$

is required. More precisely, for all $L'_{h,w}$ and J from Equation (1.8) we assume that

$$\delta_0\|v\|_J^2 \le \|L'_{h,w}v\|_{J-1}^2 \le \delta_1\|v\|_J^2, \qquad \delta_0 > 0\ \forall v \in H. \qquad (4.11)$$

Useful sufficient conditions for Equation (4.11) are found in reference 28 (Chapter 1, Section 2). These assume that $L'_{h,w} = (L'_{h,w})^*$ and all eigenvalues of $L'_{h,w}$ are such that $|\lambda(L'_{h,w})| \ge d > 0$. Conditions for Equation (4.11) lead to similar inequalities, with the operator J replaced by the operator $B \sim J$. Therefore, what we need to ensure the applicability of iterative method (2.8) and Theorem 2.1 is to show that there exists a constant, K, such that (see Equation [2.18])

$$\|L'_{h,w} - L'_{h,v}\|_{H(J)\mapsto H(J^{-1})} \le K\|v - w\|. \qquad (4.12)$$

Theorem 4.1. Let functions $a_r(x, \xi)$ satisfy the previous conditions such that all $a_r(x, \xi)$ with $r \ge 1$ do not depend on $\xi_1, \ldots, \xi_d$ and let

$$a_0(x, \xi) = \sum_{i=1}^{d} a_{0,i}(x, \xi_0)\xi_i,$$

where all $\partial a_{0,i}/\partial \xi_0$, $\partial^2 a_{0,i}/\partial \xi_0 \partial \xi_0$ are bounded functions. Then there exists a constant, K, such that Equation (4.12) holds for the previous operators, $L'_{h,w}$ and J.

Proof. For arbitrary vectors u, v, w, and z from the Euclidean space H, we have

$$(L'_{h,w}u - L'_{h,v}u, z) = \sum_{r=0}^{d}\sum_{j=0}^{d}\times$$

$$\left(\left(\frac{\partial a_r(x, D\hat{w})}{\partial \xi_j} - \frac{\partial a_r(x, D\hat{v})}{\partial \xi_j}\right)D_j\hat{u}, D_r\hat{z}\right)_{0,\Omega}. \qquad (4.13)$$

Using the formula for finite increments of a differentiable functional,[48,49] we rewrite Equation (4.13) as

$$(L'_{h,w}u - L'_{h,v}u, z) = \sum_{r=0}^{d}\sum_{j=0}^{d}\sum_{i=0}^{d}\times$$

$$\frac{\partial^2 a_r}{\partial \xi_j \partial \xi_i}\left(x, D(\hat{v} + \theta(\hat{w} - \hat{u}))D_i(\hat{w} - \hat{v})D_j\hat{u}, D_r\hat{z}\right)_{0,\Omega}.$$

Due to the additional conditions imposed on functions a_r, it follows that

$$|(L'_{h,w}u - L'_{h,v}u, z)| \leq K_r \sum_{r=0}^{d} \left|((\hat{w} - \hat{v}), \hat{u}, D_r\hat{z})_{0,\Omega}\right|$$

$$+ \sum_{i=1}^{d} K_{0,i} \left(\left|((\hat{w} - \hat{v})D_i\hat{u}, \hat{z})_{0,\Omega}\right| + \left|(D_i(\hat{w} - \hat{v})\hat{u}, \hat{z})_{0,\Omega}\right| \right)$$

$$+ K_{0,0} \left|((\hat{w} - \hat{v})\hat{u}, \hat{z})_{0,\Omega}\right|. \tag{4.14}$$

Each addend in the right-hand side of Equation (4.14) can be rewritten either in the form $X \equiv |(u_1 u_2, D_i u_3)_{0,\Omega}|$ or in the form $Y \equiv |(u_1 u_2, u_3)_{0,\Omega}|$. Using the Holder inequality,

$$|(u_1 u_2, u_3)_{0,\Omega}| \leq |u_1|_{(4)} |u_2|_{(4)} |u_3|_{(2)},$$

with $\qquad |u|_{(4)} \equiv \|u\|_{L_4(\Omega)} \equiv (1, u^4)_{0,\Omega}^{1/4},$

$$|u|_{(2)} \equiv \|u\|_{L_2(\Omega)} \equiv (1, u^2)_{0,\Omega}^{1/2},$$

and the Sobolev imbedding theorem,[28,48,55,59] $|u|_{(4)} \leq K'|u|_{1,\Omega}$ (note that $d \leq 3$ and recall that $|u|_{1,\Omega} = \|u\|_G$), we can conclude that Equation (4.14) leads to

$$|(L'_{h,w}u - L'_{h,v}u, z)| \leq K \|\hat{w} - \hat{v}\| \|\hat{u}\| \|\hat{z}\|,$$

which together with Equation (1.9) yields Equation (4.12).

In much the same way, similar estimates can be obtained for elliptic systems with

$$u \equiv [u_1(x), \ldots, u_k(x)] \in G \in \left(W_2^1(\Omega)\right)^k.$$

So far we have considered only the case of mildly nonlinear problems with the previously mentioned bounded derivatives. As a useful and important illustration we now consider the case of a bounded power nonlinearity characterized by the presence in $b(u; v)$ of an addend of the form

$$\left(\mathcal{P}(u), v\right)_G \equiv (uD_r u, av)_{0,\Omega} \tag{4.15}$$

with a bounded coefficient, $a \equiv a(x)$ (typical, for example, for the well-known Navier-Stokes system). The derivative $\mathcal{P}'_{h,w}$ of the corresponding projective-grid operator is such that

$$\left(\mathcal{P}'_{h,w}u, v\right)_H \equiv (a\hat{w}D_r\hat{u}, \hat{v})_{0,\Omega} + (a\hat{u}D_r\hat{w}, \hat{v})_{0,\Omega}, \tag{4.16}$$

$$\left(\mathcal{P}'_{h,w}u - \mathcal{P}'_{h,v}u, z\right) = \left(a(\hat{w} - \hat{v})D_r\hat{u}, \hat{z}\right)_{0,\Omega} + \left(a\hat{u}D_r(\hat{w} - \hat{v}), \hat{z}\right)_{0,\Omega}.$$

Therefore, an inequality of type (4.12) (with $L'_{h,w}$ replaced by $\mathcal{P}'_{h,w}$) can be obtained in the same manner. We stress only that conditions of Equation (4.12) can be valid for w from a certain bounded ball in the Euclidean space $H(J)$.

4.2   FOURTH-ORDER ELLIPTIC EQUATIONS

First of all we note that the previous results can be generalized to the case of Equation (4.6) with the quasibilinear form

$$b(u; v) \equiv \sum_{|\alpha| \leq 2} \big(a_\alpha(x, Du), D^\alpha v \big)_{0,\Omega}, \qquad (4.17)$$

with Ω being a bounded region in the plane, $\alpha \equiv [\alpha_1, \alpha_2]$, $D^\alpha \equiv D_1^{\alpha_1} D_2^{\alpha_2}$, $|\alpha| \equiv \alpha_1 + \alpha_2$, and functions u and v being elements of a Hilbert space $G \subset W_2^2(\Omega)$. There are two reasons why we do not specify the elements. The first is connected with more involved conditions. The second and most important one is that for fourth-order elliptic problems, construction of effective projective-grid approximations, and especially of effective model operators, becomes a very hard task. Therefore, we confine ourselves to the simplest case of difference approximation of the first boundary value problem in the unit square Q for the equation

$$\Delta^2 u + g(u) = f, \qquad (4.18)$$

which nevertheless is instructive especially from the point of view of selection of asymptotically optimal model operators B. More precisely, let $h \equiv (N+1)^{-1}$ be the step of a square grid with nodes $P_i \equiv [i_1 h, i_2 h]$ ($i \equiv [i_1, i_2]$ is a vector index), and $u_i \equiv u(P_i)$ denotes a value of the grid function at the node. We will also be using the notation $\Omega_h \equiv \{P_i : i_r \in [1, N], r = 1, 2\}$; $e_1 \equiv [1, 0]$, $e_2 \equiv [0, 1]$; $I_{-r} u_i \equiv u(P_i - h e_r)$; $I_r u_i \equiv u(P_i + h e_r)$; $\partial_r u_i \equiv h^{-1}[I_r u_i - u_i]$; $\bar{\partial}_r u_i \equiv h^{-1}[u_i - I_{-r} u_i]$; and

$$\Lambda_r u_i \equiv -\bar{\partial}_r \partial_r u_i = -h^{-2}[I_{-r} u_i - 2u_i + I_r u_i]. \qquad (4.19).$$

Then the original difference system takes the form

$$(\Delta_1 + \Delta_2)^2 u_i + g(u_i) = f_i, \qquad P_i \in Q_h, \qquad (4.20)$$

$$u_i = 0, \quad \text{if } P_i \text{ does not belong to } Q_h. \qquad (4.21)$$

If we eliminate all values u_i with x_i not belonging to Q_h, we obtain the operator equation

$$L_h(u) = f \qquad (4.22)$$

in the Euclidean space H with the simplest inner product. Here,

$$\left(L_h(u)\right)_i \equiv [(\Delta_1 + \Delta_2)^2 u]_i + g(u_i), \qquad x_i \in \Omega_h, \qquad (4.23)$$

provided that the grid function, $u \in H$, is extended to nodes not belonging to Q_h through conditions (4.21). We rewrite L_h from Equations (4.22) and (4.23) in the form

$$L_h = \Lambda_1 + \Lambda_2 + 2\Lambda_{1,2} + \mathcal{P}, \qquad (4.24)$$

where $\quad \Lambda_r u_i \equiv \Delta_r u_i, \quad \Lambda_{1,2} u_i \equiv \Delta_1 \Delta_2 u_i, \quad \mathcal{P}(u_i) \equiv g(u_i),$

$$x_i \in Q_h. \quad (4.25)$$

In the case of $g = 0$ and $\mathcal{P} = 0$, the operator $L_h \in \mathcal{L}^+(H)$ and a nearly optimal model operator, $B \sim L$, can be taken in the form

$$B \equiv \Lambda(I - R_k)^{-1}, \qquad (4.26)$$

where $\Lambda = \Lambda_1 + \Lambda_2$, and the reducing operator, R_k, corresponds to the application of the ADI method for the equation $\Lambda v = g$.[18,28] This means that solving a system with this model operator requires computational work, $W = O(h^{-2}|\ln h|)$. The case when $g(u)$ is a continuously differentiable function and $g'(u) \geq 0$, or $g'(u) \geq -c^2$ with a small number c^2, is relatively simple, since it leads to the derivative $L'_{h,w} \sim B$, where

$$L'_{n,w} u \equiv (\Lambda_1 + \Lambda_2 + 2\Lambda_{1,2})u + g'(w)u. \qquad (4.27)$$

Now we consider the case when $L'_{h,w}$ may have several negative eigenvalues, $\lambda(L'_{h,w})$, but $|\lambda(L'_{h,w})| \geq d > 0$. Then, again using the results of reference 28, Chapter 1, Section 2, we can obtain inequalities (4.12) with $J = B$ (see Equation [4.26]). In order to apply iterative method (2.8) and Theorem 2.1, we need again to prove estimate (4.12). It can be done by reasoning very similar to that used in the proof of Theorem 4.1, the only difference being that instead of the classical Sobolev imbedding theorem their also very well-known difference analogues must be used (see reference 28, Chapter 6). We remark that similar model operators can be applied for more general elliptic equations with variable coefficients. And what is especially important, significantly more complicated and practical problems of the theory of shells associated with strongly elliptic systems [28] can be treated in much the same manner. For example, in the case of the system of theory of shells involving displacements $u_1 \equiv u$, $u_2 \equiv v$, $u_3 \equiv w$, in the role of the model operator B a block-diagonal operator, $B\bar{u} \equiv [B_1 u_1, B_2 u_2, B_3 u_3]^T$ can be used, where $\bar{u} \equiv [u_1, u_2, u_3]^T$ operators B_1 and B_2 are spectrally equivalent to the operator $-\Delta_h$, and the operator B_3 is of type (4.26). Finally, we note that a very important class of elliptic problems associated with two- and three-dimensional Navier–Stokes systems in general regions can be solved by the previous iterative methods; the necessary details can be found in references 28 and 29.

REFERENCES

1. Astrakhantsev, G.P., Method of fictitious domains for a second-order elliptic equation with natural boundary conditions, *Comp. Math. and Math. Phys.*, v. 18, 114, 1978.

2. Axelson, O., and Gustafson, I., Preconditioning and two-level multigrid methods of arbitrary degree of approximation, *Math. Comp.*, v. 40, 219, 1983.

3. Axelson, O., and Vassilevski, P.S., Algebraic multilevel preconditioning methods, I, *Numer. Math.*, v. 56, 157, 1989.

4. Axelson, O., and Vassilevski, P.S., Algebraic multilevel preconditioning methods. II, *SIAM Jour. Numer. Anal.*, v. 27, 1569, 1990.

5. Axelson, O., and Vassilevski, P.S., A surrey of multilevel preconditioned iterative methods, *BIT*, v. 29, 769, 1989.

6. Babushka, I., Craig, A., Mandel, J., and Pitkaranta, J., Efficient preconditioning for the p-version finite element method in two dimensions, *SIAM Jour. Numer. Anal.*, v. 28, 624, 1991.

7. Bakhvalov, N.S., *Numerical Methods*, Mir Publishers, Moscow, 1977.

8. Bank, R., and Dupont, T., An optimal order process for solving finite element equations, *Math. Comp.*, v. 36, 35, 1981.

9. Bank, R., Dupont, T., and Yserentant, H., The hierarchical basis multigrid method, *Numer. Math.*, v. 52, 427, 1988.

10. Bramble, J.H., Ewing, R.E., Pasciak, J.E., and Schatz, A.H., A preconditioning technique for the efficient solution of problems with local grid refinements, *Comp. Meth. Appl. Mech. Engrng.*, v. 67, 149, 1988.

11. Bramble, J.H., and Pasciak J.E., Preconditioned iterative methods for nonselfadjoint or indefinite elliptic boundary value problems, in *Unification of Finite Element Methods*, North Holland, Amsterdam, 167, 1984.

12. Bramble, J.H., Pasciak, J.E., and Schatz, A.H., The construction of preconditioners for elliptic problems by substructuring, IV, *Math. Comp.*, v. 53, 1, 1989.

13. Bramble, J.H., Pasciak, J.E., and Xu, J., Parallel multilevel preconditioners, *Math. Comp.*, v. 55, 1, 1990.

14. Concus, P., and Golub, G.H., Use of fast direct methods for the efficient numerical solution of nonseparable elliptic equations, *SIAM Jour. Numer. Anal.*, v. 10, 1103, 1973.

15. Dryja, M., A capacitance matrix method for Dirichlet problem on polygonal region, *Numer. Math.*, v. 39, 51, 1982.

16. Dryja, M., and Widlund, O., Towards a unified theory of domain decomposition algorithms for elliptic problems, in *Third International Symposium on Domain Decomposition Methods for Partial Differential Equations*, Chan, T.F., Glowinski, J., Periaux, J., and Widlund, O.B., Eds., SIAM, Philadelphia, 3, 1989.

17. D'yakonov, E.G., The construction of iterative methods based on the use of spectrally equivalent operators, *Comp. Math. and Math. Phys.*, v. 6, 14, 1966.

18. D'yakonov, E.G., *Difference Methods for Boundary Value Problems.* v. 1, (*Stationary Problems*), Moscow State University, Moscow, 1971 (in Russian).

19. D'yakonov, E.G., On the solution of some elliptic difference equations, *JIMA*, v. 7, 1, 1971.

20. D'yakonov, E.G., On some operator inequalities and their applications, *Sov. Math. Dokl.*, v. 12, 921, 1971.

21. D'yakonov, E.G., On approximate methods for the solution of operator equations, *Sov. Math. Dokl.*, v. 12, 826, 1971.

22. D'yakonov, E.G., About an iterative method for solving discretized elliptic systems, *Compte rendue de l' Académie Bulgare des Sciences*, v. 28, 295, 1975 (in Russian).

23. D'yakonov, E.G., On the triangulations in the finite element and efficient iterative methods, in *Topics in Numerical Analysis. III*, Miller, J.J.H., Ed., Academic Press, London, 103, 1977.

24. D'yakonov, E.G., Asymptotic minimization of computational work in applying of projective-difference methods, in *Variationally-difference Methods in Mathematical Physics*, Marchuk, G.I., Ed., AN SSSR, Novosibirsk, 149, 1978 (in Russian).

25. D'yakonov, E.G., Estimates of computational work for boundary value problems with the Stokes operators, *Izv. Vuzov. Matem.*, v. 254, 46, 1983 (in Russian).

26. D'yakonov, E.G., Effective methods for solving eigenvalue problems with fourth-order elliptic operators, *Sov. Jour. Numer. Anal. Math. Modeling*, v. 1, 59, 1986.

27. D'yakonov, E.G., On iterative methods with saddle operators, *Sov. Math. Dokl.*, v. 35, 166, 1987.

28. D'yakonov, E.G., *Minimization of computational work. Asymptotically Optimal Algorithms for Elliptic Problems*, Nauka, Moscow, 1989 (in Russian).

29. D'yakonov, E.G., On some iterative methods for nonlinear grid systems, in *Computational Processes and Systems*, Marchuk, G.I., Ed., Nauka, Moscow, 95, 1991.

30. D'yakonov, E.G., On some modern approaches to constructing spectrally equivalent grid operators, in *Fourth International Symposium on Domain Decomposition Methods for Partial Differential Equations*, Glowinski, R., Kuznetsov, Yu.A., Meurant, G., Periaux, J., and Widlund, O.B., Eds, SIAM, Philadelphia, 35, 1991.

31. D'yakonov, E.G., On increasing the efficiency of grid methods for solution of elasticity problems, in *Computer Mechanics of Solids*, Moscow, v. 2, 133, 1991 (in Russian).

32. Eisenstat, S.C., Elman, H.C., and Schultz, M.H., Variational iterative methods for nonsymmetric systems of linear equations, *SIAM Jour. Numer. Anal.*, v. 20, 345, 1983.

33. Haase, G., Langer, U., and Meyer, A., Domain decomposition preconditioners with inexact subdomain solvers, *Jour. Numer. Linear Algebra with Applics.*, v. 1, 27, 1991.

34. Kuznetsov, Yu.A., Multigrid domain decomposition methods for elliptic problems, *Comp. Meth. Appl. Mech. Engrng.*, v. 75, 185, 1989.

35. Kuznetsov, Yu.A., Algebraic multigrid domain decomposition methods, *Sov. Jour. Numer. Meth. Math. Modelling*, v. 4, 351, 1989.

36. Langer, V., and Queek, W., Preconditioned Uzawa-type iterative methods for solving mixed finite element equations: Theory — Applications — Software, *Wissenschaftliche Schriften reihe der Technischen Universitat Karl-Marx Stadt*, N. 3, 1987.

37. Mandel, J., and McCormick, S., Iterative solution of elliptic equations with refinement: the two-level case, in *Domain Decomposition Methods for Partial Differential Equations. II*, Chan, T., Glowinski, R., Meurant, G.A., Periaux, J., and Widlund, O., Eds., SIAM, Philadelphia, 81, 1989.

38. Mandel, J., McCormick, S., and Ruge, J., An algebraic theory for multigrid methods for variational problems, *SIAM Jour. Numer. Anal.*, v. 25, 91, 1988.

39. Marchuk, G.I., Kuznetsov Yu.A., and Matsokin, A.M., Fictitious domain and domain decomposition methods, *Sov. Jour. Numer. Anal. Math. Modeling*, v. 1, 3, 1986.

40. Maitre, J.P., and Musy, F., *The Contraction Number of a Class of Two-Level Methods: An Exact Evaluation for Some Finite Element Subspaces and Model Problems*, Lecture Notes in Mathematics, v. 960, 535, Heidelberg: Springer-Verlag, 1982.

41. McCormick, S., Multigrid methods for variational problems: general theory for the V-cycle, *SIAM Jour. Numer. Anal.*, v. 22, 634, 1985.

42. McCormick, S., and Thomas, J., The fast adaptive composite grid (FAC) method for elliptic equations, *Math. Comp.*, v. 46, 439, 1986.

43. Nepomnyaschikh, S.V., Application of domain decomposition to elliptic problems with discontinuous coefficients, in *Fourth International Symposium on Domain Decomposition Methods for Partial Differential Equations*, Glowinski, R., Kuznetsov, Yu.A., Meurant, G., Periaux, J., and Widlund, O.B., Eds, SIAM, Philadelphia, 242, 1991.

44. Samarskii, A.A., and Nikolaev, E.S., *Numerical Methods for Grid Equations, Vol. II: Iterative methods*, Birkhauser, Basel, 1989.

45. Siganevich, G.L., On a variant of fictitious grid domain method, *Comp. Math. and Math. Phys.*, v. 28, 515, 1988 (in Russian).

46. Wachspress, E.L., *Iterative Solution of Elliptic Systems*, Prentice-Hall, New-York, 1966.

47. Gunn, J.E., The solution of elliptic difference equations by semi-explicit iterative techniques, *SIAM Jour. Numer. Anal.*, v. 2, 24, 1965.

48. Kantorovich, L.V., and Akilov, G.F., *Functional Analysis in Normed Spaces*, Pergamon, London, 1964.

49. Trenogin, V.A., *Functional Analysis*, Nauka, Moscow, 1980 (in Russian).

50. Glowinski, R., *Numerical Methods for Nonlinear Variational Problems*, Springer-Verlag, New York et al., 1983.

51. Marchuk, G.I., and Lebedev, V.I., *Numerical Methods in Theory of Neutrons Transport*, Harwood Academic Publishers, Chur, Switzerland, 1986.

52. Hageman, L.A., and Young, D.M., *Applied Iterative Methods*, Academic Press, New York et al., 1981.

53. Mittelman, H.D., and Roose, D., Eds, *Continuation Techniques and Bifurcation Problems*, Birkhauser Verlag, Basel et al., 1990.

54. Allgower, E.L., and George, K., Eds., *Computational Solution of Nonlinear Systems of Equations*, American Mathematical Society, Providence, Rhode Island, 1990.

55. Adams, R.A., *Sobolev Spaces*, Academic Press, New York, 1975.

56. D'yakonov, E.G., On the application of disintegrating difference operators, *Comp. Math. and Math. Phys.*, v. 3, 385, 1963, (in Russian).

57. Kobelkov, G.M., Efficient methods for solving elasticity theory problems, *Sov. Jour. Numer. Anal. Math. Modeling*, v. 6, 361, 1991.

58. Andreev, V.B., Stability of difference elliptic schemes with respect to the Dirichlet boundary conditions, *Comp. Math. and Math. Phys.*, v. 12, 598, 1972 (in Russian).

59. Siganevich, G.L., "Increasing Efficiency of Grid Methods for Multidimensional Elliptic Problems on the Basis of Usage of Fictitious Grid Regions", Ph.D. dissertation, Moscow Aviation Institute, 1992.

60. Ciarlet, P., *The Finite Element Method for Elliptic Problems*, North-Holland, Amsterdam, 1975.

61. Lions, J.L., *Quelques Méthodes dès Résolution des Problèmes aux Limites non Linéaires*, Dunod-Gauthier Villars, Paris, 1969.

How to Solve Stiff Systems
of Differential Equations
by Explicit Methods

V.I. Lebedev

Methods for solving large stiff systems of ordinary differential equations, difference or variational-difference methods for solving nonstationary problems in mathematical physics, and methods for parallelizing algorithms for multiprocessor computers, seen as a complex whole, give rise to yet another discussion of the efficacy of explicit difference schemes permitting, in this situation, obviously and naturally, the parallel mode of computation. This paper discusses explicit schemes with time-variable steps and their efficacy and argues that these schemes permit stable integration of nonstationary problems with a far greater mean time step than do schemes with constant time steps.

Stiff systems are discussed in the papers indicated in references 1–4. Mathematical physics nonstationary problem solution techniques are discussed in reference 5, and methods for parallelizing algorithms for solving mathematical physics problems are considered in reference 6.

Difference schemes have long been under scrutiny; their application was suggested in references 1 and 7–10. These papers, however, left two questions unanswered; without the answers, application of explicit schemes with variable step sizes would be questionable. These questions concern relationships between approximation conditions and stability, and the stability of realization of one computation cycle with variable parameters. This author offers the solution to the second problem in references 11–13. The first problem is the subject of this paper. Problems arising in the application of iterative methods for solution of implicit difference schemes are discussed in references 14–17.

The range of problems targeted in this research can be described as stiff problems basically characterized by most of certain attributes:

0-8493-8947-X/94 /$0.00 + $.50
(c) 1994 by CRC Press, Inc.

(1) They are multi-dimensional; (2) they are dissipative; (3) iterative algorithms for finding solutions under implicit schemes are relatively time-consuming or theoretically unsubstantial; (4) solutions have boundary-layer-type areas, and large areas where the solutions vary asymptotically, linearly with time or attain a steady state. Difference analogues of mixed problems for the heat conductivity equation with fission asymptotically linear in time are examples. The proposed method may certainly be applied to a broader range of problems.

1. PROBLEM STATEMENT AND ASSUMPTIONS

Let $u(t) = \big(u_1(t), \ldots, u_n(t)\big)$, $n > 1$, be a column vector with components dependent on time t. Consider the Cauchy problem: for $t_0 \leq t \leq T$ find u(t) as the solution to the problem

$$\frac{du}{dt} = f(u, t) \tag{1.1}$$

$$u|_{t=t_0} = u_0 \tag{1.2}$$

where $u_0 = (u_{10}, \ldots, u_{n0})$ is the prescribed vector.

Assumption 1.1.   The vector function $f(t) = \big(f_1(t), \ldots, f_n(t)\big)$ satisfies the sufficient smoothness conditions required for the existence and uniqueness of the solution to problems (1.1) and (1.2) and the validity of further argument if problems (1.1) and (1.2) are replaced by their finite-difference counterpart.

Let

$$J(t) = J(u, t) = \left\| \frac{\partial f_i}{\partial u_j} \right\| \tag{1.3}$$

be the Jacobian of $f(u, t)$, where $u(t)$ is the solution to problems (1.1) and (1.2). Equation (1.1) can be associated in the neighborhood of (u_0, t_0) with a linear differential equation,

$$\frac{du}{dt} = J(t_0)(u - u_0) + f(u_0, t), \tag{1.4}$$

which is Equation (1.1) linearized in the neighborhood of (u_0, t_0).

Assumption 1.2.   The solution to Cauchy problems (1.4) and (1.2) is assumed to approximate the solution to problems (1.1) and (1.2) with the prescribed accuracy, $\varepsilon > 0$ for $t_0 \leq t \leq \bar{t}$, where

$0 \leq \bar{t} \leq T$ is a t_0-dependent quantity. Let $\lambda_i(t)$, $i = 1, \ldots, n$ be the eigenvalue of the matrix $J(t)$: $\mathrm{Sp}(J(t))$.

Assumption 1.3. All $\lambda_i(t)$ are assumed to be real and to be in correspondence with a complete set of eigenvectors $\xi_i(t)$, $i = 1, \ldots, n$, and $\mathrm{Sp}(J(t))$ is assumed to provide true information on the qualitative behavior of the solution to Equation (1.1). There exists a t-independent constant, $C_0 > 0$, such that

$$\lambda_i(t) \leq C_0 \quad \text{for } t \in [t_0, T], \qquad i = 1, \ldots, n. \qquad (1.5)$$

Under these conditions, assume that

$$M = M(t) = \max_i \left(-\lambda_i(t) \right) > 0 \qquad (1.6)$$

and
$$S = (T - t_0)M(t) \gg 1. \qquad (1.7)$$

If conditions (1.5)–(1.7) are satisfied, Cauchy problems (1.1) and (1.2) are referred to as stiff. The quantity S is referred to as the stiffness coefficient, and the quantity

$$\mathrm{cou} = \frac{2}{M} \qquad (1.8)$$

is referred to as Courant; also, $0 < \mathrm{cou} \ll 1$.

The nonstationary decreasing fundamental solutions to system (1.4) for $\lambda_i < 0$ attenuate with t at a rate proportional to $-1/\lambda_i$, which is generally referred to as the local time constant of the system. The stiffness of the system is sometimes described by the interval within which the positive local 'time constants' of the system are enclosed.

It is known[2] that the step size, τ, in a simple explicit scheme of accuracy $\mathrm{O}(\tau)$ with a constant time step (the Euler method),

$$u(t_{k+1}) = u(t_k) + \tau f(u(t_k), t_k), \qquad \tau = t_{k+1} - t_k \qquad (1.9)$$

is to satisfy two requirements: approximation with a prescribed accuracy, and stability of computation with Equation (1.9). The stability condition leads to the inequality

$$\tau \leq \mathrm{cou}, \qquad (1.10)$$

which imposes stiff restrictions on time-step sizes in explicit schemes in many problems, specifically in nonstationary problems of mathematical physics.

2.   FORMULATION OF FOUR METHODS INVOLVING EXPLICIT DIFFERENCE SCHEMES WITH TIME-VARIABLE STEPS

Assumption 2.1.   Let it be known that at any point of the segment $[t_0, t_0 + B_0\tau]$, where $\tau > 0$, $t_0 + B_0\tau \leq T$ and the quantity $B_0 \geq 1$ is independent of t_0, the differential equation to be solved can be approximately replaced to a prescribed accuracy with one of the explicit difference schemes (given subsequently), with a time step equal to τ. Assume that

$$g = \frac{\tau}{\text{cou}} \qquad \big(\text{cou} = \text{cou}(t_0)\big).$$

2.1   FIRST METHOD

Let $N = N(g) \geq 1$ be an integral-valued function of g; let $\{h_i\}_i^N, h_i = h_i(g) > 0$, and $\{\gamma_i\}_i^N, \gamma = \gamma_i(g) \geq 0$, be sequences of numbers, and

$$l_{2N} = l_{2N}(g) = 2\sum_1^N h_i; \qquad t_{k+1/2} = t_k + h_{k+1},$$

$$t_{k+1} = t_{k+1/2} + h_{k+1}, \qquad k = 0, \ldots, N-1;$$

$$u_\alpha = u(t_\alpha).$$

Assumption 2.2.   Assume that

$$\gamma_i \leq C_1, \qquad i = 1, \ldots, N, \;\; \tau \leq l_{2N}(g) \leq B_0\tau, \qquad (2.1)$$

where $C_1 > 0$ is a constant independent of g and t_0.

An approximate solution u_k, $k = 1, \ldots, N$, to the Cauchy problems (1.1) and (1.2) at points $t_k \in [t_0, t_0 + B_0\tau]$ can be found using an explicit difference scheme of the form

$$y_{k+1/2} = u_k + h_{k+1}f(u_k, t_k), \qquad t_{k+1/2} = t_k + h_{k+1}$$

$$y_{k+1} = y_{k+1/2} + h_{k+1}f(y_{k+1/2}, t_{k+1/2}), \qquad t_{k+1} = t_{k+1/2} + h_{k+1}$$

$$u_{k+1} = y_{k+1} + \gamma_{k+1}h_{k+1}\big(f(u_k, t_k) - f(y_{k+1/2}, t_{k+1/2})\big),$$

$$k = 0, \ldots, N-1 \quad (2.2)$$

Computations using Equations (2.2) replace t_0 with t_N and vector u_0 with u_N in the Cauchy problems (1.1) and (1.2). Find new values of τ and cou, and use these to find values of $g, N, \{h_i\}_i^N$, and $\{\gamma_i\}_i^N$. Repeat the computations, using Equation (2.2).

A cycle of computations using Equations (2.2) produces the value of u_N at the point $t_N = t_0 + l_{2N}$. The quantity

$$v_{2N} = v_{2N}(g) = \frac{l_{2N}}{2N \, \text{cou}} \qquad (2.3)$$

is the average relative rate of variation of t per computation of the right-hand part of system (1.1) in method (2.2) in relation to its variation in method (1.9) at $\tau = \text{cou}$. If the approximation and stability conditions are satisfied, the greater v_{2N}, we can use the larger time interval to solve this problem with a prescribed number of computations (assuming implicitly that the computations are largely done in right-hand parts of Equation [1.1]).

2.2 SECOND METHOD

Let $\bar{N} = \bar{N}(g) \geq 1$ be an integral-valued function of g; let $\{\bar{h}_i\}_i^{\bar{N}}$, $\bar{h}_i = \bar{h}_i(g) > 0$ be a sequence of numbers, and

$$\bar{l}_{\bar{N}} = \bar{l}_{\bar{N}}(g) = \sum_1^{\bar{N}} \bar{h}_i; \qquad \bar{t}_0 = t, \ \ \bar{t}_{k+1} = \bar{t}_k + \bar{h}_{k+1}, \ \ u_k = u(\bar{t}_k).$$

Assumption 2.3. Assume that

$$\tau \leq \bar{l}_{\bar{N}}(g) \leq B_0 \tau \qquad (2.4)$$

An approximate solution u_k, $k = 1, \ldots, \bar{N}$, to the Cauchy problems (1.1) and (1.2) at points $\bar{t}_k \in [t_0, t_0 + B_0 \tau]$ can be found using an explicit difference scheme of the form

$$u_{k+1} = u_k + \bar{h}_{k+1} f(u_k, \bar{t}_k),$$
$$\bar{t}_{k+1} = \bar{t}_k + \bar{h}_{k+1}, \qquad k = 0, \ldots, \bar{N} - 1. \qquad (2.5)$$

Proceed with computations as described for algorithm (2.2). In this case, the average relative rate is expressed by the quantity

$$\bar{v}_{\bar{N}} = \bar{v}_{\bar{N}}(g) = \frac{\bar{l}_{\bar{N}}}{\bar{N} \, \text{cou}}. \qquad (2.6)$$

2.3 THIRD METHOD

This method involves application of the first method to find an approximate solution to linearized problems (1.4) and (1.2) as an approximate solution to linearized problems (1.1) and (1.2). The solution is sought on the segment $[t_0, t_0 + B_0\tau]$ at $t_0 + B_0\tau \leq \bar{t}$. Then replace (u_0, t_0) with (u_N, t_N) in Equations (1.4) and (1.2) and solve new problems (1.4) and (1.2).

2.4   FOURTH METHOD

This method is similar to the third method, differing only by algorithm (2.5) being used every time to find an approximate solution to problems (1.4) and (1.2).

The Cauchy problems (1.4) and (1.2) are linear problems of the form

$$\frac{du}{dt} = -Au + \phi(t), \qquad u|_{t=t_0} = u_0 \tag{2.7}$$

where $A = -J(t_0)$ is a matrix with constant coefficients, and $\phi(t)$ is a prescribed vector function. Cauchy problems of the type in Equation (2.7) can be of interest themselves; then the third method coincides with the first, and the fourth with the second method.

3.   IMPLICIT SCHEMES FOR LINEAR EQUATIONS

Assume that $p + q = 1$, $p > 0$, $q \geq 0$, $\tau_k = t_{k+1} - t_k$, and approximate problem (2.7) as

$$Bu_{k+1} = \psi_k \tag{3.1}$$

where

$$B = I + p\tau_k A, \quad B_1 = I - q\tau_k A, \quad \psi_k = B_1 u_k + \tau_k(p\phi_{k+1} + q\phi_k),$$

$$0 < \tau_k \leq \tau, \quad k = 0, 1, \ldots$$

Equation (3.1) is solved by iterative methods, as a rule. Examine a range of these methods:

$$u_{k+1} = v_{k+1}^{N_1} \tag{3.2}$$

$$(I + \tau_k C)(v_{k+1}^{n+1} - v_{k+1}^n) = -\omega_{k+1}(Bv_{k+1}^n - \psi_k),$$

$$k = 0, \ldots, N_1 - 1 \tag{3.3}$$

$$v_{k+1}^0 = \delta u_k. \tag{3.4}$$

In this method, N_1 is an integral number assigned a priori, or determined during the iterations; δ is an operator determining the initial approximation, which depends on the solution in the previous step; and

C and $\{w_k\}$ are, respectively, an operator and scalar parameters which accelerate the iteration's convergence.

4. INVESTIGATION OF STABILITY AND APPROXIMATION OF EXPLICIT DIFFERENCE SCHEMES

This section discusses aspects of approximation and stability of the proposed methods for homogeneous ($\phi \equiv 0$) linear Cauchy problem (2.7), with the matrix, A, continuous in t, whose spectrum satisfies the conditions of Assumption 1.3. $v_i = -\lambda_i$ are eigenvalues of the matrix.

Problem (2.7) has the form

$$\frac{du}{dt} = -Au, \qquad u|_{t=t_0} = u_0 \tag{4.1}$$

Expand u_0 in eigenvectors $\{\xi_i\}_i^n$: $u_0 = \sum_1^n d_i\,\xi_i$ to find that the solution to Equation (4.1) can be represented as

$$u(t) = \sum_{v_i>0} d_i\,\xi_i e^{-v_i t} + \sum_{v_i \le 0} d_i\,\xi_i e^{-v_i t}. \tag{4.2}$$

Assumption 4.1. Assume that $v_i > 0$, $i = 1, \ldots, n$, or $d_i \ne 0$ in Equation (4.2) if $v_i \le 0$.

Applying algorithm (2.2) to problem (4.1), obtain the equation for the operator of transition from u_0 to u_N,

$$u_N = Q_{2N}(A)u_0 = \sum_{i=1}^n Q_{2N}(v_i)d_i\,\xi_i, \tag{4.3}$$

where the polynomial, $Q_{2N}(\lambda)$, of degree $2N$ has the form

$$Q_{2N}(\lambda) = \prod_{k=1}^n \left[(1 - h_k\lambda)^2 - \gamma_k h_k^2 \lambda^2\right]. \tag{4.4}$$

Since
$$e^{-l_{2N}\lambda} = 1 - l_{2N}\lambda + \frac{(l_{2N}\lambda)^2}{2} - \ldots \tag{4.5}$$

and
$$Q_{2N}(\lambda) = 1 - l_{2N}\lambda + A_{2N}\lambda^2 + \ldots \tag{4.6}$$

then, comparing Equation (4.2) and Equation (4.3), find that if the inequality Equation (2.1) is valid, the condition for the approximation of problem (4.1) by the difference scheme of Equation (2.2) with accuracy $O(\tau)$ is automatically satisfied. But if $A_{2N} = l_{2N}^2/2$ in Equation (4.5), the order of the approximation is $O(\tau^2)$.

Assume that

$$R_{2N}^i(\lambda) = \prod_{k=1}^{i} \left[(1 - h_k\lambda)^2 - \gamma_k h_k^2 \lambda^2 \right]$$

$$Q_{2N}^i(\lambda) = \prod_{k=k+1}^{n} \left[(1 - h_k\lambda)^2 - \gamma_k h_k^2 \lambda^2 \right] \tag{4.7}$$

$$r_{2N}^i = \max_{\tilde{m} \le \lambda \le M} |R_N^i(\lambda)|, \quad q_{2N}^i = \max_{\tilde{m} \le \lambda \le M} |Q_N^i(\lambda)|, \tag{4.8}$$

where $\tilde{m} = \min_{v_i > 0} v_i$. Algorithm (2.2) is stable and approximates problem (1.1) and (1.2) at least with $O(\tau)$ on the segment $[t_0, t_0 + B_0\tau]$ if the parameters are chosen thereon to satisfy the conditions:

(1) condition (2.1);

(2) $Q_{2N}(0) = 1, |Q_{2N}(\lambda)| < 1$ for $0 < \lambda \le M$; \hfill (4.9)

(3) the quantities r_{2N}^i, q_{2N}^i and $\sum_{i=1}^{N} q_{2N}^i$ are uniformly bounded in $N(g)$ and i;

(4) at $\lambda < 0$ we have $1 < Q_{2N}(\lambda) \le e^{-l_{2N}\lambda}$. \hfill (4.10)

Condition (1) stipulates satisfaction of the predicted approximation condition for application of algorithm (2.2). If condition (2) is satisfied, there will be a decrease in the values of components of the errors which arise in previous applications of algorithm (2.2) and correspond to eigenvectors for $v_i > 0$ in an expansion of the type in Equation (4.2).

Condition (3) stipulates the limits of similar errors arising within the cycle of computation of $y_{1/2}, u_1, \ldots, y_{N-1/2}, u_N$. Condition (4) is necessary to prevent an increase in components of an approximate solution from exceeding the increase in the corresponding components of the accurate solution.

The transition operator from u_0 to $u_{\bar{N}}$ in the second method has the form

$$u_{\bar{N}} = \bar{Q}_{\bar{N}}(A)u_0 = \sum_{i=1}^{n} \bar{Q}_{\bar{N}}(v_i)d_i \, \xi_i, \tag{4.11}$$

where the polynomial of degree $\bar{N}$ is

$$\bar{Q}_{\bar{N}}(\lambda) = \prod_{k=1}^{n} (1 - \bar{h}_k\lambda) \tag{4.12}$$

and its expansion in the neighborhood $\lambda = 0$ has the form

$$\bar{Q}_{\bar{N}}(\lambda) = 1 - \bar{l}_{\bar{N}}\lambda + \bar{A}_{\bar{N}}\lambda^2 + \dots \qquad (4.13)$$

Comparing Equation (4.2) with Equation (4.11) and Equation (4.13) with Equation (4.5), find that if condition (2.4) is satisfied, problem (4.1) is approximated by the difference scheme, Equation (2.5), with an accuracy $O(\tau)$. But if $\bar{A}_{\bar{N}} = l_{\bar{N}}^2/2$, the approximation order is $O(\tau^2)$.

Introduce the notation

$$\bar{r}_{\bar{N}}^i = \max_{\tilde{m} \leq \lambda \leq M} \left| \prod_{k=1}^i (1 - \bar{h}_k\lambda) \right|,$$

$$\bar{q}_{\bar{N}}^i = \max_{\tilde{m} \leq \lambda \leq M} \left| \prod_{k=i+1}^{\bar{N}} (1 - \bar{h}_k\lambda) \right|. \qquad (4.14)$$

Algorithm (2.5) is said to be stable if the function $\bar{N}(g)$ and the parameters $\{\bar{h}_k\}_1^{\bar{N}}$ contained in the algorithm are chosen at all $g > 0$ to satisfy these conditions:

(1) condition (2.4);

(2) $\quad \bar{Q}_{\bar{N}}(0) = 1 \quad$ and $\quad |\bar{Q}_{\bar{N}}(\lambda)| < 1 \quad$ at $\quad 0 < \lambda \leq M;$ (4.15)

(3) the quantities $\bar{r}_{\bar{N}}^i$, $\bar{q}_{\bar{N}}^i$, and $\sum_{i=1}^{\bar{N}} \bar{q}_{\bar{N}}^i$ are uniformly bounded in $\bar{N}(g)$ and i;

(4) at $\lambda < 0$ we have $\quad 1 < \bar{Q}_{\bar{N}}(\lambda) \leq e^{-l_{\bar{N}}\lambda}.$ (4.16)

5. INVESTIGATION OF STABILITY AND APPROXIMATION OF IMPLICIT DIFFERENCE SCHEMES

The transition operator of implicit scheme (3.1) has the form

$$B^{-1}B_1 = (I + p\tau_k A)^{-1}(I - q\tau_k A) \qquad (5.1)$$

However, Equation (3.1) is generally solved by iterative methods. N_1 iterations result in another transition operator. Let us find this operator in the case where an approximate solution to Equation (3.1) is obtained by iterative methods (3.2)–(3.4).

Let $D = (I + \tau C)^{-1}B$, $\tau = \tau_k$, $\phi_k = \phi_{k+1} = 0$; then

$$u_{k+1} = v_{k+1}^{N_1} = R_{N_1}(D)\delta u_k + \big(I - R_{N_1}(D)\big)D^{-1}(I + \tau C)^{-1}B_1 u_k$$

$$= Q_{N_1}u_k \qquad (5.2)$$

where $\quad Q_{N_1} = R_{N_1}(D)\delta + \big(I - R_{N_1}(D)\big)D^{-1}(I + \tau C)^{-1}B_1$

$$= R_{N_1}(D)\delta + \big(I - R_{N_1}(D)\big)B^{-1}B_1 \qquad (5.3)$$

and
$$R_{N_1}(s) = \prod_{k=1}^{N_1} (1 - \omega_k s). \qquad (5.4)$$

Comparing Equation (5.1) and Equation (5.3), find that the operator, Q_{N_1}, differs from $B^{-1}B_1$, and its form depends on $N_1, C, \{\omega_k\}_1^{N_1}, \delta$. In case of an unsatisfactory choice of these operators, approximation may default or instability can appear. Therefore, the parameters $N_1, \{\omega_k\}_1^{N_1}$ and the operators C, δ must be in accordance with each other.

Write equations for operator Q_{N_1}:

if $\delta = 0$,
$$Q_{N_1} = Q_{N_1}^1 = (I - \tau R_{N_1}(D)) B^{-1} B_1; \qquad (5.5)$$

if $\delta = I$,
$$Q_{N_1} = Q_{N_1}^2 = (B_1 + \tau R_{N_1}(D)A) B^{-1}; \qquad (5.6)$$

if $\delta = I - \alpha \tau A$, where α is a scalar,
$$Q_{N_1} = Q_{N_1}^3 = (B_1 + \tau R_{N_1}(D))((1 - \alpha)A - \alpha p \tau A^2) B^{-1}; \qquad (5.7)$$

if $\delta = \alpha(I + \tau C)^{-1} B_1$,
$$Q_{N_1} = Q_{N_1}^4 = [\alpha R_{N_1}(D)$$
$$+ (I - R_{N_1}(D))] D^{-1}(I + \tau C)^{-1} B_1. \qquad (5.8)$$

If $C = 0$ and δ is a scalar, then
$$Q_{N_1} = \tilde{Q}_{N_1}(\lambda), \qquad (5.9)$$

where the polynomial
$$\tilde{Q}_{N_1}(\lambda) = \delta R_{N_1}(s) + (1 - R_{N_1}(s)) s^{-1}(1 - q \tau_k \lambda) \qquad (5.10)$$

$$s = 1 + p \tau \lambda. \qquad (5.11)$$

Transform this expression into
$$\tilde{Q}_{N_1}(\lambda) = 1 + (\delta - 1) R_{N_1}(s) - \tau_k \lambda s^{-1}(1 - R_{N_1}(s)). \qquad (5.12)$$

The approximation conditions imply that
$$\tilde{Q}_{N_1}(\lambda) = 1, \quad \tilde{Q}'_{N_1}(\lambda) = -\tau \qquad (5.13)$$

Computing these values of the polynomial in Equation (5.12) yields
$$\tilde{Q}_{N_1}(0) = 1 + (\delta - 1) R_{N_1}(1),$$

$$\tilde{Q}'_{N_1}(0) = \tau_k\big((\delta - 1)pR'_{N_1}(1) - 1 + R_{N_1}(1)\big). \qquad (5.14)$$

Set the quantities in Equation (5.13) equal to those in Equation (5.14) to obtain two solutions,

$$\delta = 1, \quad R_{N_1}(1) = 0; \qquad (5.15)$$

δ is arbitrary, but

$$R_{N_1}(1) = 0, \quad R'_{N_1}(1) = 0. \qquad (5.16)$$

Notice that conditions (5.15) and (5.16) establish certain relations between the parameters δ and $\{\omega_k\}_1^{N_1}$ of iterative methods (3.2)–(3.4). Conditions (5.15) seem more natural for small τ_k. Examine this case in more minute detail. For these conditions, one of ω_i is bound to be equal to one; assume that this is ω_1. Then the first iteration in method (3.3) gives

$$v_{k+1}^1 = u_k - \tau_k A u_k. \qquad (5.17)$$

Equation (5.17) implies an important conclusion. To succeed in using iterative methods (3.2)–(3.4) with $\delta = 1$, to find a solution to implicit scheme (3.1) with a step size, τ_k, one must make the first step in method (3.3) by employing scheme (5.17) with the same step size, τ_k, or, what amounts to the same thing, employ an iterative method of the type given in Equation (3.3) involving $N_1 - 1$ iterations starting with the initial approximation obtained by explicit scheme (5.17). The conditions $\omega_i = 1$, or the choice of v_{k+1}^1 by Equation (5.17), is thus a necessary condition in algorithms used to find approximate solutions in methods (3.1)–(3.4) and method (5.15).

Assume that at $N_1 \geq 2$

$$R_{N_1}(s) = (1 - s)R_{N_1-1}(s)$$

where

$$R_{N_1-1}(s) = \prod_{k=2}^{N_1}(1 - \omega_k s).$$

Then for $\delta = 1$,

$$\tilde{Q}_{N_1}(\lambda) = 1 - \tau_k \lambda s^{-1}\big(1 + p\tau_k \lambda R_{N_1-1}(s)\big) \qquad (5.18)$$

or

$$R_{N_1-1}(s) = (p\tau\lambda^2)^{-1}\big[(1 - \tilde{Q}_{N_1}(\lambda))(1 + p\tau\lambda) - \tau\lambda)\big]. \qquad (5.19)$$

Conditions (5.17) are obviously applicable only if one is certain that the solution varies only in amplitude with time. Then it may be assumed that $\omega_1 = \omega_2 = 1$, and, for $N_1 \geq 3$ in Equation (3.3),

$$v_{k+1}^2 = u_k - \tau_k A u_k + p\tau_k^2(1 + (\delta - 1)p)A^2 u_k \qquad (5.20)$$

and
$$\tilde{Q}_{N_1}(\lambda) = 1 + (\delta - 1)p^2\tau^2\lambda^2 R_{N_1-2}(s)$$
$$+ \tau\lambda s^{-1}\left(1 - p^2\tau^2\lambda^2 R_{N_1-2}(s)\right),$$

where
$$R_{N_1-2}(s) = \prod_{k=3}^{N_1}(1 - \omega_k s).$$

The stability conditions are formulated for methods (3.1)–(3.4) as constraints imposed on the polynomial, $\tilde{Q}_{N_1}(\lambda)$, similar to conditions (2)–(4) formulated previously for the polynomial, $\bar{Q}_{\bar{N}}(\lambda)$. It is important to note that not any realization of a converging iterative method, which involves a finite number of iterations, leads to a stable computation of a nonstationary problem. Also note that the optimum selection of parameters in method (3.3) providing its best convergence will not be the optimum in the general methods (3.1)–(3.4) and, as will be shown later, there exists such an estimate N^0 from below for N_1 iterations in methods (3.2)–(3.4), dependent on τ_k and cou, that there is no stable method of the type in Equations (3.1)–(3.4) at $N_1 < N^0$. It will be rewarding to examine the stability conditions in the case where Equation (3.1) is solved by an iterative method based on variational principles, for example, by the classical gradient method or the conjugate gradient method. The polynomial induced by the iterations will then be dependent on k (or u_k): $\tilde{Q}_{N_1}(\lambda) = \tilde{Q}_{N_1}(\lambda, k)$, and this polynomial is not actually bound to satisfy the second approximation condition in Equation (5.13). To satisfy this condition, the first iteration should use Equation (5.17). The stability of the method will depend on the behavior of the polynomials $\tilde{Q}_{N_i}(\lambda, k)$ on the spectrum of the problem at different k_1 and k_2 ($k_1 > k_2$).

Notice that at
$$Q_{2N}(\lambda) = \bar{Q}_{\bar{N}}(\lambda) = \tilde{Q}_{N_1}(\lambda) \tag{5.21}$$

methods (2.2), (2.4), and (3.1)–(3.4) applied to solve problem (3.1) are equivalent. Any of these equations may be used in application of the difference scheme of the other two methods, with the values of necessary parameters found beforehand from condition (5.21), if the rounding-off error is disregarded. This provision makes it possible, in particular, to determine the parameters in Equations (3.3) in a reasonable way.

6.   SOME ESTIMATES FOR POLYNOMIALS

Let $\prod_n$ be a range of polynomials of a degree no higher than n and have the form

$$P_n = 1 - lt + \sum_{i=2}^{n} a_i t^i \tag{6.1}$$

where l and a_i $(i = 2, \ldots, n)$ are random real numbers. Then

$$P_n(0) = 1, \quad P'_n(0) = -l. \tag{6.2}$$

Any polynomial of the $\prod_n$ range can be represented as

$$P_n(t) = \prod_{i=1}^{n}(1 - \alpha_i t),$$

where the α_i are quantities inverse to the roots of this polynomial, and

$$l = l_n = \sum_{i=1}^{n} \alpha_i = -P'_n(0). \tag{6.3}$$

Let $0 < m < M$ be given numbers. Examine a polynomial belonging in the $\prod_n$ range and having the form

$$Q_n(t) = \frac{T_n(x)}{T_n(\theta)}, \tag{6.4}$$

where $T_n(x)$ is a Chebyshev polynomial of the first kind of degree n, while

$$x = \frac{M + m - 2t}{M - m} \tag{6.5}$$

and

$$\theta = \frac{M + m}{M - m} > 1. \tag{6.6}$$

Then

$$\max_{m \leq t \leq M} |Q_n(t)| = \eta = T_n^{-1}(\theta) < 1 \tag{6.7}$$

while $Q_n(m) = \eta$ and $\eta \leq |Q_n(t)| < 1$ for $t \in (0, m]$.

Assume that

$$r = \theta + (\theta^2 - 1)^{1/2}, \quad a = \theta^{-1}. \tag{6.8}$$

The equations which relate these quantities and will be applied are

$$\eta = \frac{2r^{-n}}{1 + r^{-2n}}, \quad m = \frac{1 - a}{1 + a} M \tag{6.9}$$

whence

$$r = r(\eta, n) = \left[\eta^{-1} + (\eta^{-2} - 1)^{1/2}\right]^{1/n} \tag{6.10}$$

Compute the quantity l for the polynomial $Q_n(t)$. Use Equations (6.3)–(6.9) to obtain

$$l = -Q'_n(0) = \frac{2n U_{n-1}(\theta)}{(M - m) T_n(\theta)}$$

$$= \frac{2n(1 - \eta^2)^{1/2}}{(M - m)(\theta^2 - 1)^{1/2}} = n\left(\frac{M}{m}\right)^{1/2}(1 - \eta^2)^{1/2}\frac{1}{M},$$

or
$$l = l_n = n\left(\frac{1+a}{1-a}\right)^{1/2}\frac{(1-\eta^2)^{1/2}}{M}.\tag{6.11}$$

Here $U_{n-1}(x)$ is a Chebyshev polynomial of the second kind.

The point $m_0(m_0 < m)$ nearest to m, at which $Q_n(m_0) = \eta$, is equal to

$$m_0 = \frac{1}{2}\left[M + m - (M - m)\cos\left(\frac{\pi}{n}\right)\right].\tag{6.12}$$

Computing m by Equations (6.9) and (6.10) in terms of M and η, obtain

$$m_0 = \left[\sigma(\eta, n) + (1 - \sigma(\eta, n))\sin^2\left(\frac{\pi}{2n}\right)\right]M\tag{6.13}$$

where
$$\sigma(\eta, n) = \left(\frac{r-1}{r+1}\right)^2.\tag{6.14}$$

At $n = 1$, $m_0 = M$, and $m_0 < M$ for $n > 1$. If η monotonically increases and tends toward 1, then at a fixed n

$$r(\eta, n) \to 1, \qquad \sigma(\eta, n) \to 0,\tag{6.15}$$

decreasing monotonically.

In Equation (6.11) express the quantity a in terms of η to obtain

$$l = l_n = l(\eta, n) = y(\eta, n)\frac{n}{M}\tag{6.16}$$

where
$$y(\eta, n) = \frac{r+1}{r-1}\left(1 - \eta^2\right)^{1/2}.\tag{6.17}$$

Computations show that the function $y(\eta, n)$ at $0 < \eta \le 1$ is a continuously increasing function of η, and

$$y(1, n) = 2n.\tag{6.18}$$

Therefore
$$\max_{0<\eta\le1} l(\eta, n) = \frac{2n^2}{M}.\tag{6.19}$$

Lemma 6.1 is valid.

Lemma 6.1. For any $n \ge 1$, $M > 0$ and $0 < \eta \le 1$ among the polynomials of $P_n(t) \in \prod_n$ satisfying the additional condition

$$\max_{m_0 \le t \le M}|P_n(t)| \le \eta\tag{6.20}$$

where m_0 is determined by Equation (6.13), the value of the coefficient l is the greatest in the polynomial $Q_n(t)$ of the form in

Equations (6.4) and (6.5) in the representation of Equation (4.1), equal to l_n and determined by Equation (6.11). Such a polynomial is unique.

Proof. Pursue the proof of Lemma 6.1 by contradiction. First, notice that the polynomial $Q_n(t)$ satisfies the hypotheses of Lemma 6.1, while

$$Q_n(t_i) = (-1)^i \eta$$

at the points

$$t_i = \frac{1}{2} \left[M + m - (M - m) \cos\left(\frac{\pi i}{n}\right) \right], \qquad i = 0, \cdots, n.$$

Assume that there exists a polynomial, $P_n(t) \in \prod_n$, in the form of Equation (6.1) different from the polynomial $Q_n(t)$ and satisfying condition (6.20), for which $l \geq l_n$. Form the polynomial

$$v(t) = Q_n(t) - P_n(t).$$

Condition (6.20) implies that for $i = 0, \cdots, n$

$$v(t_i) \leq 0 \quad \text{if } i \text{ is odd,}$$

$$v(t_i) \geq 0 \quad \text{if } i \text{ is even,}$$

and also

$$v(0) = 0, \quad v'(0) = l_n - l \leq 0.$$

The conditions imply that the polynomial, $v(t)$, of degree n has more than $n + 1$ zeros, that is, $Q_n(t) = P_n(t)$. This contradiction proves the validity of the lemma. The proof of the existence of $n + 1$ zeros in the polynomial $v(t)$ falls into two versions. (1) Let all inequalities of the lemma be strict; then, according to the Rolle theorem, the polynomial $v(t)$ has zero in every interval (t_i, t_{i+1}), $i = 1, \ldots, n-1$, and the number of intervals is equal to $n - 1$. Besides, $v(0) = 0$ and/or $v'(0) = 0$, that is, the point $t = 0$ is a double root, or there exists another root within the interval $(0, t_1)$ at $v'(0) < 0$. (2) In the common case, the roots are computed similarly with regard to the fact that if $v(t_i) = 0$, $1 \leq i \leq n-1$, for a certain i the point t_i is a double root for $v(t)$.

At $\eta = 1$ and with Equation (6.19) taken into account, the lemma implies Corollary 6.1, which is a paraphrased version of the well-known Markov theorem[18] as regards estimating a derivative of a polynomial of degree n.

Corollary 6.1. The polynomial

$$Q_n(t) = T_n \left(\frac{1 - 2t}{M} \right) = (-1)^n T_{2n} \left(\left(\frac{t}{M} \right)^{1/2} \right), \qquad (6.21)$$

among all the polynomials of the $\prod_n$ range satisfying the condition

$$\max_{0 \le t \le M} |P_n(t)| \le 1, \qquad (6.22)$$

has the least derivative equal to $-2n^2/M$ at the point $t = 0$.

Corollary 6.2. At any n and with $M > 0$ there exists no polynomial of the $\prod_n$ range satisfying condition (6.21), whose derivative would be less than $-2n^2/M$ at the point $t = 0$.

Corollary 6.3. Among polynomials of a degree below n, which in the neighborhood of the point $t = 1$ are of the form $P_n(t) = 1 - t + O(t^2)$, there exists a polynomial for which the quantity r in the inequality $|P_n(t)| \le 1$ for $0 \le t \le r$ is the largest one. The solution which would realize the extreme value of r is the polynomial

$$Q_n(t) = (-1)^n T_{2n} \left(\frac{(t/2)^{1/2}}{n} \right), \qquad (6.23)$$

while $r = 2n^2$.

Corollaries 6.1, 6.2, and 6.3 are equivalent. Lemma 6.1 makes the result of Markov's work[18] more precise due to a specific location of the point at which the derivative is computed. The condition $m_0 \le t \le M$ in inequality Equation (6.20) is weaker than the condition $0 \le t \le M$ suggested in reference 6, since $0 < m_0 \le M$.

Corollary 6.4. At any $M > 0$ and l there is no polynomial of degree n of the type given in Equation (6.1), satisfying condition (6.21) at $n < (Ml/2)^{1/2}$.

Lemma 6.2. At any $n > 1$ and $\theta > 1$ the function $Q_n(x)$ (see Equation [6.4]) and $\exp\left(nU_{n-1}(\theta)/T_n(\theta)(x - \theta)\right)$ are tangent to each other at the point $x = \theta$, and

$$0 < Q_n(x) \le \exp \left(\frac{nU_{n-1}(\theta)}{T_n(\theta)} (x - \theta) \right) \qquad (6.24)$$

where $x > \cos\left(\pi/(2n)\right)$.

The equality of the values and their derivatives of these two functions at the point $x = \theta$ may be verified immediately; the left-hand inequality is obvious. Examine the relations of these functions,

$$v(x) = Q_n(x) \exp\left(\frac{-nU_{n-1}(\theta)}{T_n(\theta)}(x - \theta)\right).$$

Lemma 6.2 will be proved if we show that $v'_n(x) > 0$ at $\cos\left(\pi/(2n)\right) < x < 0$, and $v'_n(x) < 0$ for $\theta < x$. Indeed,

$$v'_n(x) = \frac{-n}{T_n(\theta)} \exp\left(\frac{-nU_{n-1}(\theta)}{T_n(\theta)}(x - \theta)\right) F(x, \theta)$$

where[9] $F(x, \theta) = T_n(\theta)U_{n-1}(x) - T_n(x)U_{n-1}(\theta)$

$$= 2(\theta - x)\left(\sum_{j=1}^{n-1} U_j(\theta)U_j(x) + \frac{1}{2}U_{n-1}(\theta)U_{n-1}(x)\right),$$

(if $n = 1$, the sum is lacking in this expression) and we find that sign $F(x, \theta) = \text{sign}\,(\theta - x)$.

Similar aspects are of interest for polynomials of the form

$$P_n(t) = 1 - lt + \frac{1}{2}(lt)^2 + \sum_{i=3}^{n} a_i t^i. \tag{6.25}$$

Explicit difference schemes generated by polynomials of this type approximate a differential problem with accuracy $O(\tau^2)$. Examine several polynomials of this type; it is sufficient to present them at $l = 1$ and $n = 3$,

$$\bar{P}_3(t) = 1 - t + \frac{t^2}{2} - \frac{t^3}{6} \quad \text{or} \quad P_3(t) = 1 - t + \frac{t^2}{2} - \frac{t^3}{16}. \tag{6.26}$$

The latter polynomial has one real root, $t_1 = 5.6786$, and

$$\max_{0 \le t \le 6.2608} |P_3(t)| \le 1.$$

Rewrite Equation (6.25) as

$$P_n(t) = \prod_{i=1}^{n}(1 - \bar{\gamma}_i t),$$

and denote

$$\bar{\sigma}_1 = \sum_{i=1}^{n} \bar{\gamma}_i, \qquad \bar{\sigma}_1 = \Sigma\bar{\gamma}_i\bar{\gamma}_k.$$

Then $l = \sigma_1$, $l^2/2 = \bar{\sigma}_2$, that is,

$$2\bar{\sigma}_2 = \bar{\sigma}_1^2, \quad \text{or} \quad \sum_{i=1}^{n} \bar{\gamma}_i^2 = 0. \tag{6.27}$$

This means that polynomial Equation (6.25) has at least one pair of complex-adjoined roots, thus can be represented as

$$P_n(t) = \left(1 - \bar{a}t + \bar{b}t^2\right) \prod_{i=1}^{n-2}(1 - \bar{\gamma}_i t) \tag{6.28}$$

where $\bar{a} = \bar{\gamma}_n + \bar{\gamma}_{n-1}$, $\bar{b} = \bar{\gamma}_n \bar{\gamma}_{n-1}$, with $(1 - \bar{a}t + \bar{b}t^2)$ always positive.

Assume that $0 < m < M$. Now describe the algorithm of the construction of the polynomial $P_n(t)$, satisfying the following conditions:

(1) Equation (6.27);
(2) $|P_n(t)| < 1$ if $0 < t \leq M$;
(3) $|P_n(t)| \leq \eta$ if $m \leq t \leq M$, where $0 < \eta < 1$, $m = m(M, n, \eta)$.

First substitute x:

$$x = \frac{M + m - 2t}{M - m},$$

assuming that $\delta = (M + m)/(M - m)$. Then $\delta > 1$, $-1 \leq x \leq 1$ if $m \leq t \leq M$, and polynomial $P_n(t)$ will be

$$Q_n(x) = q(x)S_{n-2}(x), \tag{6.29}$$

where

$$q(x) = 1 - a(\delta - x) + b(\delta - x)^2, \tag{6.30}$$

$$S_{n-2}(x) = \prod_{i=1}^{n-2}\left(1 - \gamma_i(\delta - x)\right). \tag{6.31}$$

Let λ_i be the roots of $Q_n(x)$; then $\gamma_i = (\delta - \lambda_i)^{-1}$, $a = \gamma_n + \gamma_{n-1}$, $b = \gamma_n \gamma_{n-1}$, and condition (6.27) will be

$$a^2 - 2b + \sum_{i=1}^{n-2}(\delta - \lambda_i)^{-2} = 0. \tag{6.32}$$

Let us formulate a problem.

Problem 1. Find polynomial $Q_n(x)$ of degree n of the type in Equation (6.29), the least deviating from 0 on the segment $[-1, 1]$ and satisfying conditions

$$Q_n''(\delta) = \left(Q_n'(\delta)\right)^2, \tag{6.33}$$

$$Q_N(\xi) = \max_{-1 \leq t \leq 1} |Q_n(x)|, \tag{6.34}$$

where $1 < \xi < \delta$ is the point of local maximum of $Q_n(t)$ in $(1, \delta)$.[†]
To solve this problem examine Problem 2.

Problem 2. Find for given polynomial $q(x) > 0$ as in Equation (6.30) the polynomial $Q_n(x)$ of degree n of the type in Equation (6.29), the least deviating from 0 on the segment $[-1, 1]$.

Problem 2 can be solved with sufficient accuracy using the asymptotic methods developed by S.N. Bernstein.[19,20] Assuming that $\lambda_k = \cos\theta_k$, determine θ_k as the roots of a system of equations,

$$(n - 2)\theta_k = \frac{(2k - 1)}{2}\pi - \psi(\theta_k), \tag{6.35}$$

where
$$\psi(\theta) = \arctan \Phi(\theta) - 2\theta, \qquad \Phi(\theta) = \frac{\Phi_1(\theta)}{\Phi_2(\theta)}, \tag{6.36}$$

$$\Phi_1(\theta) = \left[(\lambda_1^2 - 1)^{1/2}(\lambda_2\cos\theta - 1) + (\lambda_2^2 - 1)^{1/2}(\lambda_1\cos\theta - 1)\right]\sin\theta,$$

$$\Phi_2(\theta) = \left[(\lambda_1\cos\theta - 1)(\lambda_2\cos\theta - 1) - (\lambda_1^2 - 1)^{1/2}(\lambda_2^2 - 1)^{1/2}\right]\sin^2\theta.$$

The positive part of arctan passing through 0 is used in these equations.

The solution of system (6.35) for assigned a, b is found by the iterative method,

$$(n - 2)\theta_k^{i+1} = \frac{(2k - 1)}{2}\pi - \psi(\theta_k^i), \qquad k = 1, \ldots, n - 2, \ i = 0, 1, \ldots \tag{6.37}$$

at assigned θ_k^0.

In order to solve Problem 1, construct an outer cycle of iteration: after the cycle of Equation (6.37), determine new values of the parameters a, b in Equation (6.30) from conditions (6.34) and (6.35). Then solve system (6.36) by method (6.38) again.

A.A. Medovikov calculated the parameters of polynomial Equation (6.24) of up to the 81st order at $M = 1$ and m selected such that η is equal to 0.97. The parameters of several polynomials at

$$n = 27:$$

$l = 593.26;$	$a = 437.8552;$		$b = 98015.75;$	
γ_i : 55.72641 ;	26.29250 ;	15.55653 ;	10.37757 ;	7.470329;
5.672551;	4.482981;	3.655411;	3.057079;	2.611170;
2.270698;	2.005603;	1.795937;	1.628043;	1.492332;
1.381920;	1.291779;	1.218181;	1.158337;	1.110142;
1.072011;	1.042752;	1.021517;	1.007676;	1.000849.

[†]The solution of Problem 1 will be the Zolotarev polynomial.

Examine now a case of a pure imaginary spectrum. Let the spectrum of matrix A lay within the segment $[-iM, iM]$. Represent the polynomial $P_n(t) \in \prod_n$ as

$$P_n(t) = R_n(-t^2) - l\,t\,Q_n(-t^2) \tag{6.38}$$

where

$$R_n(\lambda) = \prod_{k=1}^{n_0}(1 - \beta_k\lambda),$$

$$Q_n(\lambda) = \prod_{k=1}^{n_1}(1 - \gamma_k\lambda), \tag{6.39}$$

$$n_0 = \left[\frac{n}{2}\right], \qquad n_1 = \left[\frac{n-1}{2}\right].$$

Conditions (6.2) hold for this polynomial if $t = 0$. Now calculate the square of the modulus of the polynomial of Equation (6.38) for pure imaginary values of t: assume that $t = ib$ (b being a real number); then, if $-t^2 = z$, where $z = b^2$ and $z \geq 0$, obtain

$$|P_n(t)|^2 = R_n^2(z) + l^2 z Q_n^2(z).$$

The stability condition is

$$R_n^2(z) + l^2 z Q_n^2(z) \leq 1 \tag{6.40}$$

if $0 \leq z \leq M$. The approximation conditions in the neighborhood of 0 will be: the polynomials $R_n(\beta^2)$, $lbQ_n(\beta^2)$ must be approximations of the functions $\cos l\beta$, $\sin l\beta$, and $|P_n(t)|^2$ if $t \in [-iM, iM]$. Examine first the case where n is odd at $n = 2m + 1$, $m \geq 1$, and

$$Q_{2m+1}(t) = T_m(1 + 2t^2) - 2t(1 + t^2)U_{m-1}(1 + 2t^2).$$

For this,

$$Q_{2m+1}(0) = 1, \quad Q_{2m+1}(t) = -2m,$$

$$Q_{2m+1}(t) = 1 - 2mt + \frac{(2mt)^2}{2} - \cdots, \tag{6.41}$$

that is, $Q_{2m+1}(\tau/(2m))$ approximates e^{-t} with an accuracy $O(t^3)$. At $-t^2 = z(0 \leq z \leq 1)$,

$$|Q_{2m+1}(t)|^2 = T_m^2(1 - 2z) + 4z(1 - z)^2 U_{m-1}^2(1 - 2z),$$

or, after substituting $1 - 2z = x(-1 \leq x \leq 1)$, obtain

$$|Q_{2m+1}(t)|^2 = T_m^2(x) + (1 - x)(1 + x)^2\frac{U_{m-1}^2(x)}{2}$$

$$= 1 - (1 - x)^2(1 + x)\frac{U_{m-1}^2(x)}{2} \leq 1$$

if $t \in [-i, i]$. Therefore, the polynomial

$$P_{2m+1}(t) = Q_{2m+1}\left(\frac{t}{M}\right) \tag{6.42}$$

satisfies conditions (6.40); for this,

$$P'_{2m+1}(0) = \frac{-2m}{M}, \tag{6.43}$$

and, since $U_{m-1}(1) = m$, $l = 2m/M$. Then

$$v_{2m+1} = \frac{l}{2m+1} = \frac{1}{2+m^{-1}}\left(\frac{2}{M}\right) \tag{6.44}$$

So polynomial Equation (6.42) is the optimum.[10] The polynomial of degree four,

$$Q_4(t) = 1 - t + \frac{t^2}{2} - \frac{t^3}{6} + \frac{t^4}{24}, \tag{6.45}$$

approximating the function e^{-t} with an accuracy of $O(t^5)$, is the optimum polynomial, that is, a polynomial with the largest stability segment on an imaginary axis.[10] Equation (6.40) will have the form

$$\left(1 - \frac{x}{2} + \frac{z^2}{24}\right)^2 + z\left(1 - \frac{z}{6}\right)^2 = 1 - \left(1 - \frac{z}{8}\right)\frac{z^3}{72} \leq 1,$$

which is valid if $0 \leq z \leq 8$. Therefore, we can take

$$P_4(t) = Q_4\left(2(2)^{1/2}\frac{t}{M}\right), \tag{6.46}$$

and then

$$v_4 = \frac{(2)^{1/2}}{4}\left(\frac{2}{M}\right). \tag{6.47}$$

The stability boundary of $Q_4(t)$ continues to $t = 2.7853$ along the real axis. The polynomial $Q_4(t)$ has only complex roots. Use the Lin method to represent $Q_4(t)$ as

$$Q_4(t) = (1 - a_1 t + b_1 t^2)(1 - a_2 t + b_2 t^2), \tag{6.48}$$

where

$a_1 = 0.914746687$;
$a_2 = 0.085253313005$;
$b_1 = 0.2644626148$;
$b_2 = 0.1575521996$.

7. DETERMINATION OF PARAMETERS OF DIFFERENCE EQUATIONS

Parameters of methods (2.2), (2.4), and (3.1)–(3.4) can be found by a transition operator determined by the polynomial $Q_n(\lambda)$ for $n = 2N$, $\bar{N}$, N_1, respectively (in iterations (3.4) the parameters ω_i, $i = 2,\ldots,N_1$, are found from Equation [5.19]). Considering the results of Section 6, the polynomial $Q_n(\lambda)$ for $n > 1$ should be assumed in the form of Equations (6.4) and (6.5) for $t = \lambda$, where $m > 0$ is the parameter to be determined later. Take $\bar{h}_1 = \tau$ in Equation (2.4) at $n = 1$.

Assume that $n \geq 2$. Fixing the quantity $0 < \eta \leq 1$, Equation (6.16) yields

$$l_n = \frac{1}{2} y(\eta, n) n\text{cou}. \tag{7.1}$$

With regard to inequality Equations (2.1) and (2.4), it is necessary that quantities η and n be chosen so that $\tau \leq 1/2y(\eta, n)n\text{cou}$, or

$$g \leq \frac{1}{2} y(\eta, n) n \leq B_0 g \tag{7.2}$$

where $g = \tau/\text{cou}$.

In the extreme case (where $\eta = 1$), the inequalities in Equation (7.2) will be

$$g \leq n^2 \leq B_0 g. \tag{7.3}$$

These inequalities determine a set of permissible values of n applicable without violating the approximation conditions (2.1) and (2.4). Assume n as an integral-valued function of g: $n = n(g)$.

Assume, still as a common case, that $g > 1$ in Equations (7.2) and (7.3), because otherwise $h_1 = \tau$, $\gamma_1 = -1$, and $N = 1$ in method (2.2); $\bar{h}_1 = \tau$ and $\bar{N} = 1$ in method (2.5); and $\tau_k = \tau$ and $N_1 = 1$ in methods (3.1)–(3.4).

Assume that $n \geq 2$ and μ_1 is the minimum root of $Q_n(t)$,

$$\mu_1 = \frac{1}{2}\left(M + m - (M - m)\tilde{\beta}_n\right), \qquad \tilde{\beta}_n = \cos\left(\frac{\pi}{2n}\right).$$

Since τ is a permissible step in the sense of accuracy in scheme (1.9), and method (2.5) is equal to μ_1^{-1}, satisfy the inequalities

$$\mu_1^{-1} \leq 2\sin^2\left(\frac{\pi}{4n(g)}\right)^{-1}, \qquad \mu_1^{-1} \leq \tau. \tag{7.4}$$

To rewrite Equation (7.4) in a more convenient form, assume that

$$h^0 = \frac{2}{M + m} = \frac{1}{2}(1 + a)\text{cou},$$

$$\tilde{G}(g) = \min\left\{g, \left(2\sin^2\left(\frac{\pi}{4n(g)}\right)^{-1}\right)\right\} \tag{7.5}$$

and $G(g)$ is a monotonically increasing positive function defined on $(1, \infty)$ and satisfying the inequalities

$$1 \le G(g) \le \tilde{G}(g) \tag{7.6}$$

and $G(g) \to \infty$ for $g \to \infty$. Then rewrite Equation (7.4) as

$$\mu_1^{-1} = \frac{1}{2}\,(1+a)\,\frac{\text{cou}}{1 - a\tilde{\beta}_n} \le \tilde{G}(g)\,\text{cou}$$

and establish the relation between the quantities a and g,

$$\frac{1}{2}\,\frac{1+a}{1 - a\tilde{\beta}_n} = G(g).$$

Then,

$$a = \frac{2G(g) - 1}{1 + 2\tilde{\beta}_n G(g)} \tag{7.7}$$

and conditions (5.4) imply that $0 < a \le 1$.

At $n = \bar{N}$ satisfying the inequality Equation (7.2), the parameters $\bar{h}_k$ of method (2.5) are expressed by quantities inverse to the roots of the polynomial $\bar{Q}_{\bar{N}}(t)$, in which the quantity m is determined by a prescribed g by Equations (6.9) and (7.6); and

$$\bar{h}_k = \frac{h^0}{1 - a\tilde{\beta}_k}, \qquad k = 1, \ldots, \bar{N} \tag{7.8}$$

where h^0 and a are defined by Equations (7.5) and (7.7), while

$$\bar{\beta}_k = \cos\left(\frac{2j_k - 1}{2\bar{N}\pi}\right), \qquad k = 1, \ldots, \bar{N} \tag{7.9}$$

$1 \le j_k \le \bar{N}$, $j_k \ne j_i$, for $k \ne i$ are roots of the polynomial $T_{\bar{N}}(x)$, made more orderly in a specific way through the transposition $\kappa_{\bar{N}} = (j_1, \ldots, j_{\bar{N}})$ to stabilize computations within Equation (2.5).[11–13] Then

$$\bar{l}_{\bar{N}} = \frac{1}{2}\,\bar{N}\left(\frac{1+a}{1-a}\right)^{1/2}(1 - \eta^2)^{1/2}\,\text{cou} \tag{7.10}$$

and

$$\bar{v}_{\bar{N}} = \frac{1}{2}\left(\frac{1+a}{1-a}\right)^{1/2}(1 - \eta^2)^{1/2} \tag{7.11}$$

$$\eta = T_{\bar{N}}^{-1}(a^{-1}). \tag{7.12}$$

Conditions (7.6) and Equations (7.5) and (7.8) imply that at $\tilde{h} = \max_i \bar{h}_i$

$$\tilde{h} = G(g)\,\text{cou} = \frac{G(g)}{g}\,\tau \le \tau, \tag{7.13}$$

and if $G(g) = g$ for some g, then

$$\tilde{h} = \tau. \tag{7.14}$$

Compute $A = \lim_{g \to \infty}(\bar{l}_{\bar{N}}/\tilde{h})$. At $g \to \infty$, there is $\bar{N} \to \infty$, $a \to 1$, and $\eta \to 1$,

$$\tilde{h} \sim \left(2\sin^2\left(\frac{\pi}{4\bar{N}}\right)\right)^{-1} \text{cou}.$$

Taking into account Equation (6.19), find

$$A = \lim_{\bar{N} \to \infty} 2\bar{N}^2 \sin^2\left(\frac{\pi}{4\bar{N}}\right) = \frac{\tau i^2}{8}. \tag{7.15}$$

Equation (7.15) proves the existence of methods (2.5) for sufficiently small values $1 - \eta > 0$, for which B_0 may be assumed equal, for example, to two in inequality Equation (2.1). On the other hand, it is obvious that the ratio of the minimum step to the maximum, equal to

$$\frac{1 - a\tilde{\beta}_{\bar{N}}}{1 + a\tilde{\beta}_{\bar{N}}} = \frac{1 + \tilde{\beta}_{\bar{N}}}{2\tilde{\beta}_{\bar{N}}G(g)}, \tag{7.16}$$

is small at $g \to \infty$. In the extreme case where $\eta = 1$ and $\bar{N} \to \infty$, the ratio decreases by the formula $\text{tg}^2\left(\pi/(4\bar{N})\right)$, while the step size varies from $\text{cou}/\left[2\cos^2\left(\pi/(4\bar{N})\right)\right]$ to $\text{cou}/\left[2\sin^2\left(\pi/(4\bar{N})\right)\right] \sim \tau$.

Method (2.2), the parameters for which we intend to determine, has better characteristics in this respect. If $2N = \bar{N}$, the corresponding ratio in the method is four times as asymptotically large as Equation (7.16) at $\eta = 1$ and $N \to \infty$, due to the fact that the minimum step size is approximately twice as large in Equation (2.2) as compared with method (2.5), while the maximum step size is half as large. The grid will thus be more even in t in method (2.2).

Determine parameters h_i and γ_i $(i = 1, \ldots, N)$ of method (2.2) by comparing polynomial Equations (4.4), (6.4), and (6.5) at $n = 2N$ and $t = \lambda$. Represent the latter polynomial as

$$\prod_{k=1}^{2N}\left(1 - \frac{h^0\lambda}{1 - a\bar{\beta}_k}\right) \tag{7.17}$$

where h^0 and a are determined by Equations (7.5) and (7.7), and $\bar{\beta}_k > 0$ in Equation (7.9) will be associated with a similar expression, with $-\bar{\beta}_k$ in Equation (7.17). Multiply the expressions within these parentheses by themselves and instead of Equation (7.17) obtain

$$\prod_{k=1}^{N}\left[\left(1 - \frac{h^0\lambda}{1 - \alpha\beta_k}\right)^2 - \alpha\beta_k h_k^2\lambda^2\right] \tag{7.18}$$

where
$$\alpha = a^2, \quad h_k = \frac{h^0}{1 - \alpha\beta_k}, \quad \beta_k = \bar{\beta}_k^2 \tag{7.19}$$

and β_k are positive square roots of $T_{2N}(x)$. Comparing Equations (7.18) and (4.4), choose in method (2.2) parameters equal to

$$h_k = \frac{h^0}{1 - \alpha \beta_k}, \qquad \gamma_k = \alpha \beta_k \qquad (7.20)$$

$$\beta_k = \cos^2\left(\frac{2j_k - 1}{4N\pi}\right), \qquad k = 1, \ldots, N \qquad (7.21)$$

where j_k is determined by the transposition $k_N = (j_1, \ldots, j_N)$ which provides numerical stability within Equation (2.2). Then

$$l_{2N} = N\left(\frac{1 + \alpha^{1/2}}{1 - \alpha^{1/2}}\right)^{1/2}(1 - \eta^2)^{1/2}\text{cou} \qquad (7.22)$$

$$v_{2N} = \frac{1}{2}\left(\frac{1 + \alpha^{1/2}}{1 - \alpha^{1/2}}\right)^{1/2}(1 - \eta^2)^{1/2} \qquad (7.23)$$

$$\eta = T_{2N}^{-1}(\alpha^{-1/2}). \qquad (7.24)$$

Then
$$h_k > \frac{1}{2}(1 + \alpha^{1/2})\text{cou}, \qquad k = 1, \ldots, N$$

and
$$\max_k h_k \leq \frac{1}{2}\left[\tau + \frac{(1 + \alpha^{1/2})\text{cou}}{2\left(1 + \alpha \cos\left(\frac{\pi}{4N}\right)\right)}\right].$$

Also, h_k and γ_k are monotonically increasing functions of α.

The formulas for l_n and v_n at $n = \bar{N} = 2N$ in methods (2.5) and (4.4) coincide with the parameters chosen in this section. Notice that if $\bar{N}(2N)$ is fixed in Equations (7.10)–(7.12) and (7.22)–(7.24), $\bar{l}_{\bar{N}}(a)$, $\bar{v}_{\bar{N}}(a)$, and $\eta(a)$ [$l_{2N}(\alpha)$, $v_{2N}(\alpha)$, and $\eta(\alpha)$] are monotonically increasing functions of a at $a \in [0, 1]$ (of α for $\alpha \in [0, 1]$, respectively), and their extreme values at $a = 1$ ($\alpha = 1$) are

$$l_n = n^2\text{cou}, \qquad v_n = n, \qquad \eta = 1 \qquad (7.25)$$

at $n = \bar{N}, 2N$.

Notice that algorithms (2.2) and (7.20)–(7.21), and (2.5) and Equations (7.8)–(7.9) are also stable if a part of the spectrum of the operator, A, with positive real numbers is within an ellipse drawn through the origin of the coordinates and having the foci at the points M and m.[24] Relations in Equation (7.25) estimate from above the gain (no more than n times) of algorithms (2.2) and (2.5) at variable step sizes, as compared with algorithm (1.9) at $\tau = \text{cou}$.

On the other hand, l_n in Equation (7.25) with Equation (7.3) taken into account can be rewritten as

$$l_n = n^2 \frac{\tau}{g} \leq B_0 \tau \tag{7.26}$$

This formula shows a potential loss (no more than $O(n)$ times) of algorithms (2.2) and (2.5) in solving linear problems of the type in Equation (2.7) as compared with a hypothetically ideal stable algorithm in which $O(n)$ computations of the right-hand part of Equation (2.7) are carried out in n steps equal to τ.

Examine now methods (3.1)–(3.4) under conditions (5.15). In this case, assuming $\tilde{Q}_{N_1}(\lambda) = Q_{N_1}(\lambda)$ by Equations (6.4) and (6.5), determine by Equation (5.19) the polynomial $R_{N_1-1}(s)$, where $s = 1 + p\tau_k\lambda$. Find the required parameters ω_i, $i = 2, \ldots, N$, in Equation (3.3) ($\omega_1 = 1$) as quantities inverse to the roots of the equation

$$R_{N_1-1}(s) = 0. \tag{7.27}$$

To ensure the numerical stability in the iteration cycle Equation (3.3), this set of parameters is likely to require a certain algorithm for mixing them. The roots may be complex. If this method is used, recall Corollary 6.4, which implies

> **Corollary 7.1.** If conditions (5.13) and (5.15) are satisfied, whatever polynomial $\tilde{Q}_{N_1}$ of degree N_1 is selected, there is no stable algorithm of Equations (3.1)–(3.4) with parameters determined by the polynomial $\tilde{Q}_{N_1}(\lambda)$ if
>
> $$N_1 < N^0 = \left(\frac{\tau_k}{\text{cou}} \right)^{1/2}. \tag{7.28}$$

If Equation (7.28), Equation (5.15), and the first condition in Equation (5.13) are satisfied, and for any $\tilde{Q}_{N_1}(\lambda)$ satisfying the stability condition $|\tilde{Q}_N(\lambda)| < 1$ for $0 < \lambda \leq M$, algorithms (3.1)–(3.4) with parameters determined by the polynomial $\tilde{Q}_{N_1}(\lambda)$ will necessarily violate the approximation conditions, provided inequality Equation (7.28) holds.

A similar analysis of implicit schemes was discussed in references 14, 15, 16, and 17. Since the resultant polynomial, $\tilde{Q}_{N_1}(\lambda)$, is selected as an extreme in our approach (see Corollary 6.4), condition (7.28) is weaker than similar conditions given in these references 14–17. A thorough analysis is required as usual to justify algorithms suggested for selection of parameters of methods (2.2), (2.4), and (3.1)–(3.4) for general nonlinear Cauchy problems (1.1) and (1.2). If the requirements formulated in Assumptions 1.1 and 1.2 are satisfied, justified algorithms will be represented by modifications of these methods applied to a linear equation of the type given in Equation (1.4) (see the third and fourth methods in Section 1).

Compare the efficacy of optimum explicit method (2.2) and implicit methods (3.1) and (3.4) used to solve a multidimensional mixed problem for a parabolic equation. Assume that $\Omega = \{0 \le x_i \le 1\}$ is an n-dimensional unit cube in $R^n(x)$. Find in $\Omega \times [0, T]$ where $T > 0$, the solution of problem (2.7), where

$$-Au = \sum_{i,j=1}^{n} \frac{\partial}{\partial x_i} \left(a_{ij}(x) \frac{\partial u}{\partial x_i} \right) + a(x)u \qquad (7.29)$$

is an elliptic operator, with limited relatively smooth variable coefficients defined on functions satisfying the boundary conditions

$$u|_{\partial \Omega} = 0, \qquad (7.30)$$

where $\partial \Omega$ is the boundary of the domain Ω.

Construct a uniform grid of step h in Ω, and let N be the total number of internal grid points. Then $N = (h^{-1} - 1)^n$. Substitute Equation (7.29) for a difference operator. Use the notations of Equation (2.5) for the corresponding system of differential equations of the straight lines method. Solve the system by

(1) explicit method (2.2) as proposed and
(2) the implicit method (3.1)–(3.4) at $C = -aD + bI$, where D is the Laplace difference operator.

The estimates of computational cost for both these methods will be qualitative. Therefore, assume the coefficients of matrix A such that cou $= h^2$, the approximation step size $\tau = h$, and the number of arithmetic operations for computing Au at a given u is $C_1 N (C_1 > 0)$. Compute $(I + \tau C)^{-1} v$ by one of the fastest methods, using the marching algorithm with the number of operations $C_2 N$ $(C_2 > 0)$. Estimate the ratio C_2/C_1 by the number 2 (which is realistic). Assume also that the average rate of convergence of iterations (5.2)–(5.4) is finite and equal to 1; iterations are carried out until the norm of the initial error decreases h^{-2} times $\left(h^{-2} > 1\varepsilon = h^2 \right)$.

The total number of operations per step t in the implicit scheme is

$$V_1 = 3C_1 N \ln \varepsilon^{-1} \qquad (7.31)$$

Now solve the problem on the same time interval, using one cycle of the optimal explicit method. Denote the number of steps in this method as N_1. This can be determined from Equation (6.19):

$$\tau = \text{cou}^{1/2} = N_1^2 \text{cou},$$

that is, $N_1 = \text{cou}^{-1/4}$. Therefore, the total number of operations, V_2, in the explicit method is $C_1 N N_1 = C_1 N \text{cou}^{-1/4}$.

Compute $W = V_2/V_1 = \text{cou}^{1/4}(3\ln\varepsilon^{-1})$, assuming $\varepsilon = h^2$, $h = N^{-1/n}$, cou $= N^{-2/n}$,

$$W = \frac{nN^{1/2n}}{6\ln N}.$$

The function W is monotonically increasing at a large N. At $n = 2$, $W = W_2 = N^{1/4}/(3\ln N)$; at $n = 3$, $W = W_3 = N^{1/6}/(2\ln N)$.

Di- and tridimensional problems with the number of unknown quantities in the interval $10^6 - 10^9$ are effectively dealt with and will remain so in the near future. Compute W for the bounds of this interval, that is,

$$\begin{aligned}
N = 10^6 &: W_2 = 0.76, \ W_3 = 0.36; \\
N = 10^9 &: W_2 = 2.87, \ W_3 = 0.76.
\end{aligned} \tag{7.32}$$

Equations (7.32) show the efficacy of the explicit method (2.2) applied to problems with a fairly large number of unknown quantities. These estimates are given for the usual computers and will certainly be better for parallel computer architectures.

Algorithms (2.2) and (2.5) can be extended to simultaneous equations of the form

$$p(t)\frac{du}{dt} = f(u,t), \tag{7.33}$$

where $p(t)$ is a sufficiently smooth positive function. To do so, substitute the second and fourth equations in algorithm (2.2) for

$$t_{k+1/2} = t_k + h_{k+1}p(\tilde{t}_k), \quad t_{k+1} = t_{k+1/2} + h_{k+1}p(\tilde{t}_k), \tag{7.34}$$

respectively, where $\tilde{t}_k = t_k + h_{k+1}p(t_k)$, and the second equation in Equation (2.5) for

$$\tilde{t}_k = \bar{t}_k + \frac{1}{2}h_{k+1}p(\tilde{t}_k), \quad \bar{t}_{k+1} = \bar{t}_k + \bar{h}_{k+1}p(\tilde{t}_k). \tag{7.35}$$

Choose step sizes h_{k+1} and $\bar{h}_{k+1}$ by the equations obtained.

8.   ONE REALIZATION OF STABLE ALGORITHMS

Let us construct one of the stable algorithms from Equation (2.2). Its advantage is that at any $n = n(g)$ (the number of steps), its parameters can be determined by a uniquely fixed sequence of numbers, $\{\omega_i\}$, the so-called T-sequence, e.[13] The parameters of algorithm (2.5) are constructed in a similar way.

Assume that $n = n(g) = 2^m 3^p$ $(0 \leq m \leq 14, 0 \leq p \leq 6)$; the members of this sequence are $2, 3, 4, 6, 8, 9, 12, \ldots, 23887822, 47775744$. If $0 < g \leq 0.5$, assume $N = 1$, $\gamma_1 = -1$, $h_1 = \tau$; and obtain in this case a difference scheme of the second-order approximation,

$$Q_2(\lambda) = 1 - \lambda + \frac{\lambda^2}{2} .$$

At $g > 0.5$, determine the functions $n(g)$, $G(g)$: let $g_0 = \bar{g}_0 = 0.5 < g_1 < \bar{g}_1 < \ldots < g_i < \bar{g}_i < \ldots < \infty$ be a partition $[0.5, \infty)$. Assume that

$$n(g) = N_i, \qquad \bar{g}_{i-1} < g \leq g_i \tag{8.1}$$

$$G(g) = \begin{cases} g, & \bar{g}_{i-1} < g < g_i \\ g_i, & g_i \leq g \leq \bar{g}_i. \end{cases} \tag{8.2}$$

Let $\{\omega_i\}_1^\infty$ be a T-sequence formed by the roots $\cos \omega_i \pi$, $i = 1, \ldots, 3^p$, of the Chebyshev polynomials $T_{3^p}(x)$ and by the permutation κ_{3^p}. To be precise, any segment of the finite sequence $\{\omega_i\}$ has $\{\cos \omega_i \pi\}$ as roots of $T_{3^p}(x)$ arranged by the permutation κ_{3^p}. Assume that for κ_{3^p} $\max \cos^2 \omega_i \pi = \cos \omega_{3^p} \pi$.

Select now a partition $\{g_i, \bar{g}_i\}$ such that at any $g > 0.5$

(1) inequality Equation (7.6) holds;

(2) $v_{2N}(g') \leq v_{2N}(g'')$ for $g' \leq g''$;

(3) $\eta_{2N}(g) \leq \xi_N < 1$ for $g \in (\bar{g}_i, \bar{g}_{i+1}]$, where $0 < \xi < 1$ is an appropriate prescribed sequence.

The numbers g_i are thus determined from $\eta_{n(g)}(g) = \xi$. Choose ξ_i from the interval $(0.4, 1)$ for the calculation. At these parameters, $B_0 = 2$ in inequality Equation (2.1), and the relative rate of methods (2.2) and (2.4), as compared with the classic explicit method, may reach the value 3×10^7 cou. Then $n = 47775744$, and the maximum step size may be approximately 5×10^{14} cou, $\eta = 0.44$.

9. ON PASSING BOUNDARY LAYERS, ESTIMATING QUANTITIES COU AND τ AND THE DUMKA CODE

It is not necessary to take into account factors affecting the determination of nonstationary fundamental solutions with small time constants in a number of nonstationary problems. For example, computations are carried out in the proximity of the boundary layer, where the solution varies very quickly and the predicted step size, τ, is very small, while we are interested in the value of a solution in a fairly large time interval. There is no need, therefore, to try to accurately approximate rapidly attenuating nonstationary fundamental solutions; one should suppress them, disposing of the parameters of method (2.2). The second reason for suppressing them is that the approximation of a nonstationary problem with space-differential operators by the straight lines method[23] (in τ) generates spurious "high-frequency harmonics" in expansion Equation (4.2) of an approximate solution, which do not necessarily require approximation. The suppression, strictly justified for problem (2.7), can be applied using the following algorithm: while realizing the method outlined in Section 2, carry out, after NK cycles, Equations (2.2) with the parameters in Equations (7.4)–(7.8) selected to reduce by 10^3 times Equation (4.3) components corresponding to eigenvalues of the matrix A within the interval $[\vartheta M, M]$, where $0 < \vartheta < 1$.

The examined explicit algorithms require that the cycle of iterations be preceded by determination of the quantity cou, or of a fairly accurate estimate, $\bar{\lambda}$, from above for $\max_i(\lambda_i)$, where λ_i is the eigenvalue of the Jacobian Equation (1.3) at $t = t_0$. In the common case, the estimate $\bar{\lambda}$ is obtained either by reinforcing the computational algorithm with a special iterative method, or applying the Gershgorin circles theorem, and so forth.

On completion of iterations (2.2), the estimate of the quantity τ, the next predictable step size, is determined by the value of the error, ε, of approximation du/dt in a separate difference.

The methods for solving stiff systems of equations by stable explicit schemes, as discussed previously, are realized in the DUMKA code. The program is written in FORTRAN with double precision. Its application requires a FUN-subprogram to calculate the right-hand parts of system (1.1), a COUR-subprogram to determine cou, and the main program. If $T = t_0$, $U = U_0$, and $H = H_0$, the structure of the main program may be schematically represented as

```
1   .   .   .   .   .   .   .   .   .   .   .   .   .   .   .   .   .   .

    CALL DUMKA (FUN, COUR, ST, N, T, U, H, P, CU, EPS,
        TEND, SK, SP, NP, NS, NK, LU, Z1, Z2)

    .   .   .   .   .   .   .   .   .   .   .   .   .   .   .   .   .

    IF (T.LT.TEND) GO TO 1
```

The reference to the FUN and COUR subprograms is

```
    CALL FUN (N, T,0, U, Z1)
    CALL COUR (NP, N, T, U, COU)
```

The DUMKA code carries out one cycle of calculations, calling the COUR subprogram once and FUN many times. When one cycle has been executed, DUMKA determines new values of T, U, H. The length of the cycle is set by the DUMKA code; values of the cycle vary from 2 to 47775744.

10. CASES OF STIFF SYSTEM COMPUTATIONS

Stiff problems arise in discretizing nonstationary mathematical physics problems in spatial variables. The dimension of problems does not matter much in the application of explicit schemes. In the cases discussed previously, Q denotes the total number of computations in the right-hand part of Equation (1.1) in solving problems (1.1) and (1.2) on the segment $[0, T]$, and $\tau_{\text{mean}} = T/Q$ is the real mean step.

Assume that $\tau_m = \max_{0 \leq t \leq T} v_{2N}$ cou; this is the maximum mean step per the complete cycle of computations by Equations (2.2) ($k = 0, \ldots, N - 1$). $\underline{\tau}$ and $\bar{\tau}$ are the minimum and maximum, respectively, of the predicted value, τ, in the process of computation. Computations were carried out by the DUMKA code, applying algorithms (2.2) and Equations (7.1)–(7.8) with the accuracy, ε, of local approximation, on a PC AT 386/387 with standard precision.

Case 1. Examine a mixed problem at $0 \leq x \leq \pi$, $0 \leq t \leq T$. Find the function $u(x, t)$ if

$$\frac{du}{dt} = Lu + au + f(t), \qquad u(0, t) = u(\pi, t) = 0, \; u(x, 0) = 1 \quad (10.1)$$

where

$$Lu = \frac{\partial^2 u}{\partial x^2} + b(x, t, u)\frac{\partial u}{\partial x^2}, \qquad f(t) = 1 + 0.1t, \; t \notin (6, 10);$$

$$f(t) = 0, \; t \in [6, 10] \quad (10.2)$$

Convection, diffusion, and kinetics terms are present in Equation (10.1). The differential operator, Lu, can be represented in a symmetric form with the scaling factor β,

$$Lu = \frac{1}{\beta}\frac{\partial}{\partial x}\beta\frac{\partial u}{\partial x},$$

where $\beta(x, t) = \exp(\int_0^x b(y, t, u(y, t))dy)$. Notice that, if

$$u_n(x, t) = \frac{1}{\sqrt{\beta}}\sin(nx),$$

then

$$Lu_n + \left(\frac{b_n^2}{4} + \frac{1}{2}\frac{db_n}{dx}\right)u_n = -n^2 u_n,$$

where $b_n = b(x, t, u_n)$. Introducing a uniform grid with a step size $h = \pi/(n+1)$, and substituting

$$\frac{1}{\beta} \frac{\partial}{\partial x} \beta \frac{\partial u}{\partial x},$$

for the second divided difference,

$$L_h u = h^{-2}\beta^{-1}(x, t)\left(\left(u(x+h, t) - u(x, t)\right)\beta\left(x + \frac{h}{2}, t\right) \right.$$

$$\left. - \left(u(x, t) - u(x-h, t)\right)\beta\left(x - \frac{h}{2}, t\right) \right)$$

$$= h^{-2}\left(\left(u(x+h, t) - u(x, t)\right)\gamma\left(x + \frac{h}{2}, t\right) \right.$$

$$\left. - \left(u(x, t) - u(x-h, t)\right)\gamma\left(x - \frac{h}{2}, t\right) \right) \qquad (10.3)$$

where $\gamma(z, t) = \exp(\int_x^z b\,dy)$. Assuming that the function, b, is linear between the points of the grid, obtain the final form of the approximation, Lu, where

$$\gamma\left(x + \frac{h}{2}, t\right) = \exp\left(\frac{h}{8}\left(3b(x, t, u(x, t)) + b(x+h, t, u(x+h, t))\right)\right),$$

$$\gamma\left(x - \frac{h}{2}, t\right) = \exp\left(-\frac{h}{8}\left(3b(x, t, u(x, t)) + b(x-h, t, u(x-h, t))\right)\right).$$

Let a, b be constants. Then u_n are eigenfunctions of the operators L, L_h with a real spectrum; obtain in this case a system of differential equations in t of the straight lines method of order n. The solution $u(x, t)$ to this problem has three distinct boundary layers, at $t = 0, 6, 10$. The function $u(x, t)$ is almost linear in t for large t. Select $n = 40, 60, 80$, and, respectively, $\varepsilon = 5 \times 10^{-3}$, $2, 5 \times 10^{-3}$, 1.25×10^{-3}; $T = 8 \times 10^7$, $a = -1$, $b = 5$. Table 1 gives the numerical results.

Case 2. Assuming the cube $\Omega = \{0 \le x_i \le \pi, \ i = 1, 2, 3\}$ with the boundary $\partial\Omega$, and $0 < t \le T$, find the solution $u(x, t)$ to the problem

$$\frac{du}{dt} = \triangle u - \sum_{i=1}^{3} b_i(x_i)\frac{\partial u}{\partial x_i} + au + f(t), \qquad (10.4)$$

$$u|_{x_i} = 0, \quad \frac{\partial u}{\partial n}\Big|_{x_i = \pi} = 0, \quad u|_{t=0} = 0;$$

TABLE 1
Results for the DUMKA code.

n	ε $\times 10^{-3}$	cou $\times 10^{-3}$	$\underline{\tau}$ $\times 10^{-4}$	$\bar{\tau}$ $\times 10^{7}$	τ_{mean}	τ_m	$\tau_{\text{mean}}/\text{cou}$	τ_m/cou
40	5.	2.91	3.5	3.0	371.8	403.4	127811	138680
60	2.5	1.32	1.6	3.0	256.5	270.5	194182	204779
80	1.25	0.75	1.0	3.0	204.7	214.0	272792	285174

the function $f(t)$ has the form of Equation (10.2). Construct a cubic grid with a step size $h = \pi/(m+0.5)$, and substitute the operator in the right-hand part for the difference operator, using Equation (10.3). The spectrum of the problem is real. Next, obtain a system of differential equations of order $n = m^3$. Select $m = 64$, $\varepsilon = 2. \times 10^{-3}$, $T = 6.7 \times 10^5$, and $b_1 = -3$, $b_2 = 2$, $b_3 = 1$, $a = -1$. The numerical results are

$$n = 64^3, \quad \text{cou} = 3.96 \times 10^{-4}, \quad \underline{\tau} = 1.4 \times 10^{-4}, \quad \bar{\tau} = 3.2 \times 10^5,$$

$$\tau_{\text{mean}} = 11.08 = 2.8 \times 10^4 \, \text{cou}, \quad \tau_m = 16.2 = 4.1 \times 10^4 \, \text{cou}.$$

Case 3. Examine a mixed problem for the Burgers equation in the region $0 \le x \le 1$, $t > 0$:

$$\frac{du}{dt} + u\frac{\partial u}{\partial x} = \nu\frac{\partial^2 u}{\partial x^2} \tag{10.5}$$

$$u(0, t) = v_1(t), \quad u(1, t) = v_2(t), \quad u(x, 0) = \omega(x).$$

Substitute Equation (10.5) on the grid $x_i = ih$ with a step size $h = (n + 1)^{-1}$ for a system of equations of the straight lines method,

$$\frac{du_i}{dt} = \nu h^{-2}(u_{i-1} - 2u_i + u_{i+1}) - \frac{1}{8h}(u_{i+1}^2 - u_{i-1}^2)$$

$$- \frac{1}{4h}u_i(u_{i+1} - u_{i-1}),$$

$$u_0 = v_1(t), \quad u_{n+1} = v_2(t), \quad u_i(0) = \omega(x_i)$$

Assume $n = 60$, $\varepsilon = 5 \times 10^{-3}$, $T = 87000$. Now, assuming the function $u(x, t) = (1 + \exp\{x/(2\nu) - t/(4\nu)\})^{-1}$ as the solution of the problem, solve it in two versions:

(1) $\nu = 1$; cou $= 1.34 \times 10^{-3}$, $\underline{\tau} = 2 \times 10^{-4}$, $\bar{\tau} = 32270$, $\tau_{\text{mean}} = 2.21 = 16441 \, \text{cou}$, $\tau_m = 3.02 = 22536 \, \text{cou}$;

(2) $\nu = 0.1$; cou $= 1.34 \times 10^{-3}$, $\underline{\tau} = 1.7 \times 10^{-3}$, $\bar{\tau} = 32630$, $\tau_{\text{mean}} = 7.1 = 5392 \, \text{cou}$, $\tau_m = 9.2 = 6985 \, \text{cou}$.

Case 4. Babushka's model:

$$\frac{du_1}{dt} = \frac{[a(u_3 - u_1)u_1]}{u_2}, \quad \frac{du_2}{dt} = -a(u_3 - u_1),$$

$$\frac{du_3}{dt} = \frac{[b - c(u_3 - u_5) - au_3(u_3 - u_1)]}{u_4}, \quad \frac{du_4}{dt} = -a(u_3 - u_1),$$

$$\frac{du_5}{dt} = \frac{-c}{d(u_5 - u_3)}, \tag{10.6}$$

$$u_1(0) = u_2(0) = u_3(0) = 1, \quad u_4(0) = -10, \quad u_5(0) = 0.9, \tag{10.7}$$

$$a = 100, \quad b = 0.9, \quad c = 1000, \quad d = 10.$$

Characteristically, this problem's solutions, u_1, u_5 and u_2^{-1}, u_3^{-1}, rapidly increase exponentially, $u_4 \rightarrow -9$. The spectrum of the Jacobi matrix lies in the proximity of the points $0, \pm 110/9$, and $-\gamma$, where $\gamma = \gamma(t) \rightarrow \infty$ if $t \rightarrow \infty$. The greatest difficulty in solving the system by explicit methods is that the value of $cou(t)$ rapidly decreases with time. Therefore, no explicit method with a restricted value of $g = \tau/cou$ is practicable. For implicit schemes, an extra difficulty may arise in the application of simple iterative methods, because the spectrum of the Jacobi matrix has a nonconstant sign. The goal of the numerical investigation is to examine the ability of the algorithm for integration of the problem on the segment [0,b], where $b > 1$, and to determine its productivity by the length of cycles. The relative accuracy of approximation ε was thus assumed to be $\varepsilon = 0.01$ for $\leq t \leq .76$ and $\varepsilon = 0.1$ for $t \geq 0.76$. Table 2 gives the results of the calculation.

The analysis of the results shows that the method discussed previously is unproductive because of a solution component arising from the error of approximation, for example $\exp\{-\gamma t\}$. Therefore, we developed another method of solution in the subspace. The eigenvalues of the Jacobian were calculated accurately by the Ferrary method; the right-hand (v_r) and left-hand ($\bar{v}_r$) eigenvectors corresponding to the eigenvalue $-\gamma$ were determined. These vectors may help determine the value of the coefficient a_γ in the expansion of the solution (which corresponds to vector v_r). If a_γ is small, we exclude the component $a_\gamma v_r$ from the solution and put into the program the value of the next eigenvalue $(-110/9)$ as a new lower boundary of the spectrum. This modification was applied to solve the system on the segment [0,31.92] at a relative accuracy, $\varepsilon = 0.0001$. The DUMKA code used 8960 steps, and $\gamma(31.92) = -1.74 \times 10^{308}$.

A standard test[25] was used to verify and validate the DUMKA code. The set contains six groups of stiff problems depending on the type of spectrum (real and complex) and the form of equations (linear or nonlinear).

TABLE 2
Results for Babushka's model.

t_N	cou	N
0.75	1.2×10^{-9}	3888
0.946	1.8×10^{-11}	49152
0.994	7.1×10^{-12}	93312
1.053	2.1×10^{-12}	186624
1.122	5.2×10^{-13}	442368
1.232	7.3×10^{-14}	492992
1.233	6.9×10^{-14}	248832
1.239	5.6×10^{-15}	1327104
1.249	4.4×10^{-15}	179472
1.347	1.0×10^{-15}	11443936

The DUMKA code was also used for solving multigroup diffusion equations with delaying neutrons (with dependence on temperature). The calculations confirmed that the DUMKA code can be used as a method of solving stiff systems.

The author thanks S.A. Frolova, V.A. Khamasa, I.V. Krolivets, G.I. Kurchenkova, and M.P. Shustova for developing codes using the DUMKA code.

REFERENCES

1. *Modern Numerical Methods for Ordinary Differential Equations*, Hall, G., and Watt, J.M., Eds., Clarendon Press, Oxford, UK, 1976.

2. Rakitsky, Jv.V., Ustinov, S.M., and Chernoritsky, I.G., *Numerical Methods of Solution of Stiff Systems*, Nauka, Moscow, 1979 (in Russian).

3. Stetterr, Hanz J., *Analysis of Discretization Methods for Ordinary Differential Equations*, Springer-Verlag, Berlin, 1973.

4. Byrne, G.D., and Hindmarsh, A.C., Stiff ODE solvers: a review of current and coming attractions, *Jour. Comp. Phys.*, v. 70, 1, 1987.

5. Marchuk, G.I. *Methods of Numerical Mathematics*, Springer, New York, 1982.

6. Lebedev, V.I., Bakhvalov, N.S., Agoshkov, V.I., Baburin, O.V., Knyazev, A.V. and Shutyaev, V.P., Parallel algorithms for solving certain stationary problems of mathematical physics, Dept. Numer. Math. USSR Ac. Sci., Moscow, 1984 (in Russian).

7. Van der Houwen, P.J., Explicit Runge-Kutta formulas with increased stability boundaries, *Numer. Math.*, v. 20, 149, 1972.

8. Riha, E., Optimal stability polynomials, *Compt.*, v. 9, 37, 1972.

9. Saulyev, V.K., *Integration of Parabolic-Type Equations by the Mesh Method*, State Publishing House of Phys. and Math. Literature, Moscow, 1960 (in Russian).

10. Van der Houwen, P.J., *Construction of Integration Formulas for Initial Value Problems*, North Holland, Amsterdam et al., 1977.

11. Lebedev, V.I., and Finogenov, S.I., On the order of choosing iterative parameters in the Chebyshev cyclic iterative methods, *Zh. Vychisl. Mat. Mat. Phys.*, v. 11, 425, 1971 (in Russian).

12. Lebedev, V.I., and Finogenov, S.I., The solution of the problem of ordering the parameters in Chebyshev iterative methods, *Zh. Vychisl. Mat. Mat. Phys.*, v. 13, 18, 1973 (in Russian).

13. Lebedev, V.I., and Finogenov, S.I., On the utilization of ordered Chebyshev parameters in iterative methods, *Zh. Vychisl. Mat. Mat. Phys.*, v. 16, 895, 1976 (in Russian).

14. Lokutsievsky, V.O., and Lokutsievsky, O.V., Utilization of Chebyshev parameters for the numerical solution of certain evolution problems, Preprint No. 98, Inst. Appl. Math. USSR Ac. Sci., Moscow, 1987 (in Russian).

15. Zhaoding, Y., "Some Difference Schemes for Solving the First Boundary Value Problem for Linear Differential Equations with Partial Derivatives", Ph.D. dissertation, Moscow State University, Moscow, 1958 (in Russian).

16. Zhukov, V.T., Numerical experiments in solving heat conduction equations by the method of local iterations, Preprint No. 97, Inst. Appl. Math. USSR Ac. Sci., Moscow, 1984 (in Russian).

17. Zhukov, V.T., Difference schemes of local iterations for parabolic equations, Preprint No. 173, Inst. Appl. Math. USSR Ac. Sci., Moscow, 1986 (in Russian).

18. Markov, V.A., *On Functions Least Deviating from Zero on a Fixed Interval*, Sankt-Peterburg, 1892 (in Russian).

19. Bernshtein, S.N., *On Polynomials Orthogonal on a Finite Segment*, Collected Works, Vol. 2, USSR Ac. Sci., Moscow, 1952 (in Russian).

20. Bernshtein, S.N., *On One Class of Orthogonal Polynomials*, Collected Works, Vol. 1, USSR Ac. Sci., Moscow, 1952 (in Russian).

21. Lebedev, V.I., Explicit difference schemes with time-variable steps for solving stiff systems of equations, Preprint No. 177, Dept. Numer. Math. USSR Ac. Sci., Moscow, 1987 (in Russian).

22. Franklin, J.N., Numerical stability in digital and analog computation for diffusion problems, *Jour. Math. Phys.*, v. 37, 305, 1959.

23. Lebedev, V.I., Equation and convergence of the differential difference method (the method of lines), *Vestnik MGU*, v. 10, 47, 1955 (in Russian).

24. Lebedev, V.I., *Iterative Methods of Solving Linear Equations*, Math. Analiz i Smezhnye Voprosy Mat., Nauka, Novosibirsk, 1978 (in Russian).

25. Enright, W.H., and Pryce, J.D., Two FORTRAN packages for assessing initial value methods, *ACM Trans. Math. Software*, v. 13, 1, March 1987.

On Numerical Methods of Solving the Navier–Stokes Equations in "Velocity-Pressure" Variables

G.M. Kobelkov

The majority of stationary and nonstationary boundary value problems for the Navier–Stokes equations do not have solutions which can be represented in analytic form even in domains with simple geometry. So nowadays the most often used tool for investigating the solution properties and obtaining concrete solutions with appropriate accuracy is the numerical simulation of the Navier–Stokes equations using computers. But in many cases the problem of constructing numerical methods and solving the resultant systems of linear and nonlinear equations is not a trivial one.

An often used numerical method for solving the Navier–Stokes equations in the two-dimensional case is the (ψ, ω) form of solution (stream function–vorticity variables).[1] The advantage of this form of solution is that the incompressibility equation is satisfied. In using (ψ, ω) solutions there arise a number of essential problems, for example, there is no boundary condition for the vorticity function, ω; it is impossible to extend these results to the three-dimensional case; there are also some difficulties with boundary conditions for the stream function, ψ, in multiconnected domains.

It is also worth noting that the mathematical basis of numerical methods in the (ψ, ω) form of notation as a rule is absent. Thus, to solve the same problems researchers frequently use different methods which are difficult to compare.

In the present paper we consider mathematical questions connected with constructing numerical solutions of the Navier–Stokes equations

0-8493-8947-X/94 /$0.00 + $.50
(c) 1994 by CRC Press, Inc.

in primitive "velocity-pressure" variables (so-called (u, v, p) variables). During our investigation we pay more attention to the problems of approximation and stability of finite-difference schemes and also to methods for solving them with a computer. We did not consider the problem of convergence of grid solutions to the exact solution of a differential problem, but those considerations should follow from the ones given here.

The peculiarity of all the methods considered in this paper is that the iterative methods used to solve the linear and nonlinear grid equations have a nonsymmetric transfer operator from one iteration step to another. This does not permit the application of the usual technique for investigation of convergence. We used a new approach to investigate the convergence of such iterative methods. This allows us to estimate the norm of the transfer operators.

It should also be pointed out that all the methods proposed in this paper are generalized to more complicated systems of equations containing the Stokes operator. In particularly, finite-difference methods were constructed and established for the convection problem (Boussinesq approach),[2,3] magnetic hydrodynamics equations,[4] electromagnetic equations,[5] and so on.

1. NUMERICAL METHODS FOR SOLUTION OF THE STATIONARY STOKES PROBLEM

Since our discussions do not depend on a space dimension, we consider only the two-dimensional case for simplicity of notation. Let $\Omega \in \mathbf{R}^2$ be a bounded domain with boundary $\partial\Omega$. We must find the solution $(\mathbf{u}, p)$ of the following boundary value problem (the Stokes problem)

$$-\boldsymbol{\Delta}\mathbf{u} + \nabla p = \mathbf{f}, \qquad \operatorname{div}\mathbf{u} = 0, \quad \mathbf{u}|_{\partial\Omega} = \mathbf{0}. \qquad (1.1)$$

Here, $\mathbf{u} = (u_1, u_2)$, $p \in L_2$, and $(p, 1) = \int_\Omega p\,dx = 0$. It is well known that a weak solution, $(\mathbf{u}, p) \in \mathbf{H}_0^1 \times L_2/\mathbf{R}^1$ of Equation (1.1), exists and is unique if $(\mathbf{f}, \mathbf{w})$ $(\mathbf{w} \in \mathbf{H}_0^1)$ is a bounded functional on L_2.[6]

Let us assume that we have some finite approximation of Equation (1.1) and $\mathbf{U}$ is a space which approximates $\mathbf{H}_0^1$ and P approximates $L_2/\mathbf{R}^1$. Thus $\mathbf{U}$ is a space of vector functions, $\mathbf{u}^h = (u_1^h, u_2^h)$, and P is a space of scalar functions, p^h, which are orthogonal to constants. We also assume that there are norms in these spaces which correspond to norms in $\mathbf{H}_0^1$ and $L_2/\mathbf{R}^1$.

Let us also assume that approximations of $\boldsymbol{\Delta}$, ∇, and div have the properties $\boldsymbol{\Delta}^h : \mathbf{U} \to \mathbf{U}$, $\nabla^h : P \to \mathbf{U}$, $\operatorname{div}^h : \mathbf{U} \to P$,

$$(\operatorname{div}^h\mathbf{v}, q) = -(\nabla^h q, \mathbf{v}) \quad \text{for any } \mathbf{v} \in \mathbf{U}, \ q \in P, \qquad (1.2)$$

$$\|q\|_P \le c_0 \sup_{v \in U} \frac{|(q, \mathrm{div}^h \mathbf{v})|}{\|\mathbf{v}\|_U}. \tag{1.3}$$

Here $(\ ,\ )$ is an analogue of the usual scalar product of L_2 in U or P. Approximations of Equation (1.1) of this kind can be constructed by finite-difference methods as well as by finite-element methods. We shall also assume that the expression $-(\boldsymbol{\Delta}^h \mathbf{u}^h, \mathbf{u}^h)^{1/2}$ defines norm $\|\mathbf{u}^h\|_1$ in U, which coincides with the norm of U; it is possible to assume the equivalence of these norms. We also denote $\|\cdot\|$ as a discrete analogue of the norm L_2 in U or P.

Thus, to obtain the approximate solution of Equation (1.1), we must find functions $\mathbf{u} \in U$ and $p \in P$ which are the solution of the discrete boundary value problem

$$-\boldsymbol{\Delta}^h \mathbf{u}^h + \nabla^h p^h = \mathbf{f}^h \text{ in } U, \mathrm{div}^h \mathbf{u} = 0 \text{ in } P. \tag{1.4}$$

It is clear that the solution of Equation (1.4) exists and is unique (the proof is similar to the continuous case). Let us construct iterative methods to solve Equation (1.4) by considering the fully implicit one-step iterative method,

$$\mathbf{B}\mathbf{v}_t - \boldsymbol{\Delta}^h \hat{\mathbf{v}}^h + \nabla^h \hat{q}^h = \mathbf{f}^h \quad \text{in } U, \quad \beta\tau q_t + \mathrm{div}^h \hat{\mathbf{v}} = 0 \quad \text{in } P. \tag{1.5}$$

Here we used the notations

$$\mathbf{v} = \mathbf{v}^n, \quad \hat{\mathbf{v}} = \mathbf{v}^{n+1}, \quad \mathbf{v}_t = \frac{(\hat{\mathbf{v}} - \mathbf{v})}{\tau}$$

and β and τ are iterative parameters, $\mathbf{B} : U \to U$ — some operator which will be determined later. We also must set up initial values $\mathbf{v}^0 \in U, q^0 \in P$.

In some sense this method is "the best". For instance, if we set $\beta = 0$ and $\mathbf{B} = 0$ then the method will converge in only one iteration step, but in this case this method makes no sense. Here we mean "the best" in the sense of the most rapid convergence.

The form of notation of iterative method (1.5) differs from the usual one (see, e.g., reference 7): $By_t + Ay = \phi$. Nevertheless, we will see later that the fully implicit form of notation has some advantage as concerns the construction and investigation of iterative methods.

Let us investigate the convergence of method (1.5). If $\mathbf{y} = \mathbf{v} - \mathbf{u}^h$, $r = q - p^h$, then from Equation (1.5) we obtain the boundary value problem for $(\mathbf{y}, r)$,

$$\begin{cases} \mathbf{B}\mathbf{y}_t - \boldsymbol{\Delta}^h \hat{\mathbf{y}} + \nabla^h \hat{r} = 0 & \text{in } U, \\ \beta\tau r_t + \mathrm{div}^h \hat{\mathbf{y}} = 0 & \text{in } P, \\ \mathbf{y}^0 = \mathbf{v} - \mathbf{u}^h \in U, \ r^0 = q - p^h \in P. \end{cases} \tag{1.6}$$

If $\mathbf{C} : \mathbf{U} \to \mathbf{U}$ is a symmetric nonnegative operator, then we denote $\|\mathbf{w}\|_\mathbf{C}$ as the seminorm $(\mathbf{Cw}, \mathbf{w})^{1/2}$. Let $\mathbf{B} = \mathbf{B}^* \geq 0$, then we denote $\|\mathbf{w}\|_2$ as the norm,

$$\|\mathbf{w}\|_2^2 = (\mathbf{Bw}, \mathbf{w}) + \tau\gamma\|\mathbf{w}\|_1^2.$$

Thus this norm depends on $\gamma > 0$.

Theorem 1.1. Let $\mathbf{B} = \mathbf{B}^* \geq 0$. Then iterative method (1.5) converges as geometric progression for any $\beta, \tau > 0$, and the estimate

$$\|\hat{\mathbf{y}}\|_2^2 + \beta\tau\|\hat{r}\|^2 \leq (1 + \alpha\tau)^{-1}\left(\|\mathbf{y}\|_2^2 + \beta\tau\|r\|^2\right) \qquad (1.7)$$

holds where $\alpha > 0$ is a constant which will be defined later.

Proof. Let us form the scalar product in $\mathbf{U}$ of the first part of Equation (1.6) with $2\tau\hat{\mathbf{y}}$, and the second part of the equation in P with $2\tau\hat{r}$. By adding the results we obtain

$$\|\hat{\mathbf{y}}\|_\mathbf{B}^2 - \|\mathbf{y}\|_\mathbf{B}^2 + \tau^2\|\mathbf{y}_t\|_\mathbf{B}^2 + 2\tau\|\hat{\mathbf{y}}\|_1^2 + \beta\tau\|\hat{r}\|^2$$
$$- \beta\tau\|r\|^2 + \beta\tau^3\|r_t\|^2 = 0. \qquad (1.8)$$

Here we used the property from Equation (1.2),

$$(\nabla^h\hat{r}, \hat{\mathbf{y}}) + (\mathrm{div}^h\,\hat{\mathbf{y}}, \hat{r}) = 0.$$

Since our space $\mathbf{U}$ is finitely dimensional, there exists a constant, κ_2, such that the inequality

$$\mathbf{B} \leq -\kappa_2\mathbf{\Delta}^h \qquad (1.9)$$

is true. Recall that this inequality is understood as a bilinear form, that is, $(\mathbf{Bw}, \mathbf{w}) \leq -\kappa_2(\mathbf{\Delta}^h\mathbf{w}, \mathbf{w})$ for any $\mathbf{w} \in \mathbf{U}$.

Now let us use the property from statement (1.3). Form the scalar product of both parts of the first part of Equation (1.6) with $\phi \in \mathbf{U}$. Then we have

$$\cdot|(\nabla^h\hat{r}, \phi)| \leq |(\mathbf{By}_t, \phi)| + |(\mathbf{\Delta}^h\hat{\mathbf{y}}, \phi)|. \qquad (1.10)$$

Now we divide both parts of Equation (1.10) by $\|\phi\|_1$ and take the sup over $\phi \in \mathbf{U}$ from the right-hand part. Using Equation (1.10) and the Cauchy inequality, we obtain

$$\frac{|(\nabla^h\hat{r}, \phi)|}{\|\phi\|_1} \leq \sqrt{\kappa_2}\,\|\mathbf{y}_t\|_\mathbf{B} + \|\hat{\mathbf{y}}\|_1.$$

Since the right-hand part of the last inequality does not depend on ϕ we can use Equation (1.2) and take sup by ϕ over the left-hand part of this inequality. Using the Babushka–Brezzi inequality (1.3), we have

$$\|\hat{r}\| \le c_0\big(\sqrt{\kappa_2}\,\|\mathbf{y}_t\|_{\mathbf{B}} + \|\hat{\mathbf{y}}\|_1\big).$$

From this inequality it is easy to obtain the estimate

$$\|\hat{r}\|^2 \le 2c_0^2\big(\kappa_2\|\mathbf{y}_t\|_{\mathbf{B}}^2 + \|\hat{\mathbf{y}}\|_1^2\big).$$

When both parts of the last estimate are multiplied by $\beta\tau^2\lambda$, where $\lambda > 0$ is arbitrary, and added to Equation (1.8), we obtain the estimate

$$\|\hat{\mathbf{y}}\|_{\mathbf{B}}^2 + \tau^2(1 - \beta\lambda)\|\mathbf{y}_t\|_{\mathbf{B}}^2 + 2\tau(1 - c_0^2\kappa_2\beta\lambda\tau)\|\hat{\mathbf{y}}\|_1^2$$
$$+\,\beta\tau(1 + \lambda\tau)\|\hat{r}\|^2 \le \|\mathbf{y}\|_{\mathbf{B}}^2 + \beta\tau\|r\|^2 \qquad (1.11)$$

Since $\lambda > 0$ is arbitrary, choose it in such a way that inequalities

$$1 - \beta\lambda \ge 0, \quad 1 - c_0^2\kappa_2\beta\lambda\tau \equiv \kappa > 0 \qquad (1.12)$$

hold. It is easy to see that for arbitrary positive $\beta, \kappa_2, \tau, c_0$, this can be done. In this case Equation (1.11) can be transformed to

$$\|\hat{\mathbf{y}}\|_{\mathbf{B}}^2 + 2\tau\kappa\|\hat{\mathbf{y}}\|_1^2 + (1 + \lambda\tau)\beta\tau\|\hat{r}\|^2 \le \|\mathbf{y}\|_{\mathbf{B}}^2 + \beta\tau\|r\|^2.$$

Using Equation (1.9) with this estimate, we have

$$\Big(1 + \frac{\tau\kappa}{\kappa_2}\Big)\|\hat{\mathbf{y}}\|_{\mathbf{B}}^2 + \tau\kappa\|\hat{\mathbf{y}}\|_1^2 + (1 + \lambda\tau)\beta\tau\|\hat{r}\|^2 \le \|\mathbf{y}\|_{\mathbf{B}}^2 + \beta\tau\|r\|^2$$

Assume that $\alpha = \min\{\kappa/\kappa_2, \lambda\}$. Then from the last inequality we obtain the estimate

$$\|\hat{\mathbf{y}}\|_{\mathbf{B}}^2 + \tau\kappa(1 + \alpha\tau)^{-1}\|\hat{\mathbf{y}}\|_1^2 + \beta\tau\|\hat{r}\|^2$$
$$\le (1 + \alpha\tau)^{-1}(\|\mathbf{y}\|_{\mathbf{B}}^2 + \beta\tau\|r\|^2).$$

Setting $\gamma = \kappa(1 + \alpha\tau)^{-1}$, we finally have the statement of Equation (1.7).

Thus, we have proved the convergence of iterative method (1.5), an obvious result. But there are two unknown items: first, how to realize this method of Equation (1.5), and second, what will occur if the operator $\mathbf{B}$ depends on τ?

First let us consider the point of the algorithm of Equation (1.5). It is easy to see that the second part of Equation (1.5) is explicitly solvable with respect to $\hat{q}$,

$$\hat{q} = q - \frac{1}{\beta} \operatorname{div}^h \hat{\mathbf{v}}.$$

Substituting this expression for $\hat{q}$ in the first part of Equation (1.5) yields

$$\mathbf{B}\mathbf{v}_t - \mathbf{\Delta}^h \hat{\mathbf{v}} + \nabla^h q - \frac{1}{\beta} \nabla^h \operatorname{div}^h \hat{\mathbf{v}} = \mathbf{f}^h.$$

Here $\hat{\mathbf{v}} = \mathbf{v} + \tau\mathbf{v}_t$. Thus the last statement can be expressed in an equivalent form,

$$\left(\mathbf{B} - \tau\mathbf{\Delta}^h - \frac{\tau}{\beta} \nabla^h \operatorname{div}^h\right)\mathbf{v}_t = \mathbf{\Delta}^h \mathbf{v} + \frac{1}{\beta} \nabla^h \operatorname{div}^h \mathbf{v} - \nabla^h q + \mathbf{f}^h.$$

Let us denote $\mathbf{C} = \mathbf{B} - \tau\mathbf{\Delta}^h - (\tau/\beta)\nabla^h \operatorname{div}^h$. Then we obtain an equation to find $\mathbf{v}_t$,

$$\mathbf{C}\mathbf{v_t} = \mathbf{\Delta}^h \mathbf{v} + \frac{1}{\beta} \nabla^h \operatorname{div}^h \mathbf{v} - \nabla^h q + \mathbf{f}^h,$$

and we have for calculation

$$\begin{cases} \mathbf{C}\dfrac{\mathbf{v}^{n+1} - \mathbf{v}^n}{\tau} = \mathbf{\Delta}^h \mathbf{v}^n + \dfrac{1}{\beta} \nabla^h \operatorname{div}^h \mathbf{v}^n - \nabla^h q^n + \mathbf{f}^h & \text{in } \mathbf{U}, \\[2mm] q^{n+1} = q^n - \dfrac{1}{\beta} \operatorname{div}^h \mathbf{v}^{n+1} & \text{in } P; \\[2mm] \mathbf{v}^0 \in \mathbf{U}, \ q^0 \in P; & (1.13) \end{cases}$$

In Equation (1.13) the operators $\mathbf{B}$ and $\mathbf{C}$ are connected by the relation

$$\mathbf{B} = \mathbf{C} + \tau\mathbf{\Delta}^h + \frac{\tau}{\beta} \nabla^h \operatorname{div}^h. \qquad (1.14)$$

Thus, if $(\mathbf{v}^n, q^n)$ has already been found, we must solve the first part of Equation (1.13) in $\mathbf{U}$ of the type $\mathbf{C}\mathbf{z} = \phi$ to find $\mathbf{v}^{n+1}$. Then it is possible to obtain q^{n+1} by using the explicit second part of Equation (1.13). Thus, one step of our iterative method from $(\mathbf{v}^n, q^n)$ to $(\mathbf{v}^{n+1}, q^{n+1})$ is now complete.

From this it follows that operator $\mathbf{C}$ should be "easily invertible" in some sense. This means that the equation $\mathbf{C}\mathbf{z} = \phi$ must be easily solvable by computer. Since iterative method (1.5) is convergent with $\mathbf{B} = \mathbf{B}^* \geq 0$, from Equation (1.14) it follows that the operator $\mathbf{C}$ must be symmetric and positive definite, because the operators $\mathbf{\Delta}$ and $\nabla^h\mathrm{div}^h$ are symmetric and $\mathbf{\Delta}^h < \mathbf{0}$, $\nabla^h\mathrm{div}^h \leq \mathbf{0}$.

Theorem 1.2. Let $\mathbf{C} = \mathbf{C}^* > \mathbf{0}$. Then for any $\beta > 0$ there exists $\bar{\tau} = \bar{\tau}(\beta) > 0$ such that for any $\tau \leq \bar{\tau}$, the operator $\mathbf{B} = \mathbf{C} + \tau\mathbf{\Delta}^h + (\tau/\beta)\nabla^h\mathrm{div}^h$ will be nonnegative and method (1.5) will converge as a geometric progression with arbitrary initial values $\mathbf{v}^0 \in \mathbf{U}, q^0 \in P$.

Proof. Since $\mathbf{C} = \mathbf{C}^* > \mathbf{0}$ and all the norms in $\mathbf{U}$ are equivalent (but constants of equivalence can depend on h, the mesh step size), then there are constants γ_1, γ_2 such that the statement

$$-\gamma_1\mathbf{\Delta}^h \leq \mathbf{C} \leq -\gamma_2\mathbf{\Delta}^h$$

holds. In accordance with the theorem assumption, $\mathbf{B} = \mathbf{B}^*$ and

$$\mathbf{B} = \mathbf{C} + \tau\mathbf{\Delta}^h + \frac{\tau}{\beta}\nabla^h\mathrm{div}^h \geq (-\gamma_1 + \tau)\mathbf{\Delta}^h$$

$$+ \frac{\tau}{\beta}\nabla^h\mathrm{div}^h \geq -\left(\gamma_1 - \tau - \frac{\tau}{\beta}\right)\mathbf{\Delta}^h.$$

Since $-\mathbf{\Delta}^h > \mathbf{0}$ and $\mathbf{C}$ does not depend on τ, β, it is possible to choose $\bar{\tau}$ as the root of the equation

$$\bar{\tau}\left(1 + \frac{1}{\beta}\right) = \gamma_1. \tag{1.15}$$

Then for any $\tau \leq \bar{\tau}$ we have $\mathbf{B} \geq \mathbf{0}$, and, as a result of Theorem 1.1, method (1.13) converges as a geometric progression. Thus, Theorem 1.2 has been proved.

Let us estimate the rate of convergence of this method. From Equation (1.7) it follows that the power of geometric progression $(1 + \alpha\tau)^{-1}$ will be minimal, with maximal value of $\alpha\tau$. We can take κ_2 equal to γ_2 (actually it is possible to set it as $\gamma_2 - \tau$, but in this case κ_2 depends on τ). Then from Equation (1.14) it follows that $\bar{\tau}$ depends only on β and γ_1. It is easy to see that the constant, λ, can be chosen in such a way that Equation (1.12) holds and κ does not depend on $\tau \leq \bar{\tau}$. For this purpose it is sufficient to set

$$\lambda = \min\left\{\frac{1}{\beta}, \frac{1}{(2c_0^2\kappa_2\beta\bar{\tau})}\right\}.$$

It is worth noting that in this case λ and κ also do not depend on $\tau \leq \bar{\tau}$. Then we can choose $\alpha = \min\{\kappa/\kappa_2, \lambda\}$. Thus, in accordance with Theorem 1.2 method (1.13) converges as a geometric progression. The power of this progression is equal to $(1 + \alpha\bar{\tau})^{-1/2}$, and $\alpha, \bar{\tau}$ depend only on $\gamma_1, \gamma_2, \beta, c_0$ but are independent of h. (As a matter of fact, there would be dependence on h in implicit form. In a general case κ_2 might depend on h, and as we will see later in this case we need additional boundedness for τ.)

Next let us consider some examples of constructing finite-difference schemes, which satisfy the properties of Equations (1.2) and (1.3).

Example 1.1. Let $h = 1/N$, the mesh step size, and

$$\overline{\Omega} = \{x = (x_1, x_2) : 0 \le x_i \le 1, \ i = 1, 2\}.$$

Let us denote $\overline{\Omega}_h$ as the set of nodes

$$\overline{\Omega}_h = \{(ih, jh), 0 \le i, j \le N\}.$$

We also introduce the sets

$$\overline{\Omega}_1 = \left\{ \left(\left(i - \frac{1}{2}\right)h, jh \right) : 0 \le i, \ j \le N \right\},$$

$$\overline{\Omega}_2 = \left\{ \left(ih, \left(j - \frac{1}{2}\right)h \right) : 0 \le i, \ j \le N \right\},$$

$$\Omega_3 = \left\{ \left(ih, jh \right) : 0 \le i, j \le N - 1, \ i^2 + j^2 \ne 0 \right\}.$$

Let $\mathbf{U}$ be a linear space of vector functions $\mathbf{v} = (v_1, v_2)$, defined on $\Omega_1 \times \Omega_2$, and P be a space of functions defined on Ω_3 and satisfying the property

$$\sum_{x \in \Omega_3} h^2 q(x) = 0.$$

Let us introduce scalar products in $\mathbf{U}$ and P,

$$(\mathbf{v}, \mathbf{w})_{\mathbf{U}} = \sum_{x \in \Omega_1} h^2 v_1(x) w_1(x) + \sum_{x \in \Omega_2} h^2 v_2(x) w_2(x),$$

$$(p, q)_P = \sum_{x \in \Omega_3} h^2 p(x) q(x).$$

Henceforth we will omit indices in scalar products if no ambiguity arises. If $\mathbf{v} \in \mathbf{U}$ then we denote a vector function which coincides with $\mathbf{v}$ on $\Omega_1 \times \Omega_2$ and equals zero on $\partial\Omega_1 \times \partial\Omega_2$ by $\tilde{\mathbf{v}}$.

Finite-difference analogues of ∇ and div are given by

$$\operatorname{div}^h \mathbf{v}|_{x \in \Omega_3} = \sum_{i=1}^{2} h^{-1} \left(\tilde{v}_i \left(x + \frac{h}{2} \mathbf{e}_i \right) - \tilde{v}_i \left(x - \frac{h}{2} \mathbf{e}_i \right) \right),$$

$$\nabla^h q|_{(x, x') \in \Omega_1 \times \Omega_2} = \left[h^{-1} \left(q \left(x + \frac{h}{2} \mathbf{e}_1 \right) - q \left(x - \frac{h}{2} \mathbf{e}_1 \right) \right),$$

$$h^{-1} \left(q \left(x + \frac{h}{2} \mathbf{e}_2 \right) - q \left(x - \frac{h}{2} \mathbf{e}_2 \right) \right) \right].$$

Here $\mathbf{e}_i$ is a unit vector in the x_i direction.

By checking we find that for approximations of ∇ and div the property of Equation (1.2) holds, that is,

$$(\operatorname{div}^h \mathbf{v}, q) = -(\nabla^h q, \mathbf{v}).$$

Let $\boldsymbol{\Delta}^h = (\Delta^h, \Delta^h)$, where Δ^h is the usual "five-nodes" approximation of the Laplace operator. Then, using the usual technique, we find that Equation (1.3) also holds.[8,9]

Our finite-difference scheme approximates the original boundary value problem (1.1) with $O(h^2)$. It is possible to show that if solution $(\mathbf{u}, p)$ of the differential problem is quite smooth then the estimate,

$$\|\mathbf{u} - \mathbf{u}^h\|_{\mathbf{H}_h^1} + \|p - p^h\|_{L_{2,h}} \le ch^{3/2},$$

holds.[10] We can also use iterative method (1.13) to find the solution of a discrete problem.

Let us consider the problem of choosing the operator, $\mathbf{C}$. If we set $\mathbf{C} = -\boldsymbol{\Delta}^h$, then $\gamma_1 = 1$, and, as shown above, the iterative method converges as a geometric progression; the rate of progression does not depend on h. We can also use some direct method to solve equation $-\boldsymbol{\Delta}^h \mathbf{z} = \phi$ (see, e.g., reference 11).

Now let us set $\mathbf{C} = (\mathbf{I} - \mu R_1)(\mathbf{I} - \mu R_2)$, where R_1 and R_2 are left and right matrices with respect to the matrix of the operator, $\boldsymbol{\Delta}^h$. In this case $\mathbf{Cz} = \phi$ is easily solvable, not only in a rectangular grid domain but in a domain of arbitrary shape. When setting $\mu = O(h)$ we obtain $\gamma_1 = O(h)$. For inequality $\mathbf{B} \ge \mathbf{0}$ to be satisfied it is necessary to choose $\bar{\tau}$ to satisfy Equation (1.15); thus, in this case $\bar{\tau} = O(h)$, $\kappa_2 = 1$, and the rate of convergence of Equation (1.13) is of the order $(1 + ch)^{-1}$. Note that the estimates for rates of convergence of our iterative method coincide with the corresponding ones for grid elliptic equations.

Example 1.2. Let us construct our finite-difference scheme for the same problem by

$$\overline{\Omega}_1 = \left\{ \left(ih, \left(j - \tfrac{1}{2}\right)h \right) : 0 \le i \le N, \ 0 \le j \le N + 1 \right\},$$

$$\overline{\Omega}_2 = \left\{ \left(\left(i - \tfrac{1}{2}\right)h, jh \right) : 0 \le i \le N + 1, \ 0 \le j \le N \right\},$$

$$\Omega_3 = \left\{ \left(\left(i + \tfrac{1}{2}\right)h, \left(j + \tfrac{1}{2}\right)h \right) : 0 \le i, \ j \le N - 1 \right\}.$$

Now let $\mathbf{U}$ be a linear space of vector functions defined on $\Omega_1 \times \Omega_2$, and P be a space of functions defined on Ω_3 and orthogonal to constants. If v_i is defined on $\Omega_i (i = 1, 2)$ then as before we denote the function which is defined on $\bar{\Omega}_i$ and coincides with v_i on Ω_i as $\tilde{v}_i$. If $x \in \partial\Omega_i \cap \partial\Omega$ then we set $\tilde{v}_i = 0$; if $x \in \bar{\Omega}_i$ and $x \notin \bar{\Omega}$ we set $\tilde{v}_i(x) = -v_i(x')$ where x' is the nearest node of $\bar{\Omega}_i$ to x. Now it is possible to define operators $\Delta^h, \nabla^h, \mathrm{div}^h$ as done previously.

Thus we obtain a finite-difference scheme which possesses the same properties as the former one but looks "more symmetric". This scheme is better suited for calculations, but there is a drawback with it. Let $\mathbf{C} = -\Delta^h$. In solving the equation $-\Delta^h \mathbf{z} = \phi$ we usually assume that the number of nodes at least on one side of Ω is equal to 2^n, which is connected with the use of Fast Fourier Transform, or Fast Reduction Method (FDM), to invert the Δ^h operator.

If we use FDM from Example 1.1, and the number of nodes on x_1- and x_2-axes are equal to M and N correspondingly, then the unknown functions, v_1 and v_2, are arrays of $(M-1) \times (N-1)$ unknowns. In Example 1.2 the arrays corresponding to v_1 and v_2 contain $(M-1) \times N$ and $M \times (N-1)$ unknowns. Thus, if we use some method to invert Δ^h, which needs 2^n nodes at least on one axis, then we must choose $M = 2^m$, $N = 2^n$. But in the previous example we had to choose the number of nodes as 2^n only in one direction.

Example 1.3.   Let us consider the application of the finite-element method to numerical solution of the Stokes problem, Equation (1.1).[12] For simplicity we take as before $\bar{\Omega}$ equal to the unit square. We shall refer to the pair of functions $(\mathbf{u}, p) \in \mathbf{H}_0^1 \times L_2/\mathbf{R}^1$ as a weak solution of Equation (1.1) if the identities

$$(\nabla\mathbf{u}, \nabla\mathbf{w}) - (p, \mathrm{div}\ \mathbf{w}) = (\mathbf{f}, \mathbf{w}) \quad \text{for any } \mathbf{w} \in \mathbf{H}_0^1,$$

$$(\mathrm{div}\ \mathbf{u}, 1) = \int_G \mathrm{div}\ \mathbf{u} dx = 0 \quad \text{for any } G \subset \Omega$$

hold. The set of lines $x_1 = ih, x_2 = jh$ splits $\bar{\Omega}$ into so-called elementary squares (cells). Let us split every cell into two triangles by drawing diagonals parallel to the line $y = x$. Thus we have triangulated our domain, Ω.

We denote the set of elementary triangles by T. Now let us split every elementary triangle $t \in T$ into four triangles, by drawing lines connecting the middles of the triangle sides. We denote the set of "small" triangles by T'. Let $H \subset \mathbf{H}_0^1$ be a space of continuous vector functions which are linear on every triangle from T', and

$Q \subset L_2/\mathbf{R}^1$ be a space of functions which are constant on every triangle from T.

Let us denote a pair of functions $(\mathbf{v}, q) \in H \times Q$ by a discrete weak solution if

$$
\begin{cases}
(\nabla \mathbf{v}, \nabla \mathbf{w}) - (q, \operatorname{div} \mathbf{w}) = (\mathbf{f}, \mathbf{w}) & \text{for any } \mathbf{w} \in H, \\
(\operatorname{div} \mathbf{v}, 1)_t = \int_t \operatorname{div} \mathbf{v} dx = 0 & \text{for any } t \subset T.
\end{cases} \quad (1.16)
$$

It is not difficult to demonstrate that if $\mathbf{u} \in \mathbf{H}^2 \cap \mathbf{H}_0^1$, $p \in H^1$ then the estimate

$$
\|\mathbf{u} - \mathbf{v}\|_{\mathbf{H}^1} + \|p - q\|_{L_2} \le ch
$$

is true. Since $\mathbf{v} \in H$ can be defined by its values on simplex vertices of T' in only one way, and q can be defined by its values in simplices of T also in only the one way, then choosing trial functions $\mathbf{w}$ as basic functions in H and taking into account the second part of Equation (1.16), we obtain a system of algebraic equations. The solution of this system exists and is unique; the operators div^h and ∇^h satisfy the properties of Equations (1.2) and (1.3). Thus we can use iterative method (1.13) to solve these simultaneous linear equations.

Consider iterative method (1.13), which we will call the (β, τ) method. This kind of method was proposed (without proof)[13] by N.N. Yanenko to solve nonstationary problems for the Navier–Stokes equations. Our method (1.13) differs in that it contains the term $1/\beta \nabla^h \operatorname{div}^h \mathbf{v}^n$ in the first part of Equation (1.13); the second part of Yanenko's equation had the form

$$
\varepsilon \frac{q^{n+1} - q^n}{\tau} + \operatorname{div}^h \mathbf{v}^{n+1} = 0.
$$

But it was not clear how to choose ε. As has been shown previously, the parameter ε should be equal to $\beta\tau$. In this case, $\beta = O(1)$ depends on the domain shape and should be chosen by numerical experiments on a computer.

Let us also comment on the role of the term $(1/\beta)\nabla^h \operatorname{div}^h \mathbf{v}^n$ in the first part of Equation (1.13). First, note that this term was absent in our original grid Equations (1.4) and appeared only due to our way of constructing the iterative method (fully implicit scheme). If we constructed an iterative method in the usual way,[7] then this term would be absent. This term is important when applied to numerical solutions of practical problems, since its presence accelerates the rate of convergence of the iterative method, especially in a nonlinear case.

During numerical experiments with the grid Stokes problem (Example 1.2) one can see that iterative method (1.13) with $C = -\Delta^h$ converges about 1.5 times faster if the term $(1/\beta)\nabla^h \text{div}^h \mathbf{v}^n$ is present (in both cases we produced numerical experiments with optimal parameters β, τ).

It is also interesting to clarify what occurs if we rewrite method (1.13) in the so-called "canonical form",[7]

$$\mathbf{D}\mathbf{z}_t + \mathbf{A}\mathbf{z} = \phi,$$

where $\mathbf{z} = (\mathbf{v}^n, q^n)$. Operators $\mathbf{D}$ and $\mathbf{A}$ are symmetric but not positive (negative) definite, and they are not commutative, that is, $\mathbf{DA} \neq \mathbf{AD}$.[14] Thus Equation (1.13) is a method with a nonsymmetric transfer operator. If this method is convergent, it is possible to estimate the norm of the transfer operator by $(1 + \alpha\tau)^{-1}$. But the only thing to be said about its spectrum is that all eigenvalues belong to the circle of radius $(1 + \alpha\tau)^{-1}$. From this it follows that it is impossible to investigate method (1.13) by using the usual techniques.[7] Theorem 1.2 can be strengthened in the following way: The statement $\mathbf{C} = \mathbf{C}^* > 0$ was required to hold, and it was assumed that $\mathbf{Cz} = \phi$ could be solved exactly by a computer. But this requirement can be weakened. If the equation $\mathbf{Cz} = \phi$ is solved approximately by some inner iterative process, then Theorem 1.2 will hold if we make some assumptions about the inner iterative method and the number of inner iterations.

Now let us show how to obtain an iterative method for Equation (1.4) with a symmetric transfer operator. We apply operator $(\Delta^h)^{-1}$ to the first part of Equation (1.4). Then we have

$$\mathbf{v} = (\Delta^h)^{-1}\nabla^h q - (\Delta^h)^{-1}\mathbf{f}^h.$$

(Here we mean $(\Delta^h)^{-1}\mathbf{f}^h$ to be a solution of the equation $\Delta^h \psi = \mathbf{f}^h$, $\psi \in \mathbf{U}$.) Substitute this expression for $\mathbf{v}$ into the second part of Equation (1.4). As a result we obtain an equation for q,

$$\text{div}^h (\Delta^h)^{-1}\nabla^h q = \text{div}^h (\Delta^h)^{-1}\mathbf{f}^h. \tag{1.17}$$

Let us introduce notations $A = \text{div}^h (\Delta^h)^{-1}\nabla^h$, $\text{div}^h (\Delta^h)^{-1}\mathbf{f}^h = \phi$. Theorem 1.3 holds.[15]

Theorem 1.3. Operator A is symmetric and positive definite on P, and there exists a constant, θ, independent of the h mesh step size such that the inequality

$$\theta I \leq A \leq I \tag{1.18}$$

is true.

Proof. Symmetry and positive definiteness directly follow from the property of Equation (1.2) and positive definiteness of the operator, $-\Delta^h$. Let us prove the validity of Equation (1.18). Obviously we have the sequence of equalities

$$(Aq, q)_P = \left(\mathrm{div}^h \left(\Delta^h\right)^{-1} \nabla^h q, q\right)_P = -\left(\left(\Delta^h\right)^{-1} \nabla^h q, \nabla^h q\right)_U$$

$$= \sup_{\mathbf{v} \in \mathbf{U}} \frac{\left(\nabla^h q, \mathbf{v}\right)_{\mathbf{U}}^2}{\left(-\Delta^h \mathbf{v}, \mathbf{v}\right)_{\mathbf{U}}} = \sup_{\mathbf{v} \in \mathbf{U}} \frac{\left(q, \mathrm{div}^h \mathbf{v}\right)_{\mathbf{U}}^2}{\|\mathbf{v}\|_1^2} \,.$$

By using Equation (1.3) we get the lower bound,

$$(Aq, q)_P \geq c_0^{-2} \|q\|_P^2.$$

The upper bound is almost obvious, that is,

$$(Aq, q)_P = \sup_{\mathbf{v} \in \mathbf{U}} \frac{(q, \mathrm{div}^h \mathbf{v})_P}{\|\mathbf{v}\|_1^2} \leq \|q\|_P^2.$$

Thus the theorem has been proved.

From Theorem 1.3, it follows that to find the solution we can use ordinary iterative methods such as the Richardson method, or CGM. One step of such an iterative method requires at least two solutions of the Dirichlet problem for the Poisson equation. The rate of convergence of these methods does not depend on the mesh step size.

Numerical experiments concerning the investigation of the rate of convergence of iterative methods for Equation (1.17) showed high efficiency of this approach to the numerical solution of the Stokes problem. Thus, CGM made the norm of the error function three times as small per step. To investigate the source of so fast a convergence, we investigate the spectrum of operator A. For simplicity we consider the finite-difference scheme from Example 1.2. We state that 1 is an eigenvalue of A, with multiplicity proportional to h^{-2} (the number of grid nodes in the domain), and the number of other eigenvalues is proportional to h^{-1} (the number of boundary nodes).

To prove this statement, we rewrite our original grid problem. Let us denote $\mathbf{U}_0$ as a space of vector functions defined on $\overline{\Omega}_1 \times \overline{\Omega}_2$ and satisfying the boundary conditions

$$v_i(x) = \begin{cases} 0, & x \in \partial\Omega \cap \partial\Omega_i, \\ -v_i(x'), & x \in \partial\Omega_i \backslash \partial\Omega, \ i = 1, 2; \end{cases}$$

here x' is the nearest to x node of Ω_i. We also denote P as a space of functions defined on Ω_3 and satisfying $\sum_{x \in \Omega_3} h^2 q(x) = 0$. Thus, our grid Stokes problem is

$$
\begin{cases}
-\boldsymbol{\Delta}_0^h \mathbf{v} + \nabla^h q = \mathbf{f}^h & \text{in } \Omega_1 \times \Omega_2, \\
\operatorname{div}^h \mathbf{v} = 0 & \text{in } \Omega_3, \quad \mathbf{v} \in \mathbf{U}_0, \ q \in P,
\end{cases} \tag{1.19}
$$

that is, the first part of Equation (1.19) should hold in every node $(x, x') \in \Omega_1 \times \Omega_2$, and the second part of Equation (1.19) should hold in every node $x \in \Omega_3$. The $\boldsymbol{\Delta}_0^h$ are the usual five-nodes approximation of the Laplace operator, therefore $\boldsymbol{\Delta}_0^h : \mathbf{U}_0 \to \mathbf{U}$.

We consider the Stokes problem with periodic boundary conditions as well as Equation (1.19),

$$
\begin{cases}
-\boldsymbol{\Delta} \mathbf{u} + \nabla p = \mathbf{f}, \\
\operatorname{div} \mathbf{u} = 0, \quad \mathbf{u} \cdot \mathbf{n}|_{\partial \Omega} = 0, \quad \dfrac{\partial (\mathbf{u} \cdot \boldsymbol{\tau})}{\partial \mathbf{n}} \Big|_{\partial \Omega} = 0.
\end{cases} \tag{1.20}
$$

Here $\mathbf{n}$ and $\boldsymbol{\tau}$ are normal and tangent unit vectors.

It is known that the weak solution of this problem exists and is unique:[16] if $\mathbf{f} \in \mathbf{L}_q$ then $\mathbf{u} \in \mathbf{W}_q^2$, $p \in W_q^1 / \mathbf{R}^1$. Let $\mathbf{U}_p$ be a space of vector functions defined on $\bar{\Omega}_1 \times \bar{\Omega}_2$ and satisfying boundary conditions

$$
v_i(x) = \begin{cases}
0, & x \in \partial \Omega \cap \partial \Omega_i, \\
v_i(x'), & x \in \partial \Omega_i \backslash \partial \Omega, \ i = 1, 2.
\end{cases} \tag{1.21}
$$

Here $x' \in \Omega_i$ is the nearest to x node. Now, corresponding to the grid problem of Equation (1.20) is

$$
\begin{cases}
-\boldsymbol{\Delta}_p^h \mathbf{v} + \nabla^h q = \mathbf{f}^h & \text{in } \Omega_1 \times \Omega_2, \\
\operatorname{div}^h \mathbf{v} = 0 & \text{in } \Omega_3, \quad \mathbf{v} \in \mathbf{U}_p, \ q \in P;
\end{cases} \tag{1.22}
$$

here $\boldsymbol{\Delta}_p^h$ is the usual five-nodes approximation of the Laplace operator, thus

$$
\boldsymbol{\Delta}_p^h : \mathbf{U}_p \to \mathbf{U}.
$$

Lemma 1.1. For an arbitrary function $q \in P$, the equality

$$
\operatorname{div}^h \left(\boldsymbol{\Delta}_p^h \right)^{-1} \nabla q = q \tag{1.23}
$$

holds, that is, $\operatorname{div}^h \left(\boldsymbol{\Delta}_p^h \right)^{-1} \nabla = I$ with I being an identity operator.

Proof. Let us denote $\left(\boldsymbol{\Delta}_p^h \right)^{-1} \nabla q = \mathbf{v}$. Then we can rewrite Equation (1.23) as

$$\begin{cases} -\boldsymbol{\Delta}_p^h \mathbf{v} + \nabla h = \mathbf{0} & \text{in } \Omega_1 \times \Omega_2, \\ \operatorname{div}^h \mathbf{v} = q & \text{in } \Omega_3, \quad \mathbf{v} \in \mathbf{U}_p, \ q \in P. \end{cases} \quad (1.24)$$

Let us continue q evenly from Ω_3 onto squares $\{-1 \le x_1 \le 0, 0 \le x_2 \le 1\}$ and $\{0 \le x_1 \le 1, -1 \le x_2 \le 0\}$. Then we shall continue our function evenly from $\{-1 \le x_1 \le 0, 0 \le x_2 \le 1\}$ onto $\{-1 \le x_1 \le 0, -1 \le x_2 \le 0\}$. Now our continued function, $\tilde{q}$, is defined on the square $\{-1 \le x_1 \le 1, -1 \le x_2 \le 1\}$. If we continue v_i oddly with respect to the x_i-axis and evenly with respect to the x_{3-i}-axis, then for the continued functions $\tilde{\mathbf{v}}, \tilde{q}$ Equations (1.24) hold due to Equation (1.21) in all the nodes where they make sense. Now it is possible to continue functions $\tilde{\mathbf{v}}$ and $\tilde{q}$ periodically to the whole plane with period 2 (we shall denote them as before $\tilde{\mathbf{v}}, \tilde{q}$), and for these functions Equations (1.24) will hold on the whole grid plane. This means that problem (1.24) can be continued periodically and Fourier analysis can be used to investigate it.

Next expand functions $\tilde{v}_1, \tilde{v}_2, \tilde{q}$ into a discrete Fourier series with corresponding systems of functions $\{\sin \pi k_1 x_1 \cos \pi k_2 x_2\}$, $\{\cos \pi k_1 x_1 \sin \pi k_2 x_2\}$, $\{\cos \pi k_1 x_1 \cos \pi k_2 x_2\}$. By substituting these expansions in Equation (1.24), we find that for any $q \in P$ the system of Equations (1.24) has a unique solution. Thus, Lemma 1.1 is proved.

Let us investigate the spectrum of $A = \operatorname{div}^h \left(\boldsymbol{\Delta}_0^h\right)^{-1} \nabla^h$. Estimate all the λ such that

$$Aq = \lambda q, \qquad q \in P. \quad (1.25)$$

Let λ be an eigenvalue of A. As previously, we denote vector function $\left(\boldsymbol{\Delta}_0^h\right)^{-1} \nabla q$ as $\mathbf{v}$. Then Equation (1.25) is equivalent to the system

$$\begin{cases} -\boldsymbol{\Delta}_0^h \mathbf{v} + \nabla^h q = \mathbf{0} & \text{in } \Omega_1 \times \Omega_2, \\ \operatorname{div}^h \mathbf{v} = \lambda q & \text{in } \Omega_3, \quad \mathbf{v} \in \mathbf{U}_0, \ q \in P, \end{cases} \quad (1.26)$$

Since from the second part of Equation (1.26) we have $q = (1/\lambda)\operatorname{div}^h \mathbf{v}$, substituting this expression into the first part of the equation yields

$$-\boldsymbol{\Delta}_0^h \mathbf{v} = \nabla^h \frac{1}{\lambda} \operatorname{div}^h \mathbf{v}, \qquad \mathbf{v} = \left(\boldsymbol{\Delta}_0^h\right)^{-1} \nabla q, \ q \in P. \quad (1.27)$$

Thus we proved the following lemma.

Lemma 1.2. If λ is an eigenvalue of Equation (1.25), then $1/\lambda$ is an eigenvalue of Equation (1.27).

It is obvious that the inverse statement also holds. If a vector function, $\mathbf{v}$, presented as $\mathbf{v} = \left(\Delta_0^h\right)^{-1}\nabla q (q \in P)$, is an eigenfunction of Equation (1.27) with corresponding eigenvalue $1/\lambda$, then by making the inverse transformation we find λ to be an eigenvalue of Equation (1.25). Thus, we find that the eigenvalue problem

$$-\Delta_p^h\mathbf{v} = \frac{1}{\lambda}\nabla\text{div}^h\mathbf{v}, \qquad \mathbf{v} = \left(\Delta_p^h\right)^{-1}\nabla q, \ \ q \in P$$

has the only eigenvalue which is equal to 1.

Let $\mathbf{V}_0$ be a space of vector functions which are represented as $\mathbf{v} = \left(\Delta_0^h\right)^{-1}\nabla q$, $(q \in P)$, and $\mathbf{V}_p$ be a space of vector functions represented as $\mathbf{v} = (\Delta_p^h)^{-1}\nabla q$, $(q \in P)$. If $\mathbf{v}$ is defined on $\bar{\Omega}_1 \times \bar{\Omega}_2$ then we use notation D^h for the grid operator

$$D^h\mathbf{v}(x) = \frac{v_1\left(x + \frac{h}{2}\,\mathbf{e}_2\right) - v_1\left(x - \frac{h}{2}\,\mathbf{e}_2\right)}{h}$$

$$- \frac{v_2\left(x + \frac{h}{2}\,\mathbf{e}_1\right) - v_2\left(x - \frac{h}{2}\,\mathbf{e}_1\right)}{h} \qquad (1.28)$$

We assume that x in Equation (1.28) is such that all the expressions in the right-hand part of Equation (1.28) make sense.

Space $\mathbf{V}$ of the vector functions defined on $\bar{\Omega}_1 \times \bar{\Omega}_2$ is equal to zero on neighboring boundary nodes (i.e., nodes from $\Omega_1 \times \Omega_2$ such that the distance from every node to boundary $\partial\Omega$ is not greater than h) and satisfies the condition $D^h\mathbf{v} = 0$ in all nodes where this expression is true. Then for any $\mathbf{v} \in \mathbf{V}$ there exists $q \in P$ such that $\Delta_0^h\mathbf{v} = \nabla^h q$.[17]

Since for any $\mathbf{v} \in \mathbf{V}$ the equality $\Delta_0^h\mathbf{v} = \Delta_p^h\mathbf{v}$ holds, an arbitrary function, $\mathbf{v} \in \mathbf{V}$, can be represented as

$$\mathbf{v} = \left(\Delta_0^h\right)^{-1}\nabla^h q = \left(\Delta_p^h\right)^{-1}\nabla^h q, \qquad q \in P. \qquad (1.29)$$

From this it follows that $\mathbf{V} \subset (\mathbf{U}_0 \cap \mathbf{U}_p)$. It is also easy to see that $\dim \mathbf{V}_0' = O(h^{-1})$, and $\dim \mathbf{V}_p' = O(h^{-1})$; here, $\mathbf{V}_0'$ and $\mathbf{V}_p'$ are orthogonal complements to $\mathbf{V}$ in $\mathbf{U}_0$ and $\mathbf{U}_p$, respectively. Thus we proved the Theorem 1.4.

Theorem 1.4. Operator $A = \text{div}^h\left(\Delta_0^h\right)^{-1}\nabla$ has eigenvalue 1 with multiplicity of $O(h^{-2})$; the number of remaining eigenvalues equals $O(h^{-1})$.

This result explains why only a few iterations are necessary in Equation (1.17).

2. METHODS FOR SOLVING NONSTATIONARY STOKES PROBLEM

In this section we briefly consider numerical methods for solving nonstationary Stokes problem. First we state the nonstationary Stokes problem to be considered. Find functions $\mathbf{u}(x,t), p(x,t)$ defined on Q_T, where $\bar{Q}_T = \bar{\Omega} \times [0,T]$, the weak solutions of Equations (2.1) which satisfy boundary and initial conditions:[6]

$$\begin{cases} \dfrac{\partial \mathbf{u}}{\partial t} = \nu \mathbf{\Delta u} - \nabla p + \mathbf{f}, \\ \operatorname{div} \mathbf{u} = 0, \quad \mathbf{u}|_{\partial\Omega \times [0,T]} = \mathbf{0}, \quad \mathbf{u}(x,0) = \mathbf{u}_0(x). \end{cases} \quad (2.1)$$

We assume that the solution, $(\mathbf{u}, p)$, is as smooth as required for our considerations.

There are two obvious possibilities for solving this problem. First, modify the incompressibility equation, changing it by a weaker requirement (e.g., $\|\operatorname{div} \mathbf{u}(t)\|_{L_2(Q_T)} \leq C_T \varepsilon$). To do this, in Equation (2.1) use slight incompressibility instead of the incompressibility equation;[13] thus we must change the second part of Equation (2.1) to

$$\varepsilon p_\varepsilon + \operatorname{div} \mathbf{u}_\varepsilon = 0.$$

Approximations of Equation (2.1) by systems of this type we call ε-*regularizations*. In this case we obtain the mixed boundary value problem

$$\begin{cases} \dfrac{\partial \mathbf{u}_\varepsilon}{\partial t} = \nu \mathbf{\Delta u}_\varepsilon - \nabla p_\varepsilon + \mathbf{f}, \\ \varepsilon p_\varepsilon + \operatorname{div} \mathbf{u}_\varepsilon = 0, \quad \mathbf{u}_\varepsilon|_{\partial\Omega \times [0,T]} = \mathbf{0}, \quad \mathbf{u}_\varepsilon(x,0) = \mathbf{u}_0(x). \end{cases} \quad (2.2)$$

If we substitute for p_ε in the first part of Equation (2.2) its expression from the second part of Equation (2.2) we obtain a problem with only $\mathbf{u}_\varepsilon$ as the dependent variable, that is,

$$\begin{cases} \dfrac{\partial \mathbf{u}_\varepsilon}{\partial t} = \nu \mathbf{\Delta u}_\varepsilon + \dfrac{1}{\varepsilon} \nabla \operatorname{div} \mathbf{u}_\varepsilon + \mathbf{f}, \\ \mathbf{u}_\varepsilon|_{\partial\Omega \times [0,T]} = \mathbf{0}, \quad \mathbf{u}_\varepsilon(x,0) = \mathbf{u}_0(x). \end{cases} \quad (2.3)$$

The differential operator in the right-hand part of Equation (2.3) is a well-known differential operator of the theory of elasticity equations. Thus the weak solution of this problem exists and is unique. From this statement follows the existence and uniqueness of the solution of Equation (2.2).

Furthermore, we would like uniform boundedness of norms $\|p_\varepsilon\|_{L_2(Q_T)}$ on ε. To prove this, multiply both parts of the first part of Equation (2.2) by $\mathbf{u}_\varepsilon$ and the second part of this equation by p_ε.[8] Integrating these expressions with respect to Ω and adding the results yields

$$\frac{1}{2}\frac{d}{dt}\|\mathbf{u}_\varepsilon(t)\|^2 + \nu\|\nabla\mathbf{u}_\varepsilon(t)\|^2 + \varepsilon\|p_\varepsilon(t)\|^2 = (\mathbf{f},\mathbf{u}_\varepsilon).$$

Here $\|\cdot\| = \|\cdot\|_{L_2(\Omega)}, (,) = (,)_{L_2(\Omega)}$.

Integrating this equation with respect to time from 0 to t, and estimating in an obvious way the right-hand part, we obtain

$$\frac{1}{2}\|\mathbf{u}_\varepsilon(t)\|^2 - \frac{1}{2}\|\mathbf{u}_0\|^2 + \nu\int_0^t\|\nabla\mathbf{u}_\varepsilon(t)\|^2 dt + \varepsilon\int_0^t\|p_\varepsilon(t)\|^2 dt$$

$$\leq \delta\int_0^t\|\nabla\mathbf{u}_\varepsilon(t)\|^2 dt + \frac{1}{4\delta}\int_0^t\|\mathbf{f}(t)\|^2_{-1} dt.$$

Let $\delta = \nu/2$. Then from the previous inequality

$$\|\mathbf{u}_\varepsilon(t)\|^2 + \nu\int_0^t\|\nabla\mathbf{u}_\varepsilon(t)\|^2 dt + 2\varepsilon\int_0^t\|p_\varepsilon(t)\|^2 dt$$

$$\leq \|\mathbf{u}_0\|^2 + \frac{1}{\nu}\int_0^t\|\mathbf{f}(t)\|^2_{-1} dt. \tag{2.4}$$

Thus $\|p_\varepsilon\|_{L_2(Q_T)} \leq C_T/\sqrt{\varepsilon}$.

A more precise estimate for norms p_ε holds, namely

$$\|p_\varepsilon\|_{L_2(Q_T)} \leq C.$$

To demonstrate this, multiply both parts of Equation (2.3) by $\partial\mathbf{u}_\varepsilon/\partial t$, forming the scalar product. Integrating by parts yields

$$\left\|\frac{\partial\mathbf{u}_\varepsilon}{\partial t}\right\|^2 + \frac{\nu}{2}\frac{d}{dt}\|\nabla\mathbf{u}_\varepsilon\|^2 + \frac{1}{2\varepsilon}\frac{d}{dt}\|\mathrm{div}\,\mathbf{u}_\varepsilon\|^2$$

$$= \left(\mathbf{f},\frac{\partial\mathbf{u}_\varepsilon}{\partial t}\right) \leq \delta\left\|\frac{\partial\mathbf{u}_\varepsilon}{\partial t}\right\|^2 + \frac{1}{4\delta}\|\mathbf{f}\|^2.$$

Let $\delta = 1/2$. Integrating the last inequality on t from 0 to t yields

$$\frac{1}{2}\int_0^t\left\|\frac{\partial\mathbf{u}_\varepsilon}{\partial t}\right\|^2 dt + \frac{\nu}{2}\|\nabla\mathbf{u}_\varepsilon\|^2 + \frac{1}{2\varepsilon}\|\mathrm{div}\,\mathbf{u}_\varepsilon\|^2 \leq c,$$

where c depends on $\mathbf{f}, T, \|\nabla\mathbf{u}_0\|^2$. Thus, $\|\partial\mathbf{u}_\varepsilon/\partial t\|_{L_2(Q_T)} \leq C_T$.

Now multiply the first part of Equation (2.2) by $\phi \in \mathbf{H}_0^1 \times L_2([0,T],$ producing the scalar product. Integrating by parts yields

$$|(p_\varepsilon, \operatorname{div} \phi)| \leq \nu |(\nabla \mathbf{u}_\varepsilon, \nabla \phi)| + \left|\left(\frac{\partial \mathbf{u}_\varepsilon}{\partial y}, \phi\right)\right| + |(\mathbf{f}, \phi)|$$

$$\leq \nu \|\nabla \mathbf{u}_\varepsilon\| \, \|\nabla \phi\| + \left\|\frac{\partial \mathbf{u}_\varepsilon}{\partial t}\right\| \, \|\phi\| + \|\mathbf{f}\|_{-1} \, \|\nabla \phi\|$$

$$\leq \left(\nu \|\nabla \mathbf{u}_\varepsilon\| + c\left\|\frac{\partial \mathbf{u}_\varepsilon}{\partial t}\right\| + \|\mathbf{f}\|_{-1}\right) \|\nabla \phi\|.$$

The Babushka-Brezzi inequality gives us[8,9]

$$\|q\|_{L_2} \leq c_0 \|\nabla q\|_{-1}, \qquad q \in L_2/\mathbf{R}^1,$$

and the last equation becomes

$$\|p_\varepsilon(t)\| \leq c_0(\nu \|\nabla \mathbf{u}_\varepsilon(t)\| + c\left\|\frac{\partial \mathbf{u}_\varepsilon(t)}{\partial t}\right\| + \|\mathbf{f}(t)\|_{-1}).$$

Since the norms $\|\partial \mathbf{u}_\varepsilon/\partial t\|_{L_2(Q_T)}$, $\|\nabla \mathbf{u}_\varepsilon\|_{L_2(Q_T)}$, and $(\int_0^T \|\mathbf{f}\|_{-1}^2 dt)^{1/2}$ are uniformly bounded on ε, then from this fact and the last inequality the desired estimate,

$$\|p_\varepsilon\|_{L_2(Q_T)} \leq c \qquad (2.5)$$

follows. It should be pointed out that the constant in the last inequality in general depends on T. If we apply the same procedure to Equation (2.1), then

$$\|\operatorname{div} \mathbf{u}_\varepsilon\|_{L_2(Q_T)} \leq C_T \varepsilon.$$

Now let us investigate the question of the error between $\mathbf{u}_\varepsilon$ and $\mathbf{u}$. We have the boundary value problem for error functions $\mathbf{v}_\varepsilon = \mathbf{u}_\varepsilon - \mathbf{u}$, $q_\varepsilon = p_\varepsilon - p$,

$$\begin{cases} \dfrac{\partial \mathbf{v}_\varepsilon}{\partial t} = \nu \Delta \mathbf{v}_\varepsilon - \nabla q_\varepsilon, & \varepsilon q_\varepsilon + \varepsilon p + \operatorname{div} \mathbf{v}_\varepsilon = 0, \\[2mm] \mathbf{v}_\varepsilon|_{\partial\Omega \times [0,T]} = 0, & \mathbf{v}_\varepsilon(x,0) = 0. \end{cases} \qquad (2.6)$$

Form a scalar product of the first part of Equation (2.6) with $\mathbf{v}_\varepsilon$ and of the second part of this equation with q_ε. Adding the results yields

$$\frac{1}{2}\frac{d}{dt}\|\mathbf{v}_\varepsilon(t)\|^2 + \nu\|\nabla \mathbf{v}_\varepsilon(t)\|^2 + \varepsilon\|q_\varepsilon(t)\|^2 = -\varepsilon(p, q_\varepsilon).$$

Using the ε inequality transforms the last statement into

$$\frac{1}{2}\frac{d}{dt}\|\mathbf{v}_\varepsilon(t)\|^2 + \nu\|\nabla\mathbf{v}_\varepsilon(t)\|^2 + \frac{\varepsilon}{2}\|q_\varepsilon(t)\|^2 \le \frac{\varepsilon}{2}\|p(t)\|^2.$$

Integrating this inequality from 0 to t yields

$$\max_{0\le t\le T}\|\mathbf{v}_\varepsilon(\mathbf{t})\| + \sqrt{\nu}\,\|\nabla\mathbf{v}_\varepsilon\|_{L_2(Q_T)} \le C_T\sqrt{\varepsilon}, \qquad t\in[0,T].$$

We can investigate other ε-regularizations (see, e.g., reference 18) of Equation (2.1) as well as Equation (2.2). In all these cases $\mathbf{u}_\varepsilon \to \mathbf{u}$ and $p_\varepsilon \to p$. We only consider methods of solving Equation (2.2) and problems arising in applying these methods, because they are common for all ε-regularizations of the Stokes problem.

First let us construct grids and approximations of the operators $\boldsymbol{\Delta}$, ∇, div. As previously, we set $\bar{\Omega}$ as a unit square and $h = 1/N$. We introduce the sets of nodes,

$$\bar{\Omega}_1 = \Big\{x = (x_1,x_2) : x_1 = \Big(i - \frac{1}{2}\Big)h,\ x_2 = jh,$$
$$0 \le i \le N+1, 0 \le j \le N\Big\},$$

$$\bar{\Omega}_2 = \Big\{x = (x_1,x_2) : x_1 = ih,\ x_2 = \Big(j - \frac{1}{2}\Big)h,$$
$$0 \le i \le N, 0 \le j \le N+1\Big\},$$

$$\Omega_3 = \Big\{x = (x_1,x_2) : x_1 = ih,\ x_2 = jh,$$
$$0 \le i,j \le N\Big\}\backslash\{(0,0),(0,1),(1,0),(1,1)\}.$$

Let $\mathbf{U}$ be a linear space of vector functions defined on $\Omega_1 \times \Omega_2$, and let P be a space of scalar functions defined on Ω_3 and orthogonal to constants. If $\mathbf{v} \in \mathbf{U}$ then we shall denote $\tilde{\mathbf{v}}$ as a function, with $\mathbf{v}$ on $\Omega_1 \times \Omega_2$ and equal to zero on $\partial\Omega_1 \times \partial\Omega_2$. Now we can define approximations of $\boldsymbol{\Delta}, \nabla$, div in the same way as was done in Example 1.1.

For definiteness let us consider the finite-difference scheme of the second-order approximation over time. The equations containing two time layers, $(\mathbf{u}_\varepsilon^h, p_\varepsilon^h)$ and $(\hat{\mathbf{u}}_\varepsilon^h, \hat{p}_\varepsilon^h)$, will be

$$\mathbf{u}_{\varepsilon t}^h = \frac{\nu}{2}\,\boldsymbol{\Delta}^h\big(\hat{\mathbf{u}}_\varepsilon^h + \mathbf{u}_\varepsilon^h\big) - \frac{1}{2}\nabla^h\big(\hat{p}_\varepsilon^h + p_\varepsilon^h\big) + \frac{1}{2}\big(\hat{\mathbf{f}}^h + \mathbf{f}^h\big),$$

$$\varepsilon\big(\hat{p}_\varepsilon^h + p_\varepsilon^h\big) + \text{div}^h\big(\hat{\mathbf{u}}_\varepsilon^h + \mathbf{u}_\varepsilon^h\big) = 0. \tag{2.7}$$

It is easy to show that this finite-difference scheme is stable and is of second-order approximation, $O(h^2 + \tau^2)$. To clarify any problems arising during the realization of Equation (2.7), from the second part of Equation (2.7) we have

$$\hat{p}_\varepsilon^h = -p_\varepsilon^h - \frac{1}{\varepsilon} \mathrm{div}^h \left(\hat{\mathbf{u}}_\varepsilon^h + \mathbf{u}_\varepsilon^h \right).$$

Substituting this expression into the first part of Equation (2.7) yields

$$\mathbf{u}_{\varepsilon t}^h = \frac{\nu}{2} \, \boldsymbol{\Delta}^h \left(\hat{\mathbf{u}}_\varepsilon^h + \mathbf{u}_\varepsilon^h \right) + \frac{1}{2\varepsilon} \, \nabla^h \mathrm{div}^h \left(\hat{\mathbf{u}}_\varepsilon^h + \mathbf{u}_\varepsilon^h \right) + \frac{1}{2} \left(\hat{\mathbf{f}}^h + \mathbf{f}^h \right).$$

Thus, the unknown function, $\hat{\mathbf{u}}_\varepsilon^h$, is the solution of the grid boundary value problem,

$$\mathbf{A}\hat{\mathbf{u}}_\varepsilon^h \equiv \left(\mathbf{I} - \frac{\nu \tau}{2} \, \boldsymbol{\Delta}^h - \frac{\tau}{2\varepsilon} \, \nabla^h \mathrm{div}^h \right) \hat{\mathbf{u}}_\varepsilon^h = \phi^h, \qquad \mathbf{u}_\varepsilon^h \in \mathbf{U}, \quad (2.8)$$

where $\qquad \phi^h = \mathbf{u}_\varepsilon^h + \frac{\tau \nu}{2} \, \boldsymbol{\Delta}^h \mathbf{u}_\varepsilon^h + \frac{\tau}{2\varepsilon} \, \nabla^h \mathrm{div}^h \mathbf{u}_\varepsilon^h + \frac{\tau}{2} \left(\hat{\mathbf{f}}^h + \mathbf{f}^h \right).$

The operator $\mathbf{A}$ is symmetric and positive definite. It is not difficult to show (using Equation [1.3]) that its eigenvalues belong to the interval $[1 + c_1 \nu \tau, 1 + c_2 (\nu \tau)/h^2 + (c_3 \tau)/\varepsilon h^2]$. Therefore, the rate of convergence of the classical iterative methods of solving Equation (2.8) depends on ε. Thus, we have two opposing requirements. From the first one we must choose ε fairly small to provide a short distance between $\mathbf{u}_\varepsilon^h$ and $\mathbf{u}^h$. On the other hand, the rate of convergence of classical iterative methods for solving Equation (2.8) deteriorates with small ε — the number of iterations is proportional to $\varepsilon^{-1} |\ln \delta|$ if we want to get the solution with accuracy δ.

New original iterative methods were proposed to solve equations similar to Equation (2.8).[15,19] The rate of convergence of these methods does not depend on ε, but in our case the rate of convergence of these methods depends on τ. Since the solution of a nonstationary problem changes smoothly with the time variable, we have an appropriate initial value to solve Equation (2.8);[7] this coincides with the functions $\mathbf{u}_\varepsilon^h$ and p_ε^h for the previous time step. Thus it is sufficient to make only a few iterations for every time step. (As a rule in practice, it is sufficient to make no more than four iterations per time step.)

If we want a more precise solution (see, e.g., reference 20) we can use asymptotic methods to expand $\mathbf{u}_\varepsilon$. It is not difficult to show that if $\partial \Omega$ and $\mathbf{u}, \mathbf{f}$ are smooth enough then the expansion,

$$\mathbf{u}_\varepsilon = \mathbf{u} + \varepsilon \mathbf{u}_1 + \varepsilon^2 \mathbf{u}_2 + \cdots + \varepsilon^{n-1} \mathbf{u}_{n-1} + O(\varepsilon^n), \qquad (2.9)$$

holds where $\|\nabla \mathbf{u}_i\|_{L_2(Q_T)} \leq c$.

Thus we must solve problem (2.8) with $\varepsilon = \varepsilon_1, \ldots, \varepsilon_n$. From Equation (2.9) we have

$$\mathbf{u}_{\varepsilon_i} = \mathbf{u} + \varepsilon_i \mathbf{u}_1 + \varepsilon_i^2 \mathbf{u}_2 + \cdots, \qquad i = 1, 2, \ldots, n. \qquad (2.10)$$

We shall find the approximation to $\mathbf{u}$ from Equation (2.10) in such a way that the difference between $\mathbf{u}$ and $\mathbf{u}_\varepsilon$ must have as high an order with ε as possible. Multiply the ith equation of Equation (2.10) by a parameter, α_i, and try to find these parameters in such a way that the statement

$$\mathbf{u} = \sum_{i=1}^{n} \alpha_i \mathbf{u}_{\varepsilon_i} + O(\varepsilon^n), \qquad \varepsilon = \max\{\varepsilon_i\}. \qquad (2.11)$$

will hold. In this case the equalities

$$\sum_{i=1}^{n} \varepsilon_i^k \alpha_i = \delta_0^k, \qquad k = 0, 1, \ldots, n-1; \qquad (2.12)$$

should be true, where here δ_i^k is the Kroneker symbol. Solving this system with respect to α_i, we obtain coefficients of Equation (2.11). Thus we have found the necessary approach.

It is worth noting that as a rule this system of equations is ill conditioned. For instance, if we take $\varepsilon_i = \varepsilon/i$, then the determinant of this system is a *Wandermond determinant*, and there arise round-off errors even in the case $n = 4$. If $n = 6$, the results obtained will be too far from the correct ones. Nevertheless, this technique allows one to develop a more precise approach to the solution of the Stokes problem if the number of terms in Equation (2.11) is not too large.

Another way to solve Equation (2.1) is to construct a finite-difference scheme immediately for the original problem. First let us consider the analogue of the explicit finite-difference scheme for Equation (2.1). Equations connecting two values $(\mathbf{v}, q)$ and $(\hat{\mathbf{v}}, \hat{q})$ on neighboring time steps are

$$\begin{cases} \mathbf{v}_t = \nu \mathbf{\Delta}^h \mathbf{v} - \nabla^h \hat{q} + \mathbf{f}^h & \text{in } \mathbf{U}, \\ \operatorname{div}^h \hat{\mathbf{v}} = 0 & \text{in } P, \\ \mathbf{v}(x, 0) = \mathbf{v}_0(x) \in \mathbf{U}. \end{cases} \qquad (2.13)$$

Let us investigate the stability of Equation (2.13). Forming the scalar product of the first part of Equation (2.13) with $2\tau \hat{\mathbf{v}}$, we obtain

$$\|\hat{\mathbf{v}}\|^2 + \tau^2 \|\mathbf{v}_t\|^2 - \|\mathbf{v}\|^2 - 2\tau\nu(\mathbf{\Delta}^h \mathbf{v}, \hat{\mathbf{v}}) = 2\tau(\mathbf{f}^h, \hat{\mathbf{v}}). \qquad (2.14)$$

Scalar products in Equation (2.14) will be estimated by

$$2\tau\nu|(\boldsymbol{\Delta}^h\mathbf{v}, \hat{\mathbf{v}})| \geq 2\tau\nu\|\hat{\mathbf{v}}\|_1^2 - 2\tau^2\nu|(\boldsymbol{\Delta}^h\mathbf{v}_t, \hat{\mathbf{v}})|$$

$$\geq 2\tau\nu\|\hat{\mathbf{v}}\|_1^2 - \frac{\tau^2\nu}{2\varepsilon}\|\hat{\mathbf{v}}\|_1^2 - 2\tau^2\nu\varepsilon\|\mathbf{v}_t\|_1^2,$$

$$2\tau|(\mathbf{f}^h, \hat{\mathbf{v}})| \leq \tau\varepsilon_1\|\hat{\mathbf{v}}\|_1^2 + \frac{\tau}{\varepsilon_1}\|\mathbf{f}^h\|_{-1}^2.$$

Then from Equation (2.14) we obtain the inequality

$$\|\hat{\mathbf{v}}\|^2 + \tau^2\left(1 - \frac{c_4\nu\varepsilon}{h^2}\right)\|\mathbf{v}_t\|^2$$

$$+ \tau\left(2\nu - \frac{\tau\nu}{2\varepsilon} - \varepsilon_1\right)\|\hat{\mathbf{v}}\|_1^2 \leq \|\mathbf{v}\|^2 + \frac{\tau}{\varepsilon_1}\|\mathbf{f}^h\|_{-1}^2. \quad (2.15)$$

Let us assume $\varepsilon = h^2/(c_4\nu)$ and $\varepsilon_1 = \nu$. In this case, from inequality (2.15) for arbitrary $\tau : \tau \leq \bar{\tau} = 2h^2/(c_4\nu)$ the validity of the statement

$$\|\hat{\mathbf{v}}\|^2 \leq \|\mathbf{v}\|^2 + \frac{\tau}{\nu}\|\mathbf{f}^h\|_{-1}^2$$

follows. This expression connects the norms of $\mathbf{v}$ on two neighboring time steps. From this inequality we obviously get the final estimate,

$$\|\mathbf{v}(t)\|^2 \leq \|\mathbf{v}_0\|^2 + \frac{1}{\nu}\sum_{k=0}^{n-1}\tau\|\mathbf{f}^h(k\tau)\|_{-1}^2, \qquad t = n\tau \leq T. \quad (2.16)$$

Therefore, the explicit finite-difference scheme, Equation (2.13), is stable, with

$$\tau \leq \frac{ch^2}{\nu},$$

that is, this scheme is conditionally stable. It is possible to show that this requirement is not only sufficient but necessary as well.

Let us consider an algorithm for realizing Equation (2.13). From the first part of Equation (2.13) we have

$$\hat{\mathbf{v}} = \mathbf{v} + \tau\nu\boldsymbol{\Delta}^h\mathbf{v} - \tau\nabla^h\hat{q} + \tau\mathbf{f}^h. \quad (2.17)$$

Substituting this expression for $\hat{\mathbf{v}}$ into the second part of Equation (2.13) yields

$$\Delta^h\hat{q} = \phi, \quad (2.18)$$

where $\qquad \phi = \nu\mathrm{div}^h\boldsymbol{\Delta}^h\mathbf{v} + \mathrm{div}^h\mathbf{f}^h + \frac{1}{\tau}\mathrm{div}^h\mathbf{v}.$

In addition, from Equation (2.17) we have boundary conditions which correspond to the finite-difference analogue of the Neimann problem. During the realization of the explicit finite-difference scheme, Equation (2.13), we find $\hat{\mathbf{v}}$ by the explicit Equation (2.17), using known $\mathbf{v}, \hat{q}$; we also must solve the finite-difference analogue of the Neimann problem, Equation (2.18), to find $\hat{q}$ ($\mathbf{v}, q$ are known).

It is easy to show that Equations (2.17) and (2.18) coincide with the well-known finite-difference scheme,[21] but in contrast to the cited paper, in our case grids and approximations of $\mathbf{\Delta}, \nabla, \text{div}$ were constructed in such a way that we had no problems with the boundary conditions for $\hat{q}$.

Let us also consider a fully implicit finite-difference scheme to solve Equation (2.1):[22]

$$\begin{cases} \mathbf{v}_t = \nu \mathbf{\Delta}^h \hat{\mathbf{v}} - \nabla^h \hat{q} + \mathbf{f}^h & \text{in } \mathbf{U}, \\ \text{div}^h \hat{\mathbf{v}} = 0 & \text{in } P, \\ \mathbf{v}(x,0) = \mathbf{v}_0(x) \in \mathbf{U}. \end{cases} \qquad (2.19)$$

This finite-difference scheme approximates our original problem (2.1) with the second-order $O(h^2 + \tau)$, with respect to the space variables as well as to Equation (2.13). Let us investigate the stability of this scheme. Forming scalar products of the first part of Equation (2.19) with $2\tau \hat{\mathbf{v}}$, and the second part of Equation (2.19) with $2\tau \hat{q}$, and adding the results, we obtain

$$\|\hat{\mathbf{v}}\|^2 + \tau^2 \|\mathbf{v}_t\|^2 - \|\mathbf{v}\|^2 + 2\tau\nu \|\hat{\mathbf{v}}\|_1^2 = 2\tau(\mathbf{f}^h, \hat{\mathbf{v}}).$$

From this we have the estimate

$$\|\hat{\mathbf{v}}\|^2 + \tau^2 \|\mathbf{v}_t\|^2 + \tau\nu \|\hat{\mathbf{v}}\|_1^2 \leq \|\mathbf{v}\|^2 + \frac{\tau}{\nu} \|\mathbf{f}^h\|_{-1}^2, \qquad (2.20)$$

which determines the stability of our scheme with respect to the right-hand part and the initial values.

To prove the stability of Equation (2.19), we must find an a priori estimate for pressure. We do this in the same way as in the case of investigating the stationary problem. Form a scalar product of the first part of Equation (2.19), with $\phi \in \mathbf{U}$. Then we have the sequence of obvious inequalities,

$$|(\hat{q}, \text{div}^h \phi)| \leq |(\mathbf{v}_t, \phi)| + \nu |(\mathbf{\Delta}^h \hat{\mathbf{v}}, \phi)| + |(\mathbf{f}^h, \phi)|$$

$$\leq c_5 \|\mathbf{v}_t\| \, \|\phi\|_1 + \nu \|\hat{\mathbf{v}}\|_1 \, \|\phi\|_1 + \|\mathbf{f}^h\|_{-1} \, \|\phi\|_1.$$

From this and condition (1.3), it follows that

$$\|\hat{q}\| \leq c_0 \big(c_5 \|\mathbf{v}_t\| + \nu \|\hat{\mathbf{v}}\|_1 + \|\mathbf{f}^h\|_{-1} \big),$$

and finally we have

$$\|\hat{q}\|^2 \le 3c_0^2\big(c_5^2\|\mathbf{v}_t\|^2 + \nu^2\|\hat{\mathbf{v}}\|_1^2 + \|\mathbf{f}^h\|_{-1}^2\big).$$

Multiplying both parts of this inequality by $\lambda\tau^2$, where λ is an arbitrary positive number, and adding this with statement (2.20) yields

$$\|\hat{\mathbf{v}}\|^2 + \tau^2\big(1 - 3c_0^2c_5^2\lambda\big)\|\mathbf{v}_t\|^2 + 2\tau\nu\left(1 - \frac{3c_0^2\nu\lambda\tau}{2}\right)\|\hat{\mathbf{v}}\|_1^2$$

$$+ \lambda\tau^2\|\hat{q}\|^2 \le \|\mathbf{v}\|^2 + \tau\big(1 + 3c_0^2\lambda\tau\big)\|\mathbf{f}^h\|_{-1}^2. \qquad (2.21)$$

Let us assume $\lambda = (3c_0^2c_5^2)^{-1}, \bar{\tau} = c_5^2/\nu$. Then for any $\tau \le \bar{\tau}$ from inequality (2.21) there follows a statement,

$$\|\hat{\mathbf{v}}\|^2 + \tau\nu\|\hat{\mathbf{v}}\|_1^2 + \lambda\tau\|\hat{q}\|^2 \le \|\mathbf{v}\|^2 + \tau\big(1 + 3c_0^2\lambda\tau\big)\|\mathbf{f}^h\|_{-1}^2.$$

In particular, from this we obtain the estimate

$$\|\mathbf{v}^n\|^2 + \frac{\tau\nu\gamma}{4}\|\mathbf{v}^n\|_1^2 + \lambda\gamma\tau^2\|q^n\|^2 \le \gamma^n\|\mathbf{v}_0\|^2$$

$$+ \alpha\gamma\sum_{k=1}^{n}\tau\gamma^{n-k+1}\|\mathbf{f}^h(k\tau)\|_{-1}^2, \qquad (2.22)$$

where $\alpha = 3c_0^2\lambda\tau$, $\gamma = (1 + c_6\tau\nu)^{-1} < 1$, $\mathbf{v}^n = \mathbf{v}(n\tau)$. Inequality (2.22) means the stability of the finite-difference scheme with respect to the initial value and the right-hand part.

Let us discuss the realization of Equation (2.19). From the first part of the equation we have

$$\hat{\mathbf{v}} = (\mathbf{I} - \tau\nu\mathbf{\Delta}^h)^{-1}(\mathbf{v} - \tau\nabla^h\hat{q} + \tau\mathbf{f}^h).$$

By substituting this expression into the second part of Equation (2.19) we obtain

$$A\hat{q} \equiv -\text{div}^h(\mathbf{I} - \tau\nu\mathbf{\Delta}^h)^{-1}\nabla^h\hat{q} = \phi, \qquad (2.23)$$

where $\phi = -(1/\tau)\text{div}^h(\mathbf{I} - \tau\nu\mathbf{\Delta}^h)^{-1}\mathbf{v} - \text{div}^h(\mathbf{I} - \tau\nu\mathbf{\Delta}^h)^{-1}\mathbf{f}^h$. Operator A is symmetric and positive definite. Thus, it is possible to use CGM to solve Equation (2.23). We must also take into account that we have "good" initial approximation during the solution of Equation (2.23) which is equal to q. After finding the approximation $\tilde{q}$ to $\hat{q}$, one can find $\hat{\mathbf{v}}$ from the equation

$$(\mathbf{I} - \tau\nu\mathbf{\Delta}^h)\hat{\mathbf{v}} = \mathbf{v} - \nabla^h\tilde{q} + \mathbf{f}^h.$$

The error function in the equation $\operatorname{div}^h \hat{\mathbf{v}} = 0$ depends on how accurately we solve Equation (2.23).[22]

3. NUMERICAL METHODS FOR NONLINEAR EQUATIONS

For simplicity we again assume $\bar{\Omega} = \{x = (x_1, x_2) : 0 \leq x_i \leq 1, i = 1, 2\}$. We shall find a pair of functions $(\mathbf{u}, p) \in \mathbf{H}_0^1 \times L_2/\mathbf{R}$, which is the weak solution of the system of differential equations,[6]

$$\begin{cases} -\nu \Delta \mathbf{u} + \nabla p + u_k \mathbf{u}_{x_k} = \mathbf{F}, \\ \operatorname{div} \mathbf{u} = 0. \end{cases} \tag{3.1}$$

Here $\nu > 0$ is a coefficient of kinematic viscosity, $\mathbf{u} = (u_1, u_2)$.

First consider an approximation of Equation (3.1). For certainty we introduce grids $\bar{\Omega}_1, \bar{\Omega}_2, \Omega_3$ as in Example 1.2. We approximate the linear part of Equation (3.1) as was done previously. As for the approximation of nonlinear terms, note that in a differential case the statement

$$(u_k \mathbf{v}_{x_k}, \mathbf{v}) \equiv \int_\Omega u_k v^j_{x_k} v^j \, dx = 0 \tag{3.2}$$

holds and is true for arbitrary $\mathbf{u}, \mathbf{v} \in \mathbf{H}_0^1$ such that $\operatorname{div} \mathbf{u} = 0$. Thus we must approximate the nonlinear terms in Equation (3.1) in such a way that the finite-dimensional analogue of this statement holds also. To do this, let $\mathbf{N}(\mathbf{u}^h, \mathbf{v}^h)$ be an approximation of the nonlinear terms $u_k \mathbf{v}_{x_k}$. We introduce it in a natural way,[3,23]

$$N_1(\mathbf{u}^h, \mathbf{v}^h)|_{x \in \Omega_1} =$$

$$\frac{1}{4h} \left[\left(u_1^h(x - h\mathbf{e}_1) + u_1^h(x) \right) \left(v_1^h(x) - v_1^h(x - h\mathbf{e}_1) \right) \right.$$

$$\left. + \left(u_1^h(x) + u_1^h(x + h\mathbf{e}_1) \right) \left(v_1^h(x + h\mathbf{e}_1) - v_1^h(x) \right) \right]$$

$$+ \frac{1}{4h} \left[\left(u_2^h \left(x + \frac{h}{2}(\mathbf{e}_1 + \mathbf{e}_2) \right) + u_2^h \left(x + \frac{h}{2}(\mathbf{e}_1 - \mathbf{e}_2) \right) \right) \times \right.$$

$$\left(v_1^h(x + h\mathbf{e}_1) - v_1^h(x) \right)$$

$$+ \left(u_2^h \left(x + \frac{h}{2}(\mathbf{e}_1 - \mathbf{e}_2) \right) + u_2^h \left(x - \frac{h}{2}(\mathbf{e}_1 + \mathbf{e}_2) \right) \right) \times$$

$$\left. \left(v_1^h(x) - v_1^h(x - h\mathbf{e}_1) \right) \right].$$

The approximating expression for $N_2(\mathbf{u}^h, \mathbf{v}^h)$ can be written in the same way.

Thus we obtained the finite-difference analogue of Equation (3.1). Next we find functions $\mathbf{u}^h \in \mathbf{U}$, $p^h \in P$, which satisfy the grid boundary value problem,

$$\begin{cases} -\nu \mathbf{\Delta}^h \mathbf{u}^h + \nabla^h p^h + \mathbf{N}(\mathbf{u}^h, \mathbf{u}^h) = \mathbf{F}^h, \\ \operatorname{div}^h \mathbf{u}^h = 0. \end{cases} \tag{3.3}$$

Later we will need an approximation of the term $(1/2)\operatorname{div}\mathbf{u}\cdot\mathbf{v}$. It is connected with the fact that in almost all our iterative methods for Equation (3.3) functions $\mathbf{u}^n$, approximating $\mathbf{u}$, do not belong to the subspace of solenoidal functions (i.e., functions from $\mathbf{U}$ which satisfy the condition $\operatorname{div}^h \mathbf{u}^h = 0$). Thus, the condition $(\mathbf{N}(\mathbf{v}, \mathbf{w}), \mathbf{w}) = 0$ in a general case is not valid with $\mathbf{v}, \mathbf{w} \in \mathbf{U}$. On the other hand,

$$\left(u_k \mathbf{v}_{x_k} + \frac{1}{2} \operatorname{div}\mathbf{u}\cdot\mathbf{v}, \mathbf{v} \right) = 0 \quad \text{for any } \mathbf{u}, \mathbf{v} \in \mathbf{H}_0^1. \tag{3.4}$$

This statement also holds when $\mathbf{u}$ is not solenoidal. If $\operatorname{div}\mathbf{u} = 0$, then the second term in Equation (3.4) vanishes and Equation (3.4) coincides with Equation (3.2).

Constructing an approximation, $\mathbf{K}(\mathbf{u}^h, \mathbf{v}^h)$ of the expression $(1/2)\operatorname{div}\mathbf{u}\cdot\mathbf{v}$, in such a way that the finite-difference analogue of Equation (3.4) is valid yields

$$K_j(\mathbf{u}^h, \mathbf{v}^h) = \frac{1}{4}\left[\operatorname{div}^h \mathbf{u}^h \left(x + \frac{h}{2}\,\mathbf{e}_j \right) + \operatorname{div}^h \mathbf{u}^h \left(x - \frac{h}{2}\,\mathbf{e}_j \right) \right] v_j(x),$$
$$j = 1, 2, x \in \Omega_j.$$

Lemma 3.1. For arbitrary functions $\mathbf{v}, \mathbf{w} \in \mathbf{U}$ the statement

$$\big(\mathbf{N}(\mathbf{v}, \mathbf{w}) + \mathbf{K}(\mathbf{v}, \mathbf{w}), \mathbf{w} \big) = 0.$$

holds. If in addition the preceding $\mathbf{v}$ satisfies the equation

$$\operatorname{div}^h \mathbf{v} = 0,$$

then $\qquad\qquad (\mathbf{N}(\mathbf{v}, \mathbf{w}), \mathbf{w}) = 0.$

The proof can be obtained directly by using a finite-difference analogue of the formula for integrating by parts. Since our problem is nonlinear, let us consider what requirements our iterative method of the solution of Equation (3.3) should satisfy. It is known that in a general case the solution of Equation (3.1) with small ν is not unique.[24]

Forming a scalar product of the first part of Equation (3.1) with $\mathbf{u}$, and taking into account the second part of Equation (3.1), we obtain the obvious estimate,

$$\nu\|\nabla\mathbf{u}\| \le \|\mathbf{f}\|_{-1}.$$

Thus, in a differential case norms of all possible solutions of Equation (3.1) are bounded by the value $\|\mathbf{f}\|_{-1}/\nu$. Also, it is known that if $\nu > \nu_0$, where ν_0 depends on $\Omega, \mathbf{f}$, then the solution of Equation (3.1) is unique.[6]

From these considerations we impose some requirements on the iterative method of solving the finite-dimensional analogue of the Navier-Stokes problem:

all iterative approximations $(\mathbf{V}^n, Q^n)$ to the solution of Equation (3.3) should be uniformly bounded by the constant, depending only on $\nu, \Omega, \mathbf{F}^h$, and initial value $(\mathbf{V}^0, Q^0)$;

if the coefficient of kinematic viscosity is not too small, then the iterative method must converge to the solution of Equation (3.3) as a geometric progression; this means there exists $\nu_0 = \nu_0(\Omega, \mathbf{F}^h)$ such that for any $\nu > \nu_0$ the iterative method converges. In this case, problem (3.3) has a unique solution.

Let us write out the algorithm of our iterative method as it was done for the linear problem. Equations connecting two neighboring approximations $(\mathbf{V}^n, Q^n) \equiv (\mathbf{V}, Q)$ and $(\mathbf{V}^{n+1}, Q^{n+1}) \equiv (\hat{\mathbf{V}}, \hat{Q})$ are

$$\begin{cases} \nu\mathbf{C}\mathbf{V}_t = \nu\mathbf{\Delta}^h\mathbf{V} + \dfrac{\nu}{\beta}\,\nabla^h\mathrm{div}^h\,\mathbf{V} - \nabla^h Q - \mathbf{N}(\mathbf{V},\mathbf{V}) - \mathbf{K}(\mathbf{V},\mathbf{V}) + \mathbf{F}^h, \\[2mm] \hat{Q} = Q - \dfrac{\nu}{\beta}\,\mathrm{div}^h\,\mathbf{V}; \end{cases} \quad (3.5)$$

here $\mathbf{C}$ is some symmetric positive definite operator from $\mathbf{U}$ into $\mathbf{U}$. The equation $\mathbf{C}\mathbf{w} = \phi$ is assumed to be "easily solvable". We assumed nonlinear terms in Equation (3.5) to be on the lower level. This was done because this case was the most difficult for the investigation and the simplest for the realization.

Let us examine the properties of iterative method (3.5). Substituting $\mathbf{V} = \nu\mathbf{v}, Q = \nu^2 q, \mathbf{F}^h = \nu^2\mathbf{f}$ in Equation (3.5), we obtain

$$\begin{cases} \mathbf{C}\mathbf{v}_t = \mathbf{\Delta}^h\mathbf{v} + \dfrac{1}{\beta}\,\nabla^h\mathrm{div}^h\,\mathbf{v} - \nabla^h q - \mathbf{N}(\mathbf{v},\mathbf{v}) - \mathbf{K}(\mathbf{v},\mathbf{v}) + \mathbf{f}, \\[2mm] \hat{q} = q - \dfrac{1}{\beta}\,\mathrm{div}^h\,\hat{\mathbf{v}}; \end{cases} \quad (3.6)$$

From the first part of Equation (3.6) it follows that $\hat{\mathbf{v}} \in \mathbf{U}$. If $q \in P$, then from the second part of Equation (3.6) it follows that $\hat{q} \in P$. Thus, Equation (3.6) puts into correspondence every pair of functions

$(\mathbf{v}, q) \in \mathbf{U} \times P$ with another pair of functions $(\hat{\mathbf{v}}, \hat{q})$ which also belongs to $\mathbf{U} \times P$.

Let us denote $\mathbf{B} = \mathbf{C} + \tau \mathbf{\Delta}^h + (\tau/\beta)\nabla^h \mathrm{div}^h$. Then from Equation (3.6) one obtains (compare with the linear case)

$$\begin{cases} \mathbf{B}\mathbf{v}_t - \mathbf{\Delta}^h \hat{\mathbf{v}} + \nabla^h \hat{q} + \mathbf{N}(\mathbf{v}, \mathbf{v}) + \mathbf{K}(\mathbf{v}, \mathbf{v}) = \mathbf{f}, \\ \beta \tau q_t + \mathrm{div}^h \hat{\mathbf{v}} = 0. \end{cases} \quad (3.7)$$

Theorem 3.1. Let $\mathbf{C} = \mathbf{C}^* > 0$. Then for arbitrary $\beta > 0, \nu > 0, \mathbf{f}, (\mathbf{v}^0, q^0) \in \mathbf{U} \times P$ there exists $\bar{\tau}, M$ such that for any $\tau \leq \bar{\tau}$ the estimate, that is,

$$\|\mathbf{v}^n\|_{\mathbf{B}}^2 + \beta \tau \|q^n\|^2 \leq M^2, n \geq 0, \quad (3.8)$$

will hold.

Proof. Let $(\mathbf{v}^0, q^0), \mathbf{f}$, and β be fixed. From the assumptions of the theorem and the definition of operator $\mathbf{B}$ it follows that there exists $\tau' = \tau'(\beta)$ such that for any $\tau \leq \tau'$ the inequalities

$$\mathbf{B} \geq \rho \mathbf{I}, \qquad -\kappa_1 \mathbf{\Delta}^h \leq \mathbf{B} \leq -\kappa_2 \mathbf{\Delta}^h.$$

hold. Constants κ_i and ρ do not depend on $\tau \leq \tau'$ but could depend on β. Assume that $\lambda = \min\{1/2, \kappa_2/(8c_0^2\beta)\}$, where c_0 is the constant from inequality Equation (1.3), and assume that the constant, M, satisfies the inequality $\lambda M^2/\kappa_2 \geq 4\|\mathbf{f}\|_{-1}^2$.

Let us also assume that estimate Equation (3.8) holds for $(\mathbf{v}, q)$. We will try to find conditions to be satisfied by constant M and parameter τ in such a way that the estimate Equation (3.8) will also be true for $(\hat{\mathbf{v}}, \hat{q})$. Now form scalar products of the first part of Equation (3.7) with $2\tau\hat{\mathbf{v}}$ in $\mathbf{U}$ and the second part of Equation (3.7) with $2\tau\hat{q}$ in P. Adding the results yields

$$\|\hat{\mathbf{v}}\|_{\mathbf{B}}^2 - \|\mathbf{v}\|_{\mathbf{B}}^2 + \tau^2 \|\mathbf{v}_t\|_{\mathbf{B}}^2 + 2\tau\|\hat{\mathbf{v}}\|_1^2 + \beta\tau\|\hat{q}\|^2 - \beta\tau\|q\|^2$$

$$+ \beta\tau^3\|q_t\|^2 + 2\tau\big(\mathbf{N}(\hat{\mathbf{v}}, \mathbf{v}), \hat{\mathbf{v}}\big) + 2\tau\big(\mathbf{K}(\mathbf{v}, \mathbf{v}), \hat{\mathbf{v}}\big) = 2\tau\big(\mathbf{f}, \hat{\mathbf{v}}\big).$$

$$(3.9)$$

Estimating scalar products in this identity in view of Lemma 3.1, we have

$$\big(\mathbf{N}(\mathbf{v}, \mathbf{v}), \hat{\mathbf{v}}\big) + \big(\mathbf{K}(\mathbf{v}, \mathbf{v}), \hat{\mathbf{v}}\big) = \big(\mathbf{N}(\mathbf{v}, \hat{\mathbf{v}}), \hat{\mathbf{v}}\big) + \big(\mathbf{K}(\mathbf{v}, \hat{\mathbf{v}}), \hat{\mathbf{v}}\big)$$

$$- \tau\big(\mathbf{N}(\mathbf{v}, \mathbf{v}_t), \hat{\mathbf{v}}\big) - \tau\big(\mathbf{K}(\mathbf{v}, \mathbf{v}_t), \hat{\mathbf{v}}\big)$$

$$= \tau\big(\mathbf{N}(\mathbf{v}, \mathbf{v}_t) + \mathbf{K}(\mathbf{v}, \mathbf{v}_t), \hat{\mathbf{v}}\big).$$

On the other hand,[6]

$$\left|\left(\mathbf{K}(\mathbf{v}, \mathbf{v}_t), \hat{\mathbf{v}}\right)\right| \leq c\|\mathbf{v}_t\|_1\|\mathbf{v}\hat{\mathbf{v}}\|$$

$$\leq \frac{c}{\sqrt{\kappa_1}}\,\|\mathbf{v}_t\|_{\mathbf{B}}\left(\|\mathbf{v}\|\,\|\hat{\mathbf{v}}\|\right)^{1/2}\left(\|\mathbf{v}\|_1\,\|\hat{\mathbf{v}}\|_1\right)^{1/2}$$

$$\leq \frac{c}{(\rho\kappa_1^3)^{1/4}}\,\|\mathbf{v}_t\|_{\mathbf{B}}\,\|\mathbf{v}\|_{\mathbf{B}}\,\|\hat{\mathbf{v}}\|_1$$

$$\leq \frac{1}{8}\,\|\mathbf{v}_t\|_{\mathbf{B}}^2 + \frac{cM^2}{\sqrt{\rho\kappa_1^3}}\,\|\hat{\mathbf{v}}\|_1^2.$$

In the same way it follows that

$$\left|\left(\mathbf{K}(\mathbf{v}, \mathbf{v}_t), \hat{\mathbf{v}}\right)\right| \leq \frac{1}{8}\,\|\mathbf{v}_t\|_{\mathbf{B}}^2 + \frac{cM^2}{\sqrt{\rho\kappa_1^3}}\,\|\hat{\mathbf{v}}\|_1^2.$$

Thus we have the final estimate for the scalar product,

$$2\tau|(\mathbf{N}(\mathbf{v}, \mathbf{v}) + \mathbf{K}(\mathbf{v}, \mathbf{v}), \hat{\mathbf{v}})| \leq \frac{\tau^2}{2}\,\|\mathbf{v}_t\|_{\mathbf{B}}^2 + \frac{c_7 M^2 \tau^2}{\sqrt{\rho\kappa_1^3}}\,\|\hat{\mathbf{v}}\|_1^2.$$

Since $2\tau|(\mathbf{f}, \hat{\mathbf{v}})| \leq \tau\|\mathbf{f}\|_{-1}^2 + \tau\|\hat{\mathbf{v}}\|_1^2$, we can rewrite Equation (3.9) as

$$\|\hat{\mathbf{v}}\|_{\mathbf{B}}^2 + \frac{\tau^2}{2}\,\|\mathbf{v}_t\|_{\mathbf{B}}^2 + (\tau - \frac{c_7 M^2 \tau^2}{\sqrt{\rho\kappa_1^3}})\|\hat{\mathbf{v}}\|_1^2 + \beta\tau\|\hat{q}\|^2$$

$$\leq \|\mathbf{v}\|_{\mathbf{B}}^2 + \beta\tau\|q\|^2 + \tau\|\mathbf{f}\|_{-1}^2. \tag{3.10}$$

Estimate the norm $\|\hat{q}\|$ from the first part of Equation (3.7). For arbitrary $\phi \in \mathbf{U}$,

$$|(\nabla^h\hat{q}, \phi)| \leq |(\mathbf{Bv_t}, \phi)| + |(-\mathbf{\Delta}^h\hat{\mathbf{v}}, \phi)| + |(\mathbf{N}(\mathbf{v}, \mathbf{v}), \phi)|$$

$$+ |(\mathbf{K}(\mathbf{v}, \mathbf{v}), \phi)| + |(\mathbf{f}, \phi)|.$$

It is possible to estimate the scalar products in the left-hand part of this inequality by

$$|(\mathbf{Bv_t}, \phi)| \leq \sqrt{\kappa_2}\,\|\mathbf{v}_t\|_{\mathbf{B}}\,\|\phi\|_1, \quad |(-\mathbf{\Delta}^h\hat{\mathbf{v}}, \phi)| \leq \|\hat{\mathbf{v}}\|_1\,\|\phi\|_1,$$

$$|(\mathbf{N}(\mathbf{v}, \mathbf{v}), \phi)| \leq c\|\mathbf{v}\|_1\,\|\mathbf{v}\phi\| \leq \frac{c}{(\rho\kappa_1^3)^{1/4}}\,\|\mathbf{v}\|_{\mathbf{B}}^2\,\|\phi\|_1,$$

$$|(\mathbf{K}(\mathbf{v}, \mathbf{v}), \phi)| \leq \frac{c}{(\rho\kappa_1^3)^{1/4}}\,\|\mathbf{v}\|_{\mathbf{B}}^2\,\|\phi\|_1, \quad |(\mathbf{f}, \phi)| \leq \|\mathbf{f}\|_{-1}\|\phi\|_1.$$

Using Equation (1.3) and the previous inequalities, one obtains

$$\|\hat{q}\| \leq c_0 \left(\sqrt{\kappa_2} \|\mathbf{v}_t\|_{\mathbf{B}} + \|\hat{\mathbf{v}}\|_1 + \frac{c}{(\rho\kappa_1^3)^{1/4}} \|\mathbf{v}\|_{\mathbf{B}}^2 + \|\mathbf{f}\|_{-1} \right).$$

From this estimate it obviously follows that

$$\|\hat{q}\|^2 \leq 4c_0^2 \left(\kappa_2 \|\mathbf{v}_t\|_{\mathbf{B}}^2 + \|\hat{\mathbf{v}}\|_1^2 + \frac{c}{\sqrt{\rho\kappa_1^3}} \|\mathbf{v}\|_{\mathbf{B}}^4 + \|\mathbf{f}\|_{-1}^2 \right).$$

Multiplying this inequality by $\lambda\beta\tau^2/\kappa_2$ and adding Equation (3.10), we obtain

$$\|\hat{\mathbf{v}}\|_{\mathbf{B}}^2 + \tau\left(1 - \frac{c_7 M^2 \tau}{\sqrt{\rho\kappa_1^3}} - \frac{4c_0^2 \lambda\beta\tau}{\kappa_2}\right)\|\hat{\mathbf{v}}\|_1^2$$

$$+ \tau^2\left(\frac{1}{2} - \frac{4c_0^2 \lambda\beta}{\kappa_2}\right)\|\mathbf{v}_t\|_{\mathbf{B}}^2 + \beta\tau\left(1 + \frac{\lambda\tau}{\kappa_2}\right)\|\hat{q}\|^2$$

$$\leq \left(1 + \frac{4c_8 c_0^2 M^2 \lambda\beta\tau^2}{\kappa_2\sqrt{\rho\kappa_1^3}}\right)\|\mathbf{v}\|_{\mathbf{B}}^2 + \beta\tau\|q\|^2$$

$$+ \left(\tau + \frac{4c_0^2 \lambda\beta\tau^2}{\kappa_2}\right)\|\mathbf{f}\|_{-1}^2. \tag{3.11}$$

Let us denote $\xi = 1/\sqrt{\rho\kappa_1^3}$, $\eta = 4c_0\lambda\beta/\kappa_2$. Using the definition of λ, we have from Equation (3.11) the estimate

$$\|\hat{\mathbf{v}}\|_{\mathbf{B}}^2 + \tau(1 - c_7\xi M^2\tau - \eta\tau)\|\hat{\mathbf{v}}\|_1^2 + \beta\tau\left(1 + \frac{\lambda\tau}{\kappa_2}\right)\|\hat{q}\|^2$$

$$\leq \left(1 + c_8\xi\eta M^2\tau^2\right)\|\mathbf{v}\|_{\mathbf{B}}^2 + \beta\tau\|q\|^2 + \left(\tau + \eta\tau^2\right)\|\mathbf{f}\|_{-1}^2.$$

If $\tau \leq \tau''$, where $\tau'' = \min\{(1-\lambda)/(c_7\xi M^2 + \eta), (1/\eta)\}$ then the preceding inequality can be reduced to the form

$$\left[1 + \tau\left(\frac{1 - c_7\xi\tau M^2 - \eta\tau}{\kappa_2}\right)\right]\|\hat{\mathbf{v}}\|_{\mathbf{B}}^2 + \beta\tau\left(1 + \frac{\lambda\tau}{\kappa_2}\right)\|\hat{q}\|^2$$

$$\leq \left(1 + c_8\xi\eta M^2\tau^2\right)\|\mathbf{v}\|_{\mathbf{B}}^2 + \beta\tau\|q\|^2 + \tau(1 + \eta\tau)\|\mathbf{f}\|_{-1}^2. \tag{3.12}$$

Note that the inequality $1 - c_7\xi\tau M^2 - \eta\tau \geq \lambda$ is valid by virtue of choosing τ''. From this and Equation (3.12) it follows that for the estimate $\|\hat{\mathbf{v}}\|_{\mathbf{B}}^2 + \beta\tau\|\hat{q}\|^2 \leq M^2$ to be valid it is sufficient that

$$(1 + c_8\xi\eta M^2\tau^2)\|\mathbf{v}\|_{\mathbf{B}}^2 + \beta\tau\|q\|^2 + (\tau + \eta\tau^2)\|\mathbf{f}\|_{-1}^2$$

$$\leq \left(1 + \lambda\tau\kappa_2\right)M^2 \qquad (3.13)$$

holds. Taking into account this estimate and the condition $\|\mathbf{v}\|_{\mathbf{B}}^2 + \beta\tau\|q\|^2 \leq M^2$, the inequality

$$c_8\xi\eta M^2\tau + (1 + \eta\tau)\|\mathbf{f}\|_{-1}^2 \leq \frac{\lambda}{\kappa_2} M^2$$

is obtained, and from this statement we have

$$\left(\frac{\lambda}{\kappa_2}\right)M^2 - c_8\xi\eta\tau M^2 \geq (1 + \eta\tau)\|\mathbf{f}\|_{-1}^2.$$

From this expression and the inequality for M, already mentioned,

$$\frac{\lambda}{\kappa_2} - c_8\xi\eta\tau M^2 \geq 0. \qquad (3.14)$$

This inequality holds if $\tau \leq \lambda/(c_8\kappa_2\xi\eta M^2)$. Thus, if we choose M such that it satisfies the inequality

$$M^2 = \max\left\{\frac{4\kappa_2\|\mathbf{f}\|_{-1}^2}{\lambda}, \|\mathbf{v}^0\|_{\mathbf{C}}^2 + \beta\|q^0\|^2\right\},$$

and choose $\bar{\tau}$ from the statement

$$\bar{\tau} = \min\left\{1, \tau', \tau'', \frac{\lambda}{c_8\kappa_2\xi\eta M^2}\right\}, \qquad (3.15)$$

then, in accordance with our discussions for any $\tau \in (0, \bar{\tau}]$, the inequality Equation (3.8) holds. The theorem is proved.

Let us clarify the dependence of $\bar{\tau}$ on the viscosity coefficient ν. For simplicity, consider the case $\mathbf{C} = -\boldsymbol{\Delta}^h$, $\|\mathbf{F}^h\|_{-1}$, $\beta = O(1)$. From these assumptions and Theorem 3.1 we have ρ, κ_i, $\lambda = O(1)$ and $\tau', \xi, \eta = O(1)$, $\tau'' = O(M^{-2})$. Since $M^{-2} = O(\nu^4)$, from Equation (3.15) it follows that $\bar{\tau} = O(\nu^4)$.

This boundedness is connected with the approximations of nonlinear terms on the lower layer of our iterative method (3.5). If we consider the iterative method

$$\begin{cases} \mathbf{B}\mathbf{v}_t - \boldsymbol{\Delta}^h\hat{\mathbf{v}} + \nabla^h\hat{q} + \mathbf{N}(\mathbf{v}, \hat{\mathbf{v}}) + \mathbf{K}(\mathbf{v}, \hat{\mathbf{v}}) = \mathbf{f}, \\ \beta\tau q_t + \operatorname{div}^h\hat{\mathbf{v}} = 0, \end{cases}$$

then the boundedness on $\bar{\tau}$ is weaker. But in this case, on every step of this iterative method we must solve a system of equations,

$$\mathbf{C}\hat{\mathbf{v}} + \tau\mathbf{N}(\mathbf{v}, \hat{\mathbf{v}}) + \tau\mathbf{K}(\mathbf{v}, \hat{\mathbf{v}}) = \phi. \tag{3.16}$$

To solve these equations we can use the inner iterative method,

$$\mu(\mathbf{w}^{k+1} - \mathbf{w}^k) + \mathbf{C}\mathbf{w}^{k+1} + \tau\mathbf{N}(\mathbf{v}, \mathbf{w}^k) + \tau\mathbf{K}(\mathbf{v}, \mathbf{w}^k) = \phi, \tag{3.17}$$

where $\mu \geq 0$ is an iterative parameter, and we can choose either $\mathbf{v}$ (the approach on a previous layer) or $\hat{\mathbf{v}}$ from Equation (3.7) as the initial value, $\mathbf{w}^0$, for this inner iterative method. Since we have a pretty good initial value for the inner iterative process it is sufficient to perform not more then four iterations (3.17) for each step.

Considering the conditions for convergence of the iterative method, Theorem 3.2 holds.

Theorem 3.2. Let all conditions of Theorem 3.1 hold. If the right-hand part, $\mathbf{F}^h$, and the viscosity coefficient, ν, satisfy the inequality

$$\left\|\mathbf{F}^h\right\|_{-1}/\nu^2 < c_9, \tag{3.18}$$

where the constant, c_9, depends only on the domain, then iterative method (3.5) converges to the solution $(\mathbf{u}^h, p^h)$ of problem (3.3) as a geometric progression. The rate of convergence of this iterative method is proportional to $(1 + c\tau)^{-1/2}$.

The proof of this theorem is obtained by using the same technique as in the previous case. It is to be pointed out that condition (3.18), providing convergence of the iterative method, coincides by order with the condition which guarantees the uniqueness of the solution of Equation (3.1).[6]

As for the numerical solution of a nonstationary problem, one of the most preferable finite-difference schemes is that which mainly coincides with the finite-difference scheme for the nonstationary Stokes problem (2.19).[22] In this numerical method we can either take approximations of nonlinear terms from the lower layer or use linearization of these terms. In the first case we encounter the problem of solving the grid Stokes problem on every time step; the methods of solution of this problem have been considered in Section 1. In another case the situation is a bit worse, because we must solve the system of equations, which is more complicated. It can be done by using some inner iterative method. Note that in the case of a large Reynolds number upwind approximation of nonlinear terms is to be used.

Using the methods discussed here, we managed to simulate a number of interesting incompressible flows. For instance, the flow in a cube cavity was presented in reference 25. We used a $65 \times 33 \times 48$ node grid (greater than 10^5 nodes) with Reynolds numbers Re $= 100, 400,$ and 1000. Another example of solving the Navier–Stokes equations is numerical simulation of flow around a car with a nonuniform grid (the number of nodes was greater than 10^5) and Re $= 5 \times 10^6$. In the latter case we were certain to solve a nonstationary problem. All the results confirmed the high efficiency of the methods proposed.

REFERENCES

1. Brailovskaya, I.U., Kuskova T.V., and Chudov L.A., Finite difference methods of solving Navier-Stokes equations, *Vychisl. Met. Program*, v. 11, No. 3, Moscow Univ., 1968 (in Russian).

2. Kobelkov, G.M., The solution of the problem of free stationary convection, *Dokl. Acad. Nauk. SSSR*, v. 255, No. 2, 277, 1980 (in Russian).

3. Kobelkov, G.M., On the numerical solution of the problem of stationary convection in original variables, Preprint No. 57, *Otdel Vychisl. Mat.*, 1983 (in Russian).

4. Chijonkov, E.V., On one system of equations of magnetic hydrodynamics, *Dokl. Acad. Nauk. SSSR*, v. 278, No. 5, 1074, 1984 (in Russian).

5. Sokolov, A.G., On the numerical solution of some stationary boundary value problems of electromagnetic hydrodynamics, Preprint No. 71, *Otdel Vychisl. Mat.*, p. 3, 1984 (in Russian).

6. Ladyjenskaya, O.A., *Mathematical Aspects of the Dynamics of a Viscous Incompressible Fluid*, Nauka, Moscow, 1970 (in Russian).

7. Samarskiy, A.A., *Theory of Difference Schemes*, Nauka, Moscow, 1977 (in Russian).

8. Kobelkov, G.M., On equivalent norms of subspaces L_2, *Analysis Mathematica*, v. 177, No. 3, 1977 (in Russian).

9. Kobelkov, G.M., and Valedinskiy, V.D., On the inequality $\|p\| \leq c\|$ grad $p\|_{-1}$ and its finite dimensional image, *Sov. Jour. Numer. Anal. Math. Modeling*, v. 1, No. 3, 189, 1986.

10. Kobelkov, G.M., On the numerical solution of the Stokes problem, *Zh. Vychisl. Mat. Mat. Phys.*, v. 15, No. 3, 786, 1975 (in Russian).

11. Samarskiy, A.A., and Nikolaev, E.S., *The Methods of Solving Grid Equations*, Nauka, Moscow, 1978 (in Russian).

12. D'yakonov, E.G., The estimates of computing work for boundary value problems with the Stokes operator, *Izvestiya Vuzov, Mat.*, No. 7, 46, 1983 (in Russian).

13. Janenko, N.N., *The Method of Splitting Steps for Solving of Multidimensional Problems of Mathematical Physics*, Nauka, Novosibirsk, 1967 (in Russian).

14. Kobelkov, G.M., On the iterative method of the stationary Navier-Stokes equations solution, *Proc. International Conf. on Variational-difference Schemes in Mat. Phys.*, Novosibirsk, 74, 1981 (in Russian).

15. Kobelkov, G.M., The iterative methods for some classes of difference schemes, *Chisl. Metody Mech. Sploshnoy Sredy*, Novosibirsk, v. 12, No. 6, 38, 1981 (in Russian).

16. Kobelkov, G.M., On the existence theorems for the elasticity theory equations, *Matem. Zametki*, v. 17, No. 4, 599, 1975 (in Russian).

17. Lebedev, V.I., Finite difference analogues of orthogonal expansions of essential differential operators and boundary value problems of mathematical physics, I, *Zh. Vychisl. Mat. Mat. Phys.*, v. 4, No. 3, 449, 1964 (in Russian).

18. Smagulov, S., On parabolic approximation of the Navier-Stokes equations, *Chisl. Metody Mech. Sploshnoy Sredy*, Novosibirsk, v. 10, No. 1, 111, 1979 (in Russian).

19. Kobelkov, G.M., On the iterative method of solving finite difference problems of elasticity theory, *Dokl. Acad. Nauk. SSSR*, v. 233, No. 5, 776, 1977 (in Russian).

20. Bakhvalov, N.S., Zidkov, N.P., and Kobelkov, G.M., *Numerical Methods*, Nauka, Moscow, 1987 (in Russian).

21. Belotserkovsky, O.M., Guschin, V.A., and Schennikov, V.V., Splitting method and its application to simulation of incompressible viscous fluid dynamics flow, *Zh. Vychisl. Mat. Mat. Phys.*, v. 15, No. 2, 177, 1975 (in Russian).

22. Kobelkov, G.M., The numerical solution of the Navier-Stokes equations with large Reynolds numbers, *Zh. Vychisl. Mat. Mat. Phys.*, v. 24, No. 2, 294, 1984 (in Russian).

23. Kobelkov, G.M., On the methods of Navier-Stokes equations solution, *Dokl. Acad. Nauk. SSSR*, v. 243, No. 4, 843, 1978 (in Russian).

24. Temam, R., *Navier-Stokes Equations*, North-Holland, Amsterdam, 1979.

25. Isakov, A.B., and Kobelkov, G.M., On a numerical solution of the problem of viscous incompressible flow in cube cavity, Preprint No. 179, *Otdel Vychisl. Mat.*, 1988 (in Russian).

Stiff Systems of Ordinary Differential Equations

R.P. Fedorenko

This paper is an attempt at a brief survey of the theory of a specific range of ordinary differential equations (ODE) known as "stiff systems" (SS). We will discuss theoretical questions only. The numerical procedure for solving SS has been applied for about 30 years and has been well developed. Gear's well-known code is an example, and now there are many similar codes. However, there is no universally accepted theoretical conception of SS. We will proceed from initial notions and definitions through to any theoretical concepts developed in recent years. The concept of SS discussed subsequently in this text is by no means complete nor the only one possible. We chose to narrow the subject in order to provide a fairly clear picture.

Let us begin with a formal definition of a SS.

Definition 1. The Cauchy problem,

$$\dot{x} = f(x), \qquad x(0) = X_0, \qquad 0 \le t \le T, \qquad x \in R^N,$$

is referred to as "stiff" if the spectrum of the matrix, $f_x(x)$, can be divided into two essentially different parts:

(1) a stiff spectrum, where eigenvalues and eigenvectors are denoted as $\Lambda_i(x)$, $\Phi_i(x)$, $(i = 1, \ldots, I)$, respectively. For the stiff spectrum,

$$\mathrm{Re}\Lambda_i(x) \le -L < 0, \qquad |\mathrm{Im}\Lambda_i(x)| < |\mathrm{Re}\Lambda_i(x)|,$$

$$i = 1, \ldots, I$$

does hold;

0-8493-8947-X/94 /$0.00 + $.50
(c) 1994 by CRC Press, Inc.

(2) a soft spectrum, where eigenvalues and eigenvectors are denoted as $\lambda_j(x)$, $\varphi_j(x)$, $j = 1, \ldots, J = N - \dot{I}$.

For this part of the spectrum,

$$|\lambda_j(x)| \leq l, \qquad j = 1, \ldots, J$$

does hold.

The Cauchy problem (Definition 1) is stiff if L/l is a large parameter.
Scaling the time, we can assume $l = 1$. The spectrum of the matrix
$f_x(x)$ depends on x, so it is more accurate to write $\dot{I}(x)$, $J(x)$. Later,
however, we confine our examination to the part of the space R^N, where
all the inequalities in Definition 1 hold independently of x.

In applications, the stiffness L may have values of $10^3, 10^6, 10^9$, or
higher. The condition $|\text{Im}\Lambda| \leq |\text{Re}\Lambda|$ is not necessary, but for simplicity
we exclude the case of $|\text{Im}\Lambda| \gg |\text{Re}\Lambda|$. It is assumed that T is large in
relation to l, but not very large: for example, $Tl \simeq 10$, 20. In relation to
L the time T is very large: $TL \approx 10^3, 10^6, 10^9, \ldots$ Because $\|f_x\| \approx \text{O}(L)$,
so $\|f_x\| \cdot T \gg 1$.

Given these conditions, we refer to the Cauchy problem as "stiff",
the system $\dot{x} = f(x)$ as a "stiff system", and the matrix $f_x(x)$ as a
"stiff" matrix. Figure 1 shows the typical spectrum of the stiff matrix
f_x. We will use the eigenvectors of the matrix f_x^T; they will be denoted
as Φ_i^*, $\varphi_j^*(x)$.

Numerical integration of SS is a problem for large step sizes. We
will denote the step size in the difference equations as τ. It must be
small in relation to the soft spectrum (i.e., $\tau \ll 1$), but it can be very
large in relation to the full spectrum: $\tau\|f_x\| \approx \tau L \gg 1$. In applications,
$\tau \simeq 10^{-1} \ 10^{-2}$, $\tau L \simeq 10^3, \ 10^6, \ldots$ Standard algorithms for numerical
solution of ODE are inapplicable in this situation, which calls for special
methods based on implicit difference schemes.

1. LINEAR STIFF SYSTEMS

A linear system with a constant stiff matrix A

$$\dot{x} = Ax, \qquad x(0) = X_0, \qquad 0 \leq t \leq T,$$

is convenient for introduction of the first notions and illustration of the
possibilities offered by implicit schemes. An accurate solution is available
immediately in terms of eigenvectors and eigenvalues,

$$x(t) = \sum_i C_i e^{\Lambda_i t} \Phi_i + \sum_j c_j e^{\lambda_j t} \varphi_j, \tag{1.1}$$

FIGURE 1. The typical spectrum for the stiff matrix, f_x.

where $C_i = (X_0, \Phi_i^*)$ and $c_j = (X_0, \varphi_j^*)$. The systems of vectors $\{\Phi_i, \varphi_j\}$ and $\{\Phi_i^*, \varphi_j^*\}$ are assumed to be biorthogonal. The first term on the right-hand part of Equation (1.1) is a function decreasing as $\exp(-Lt)$. After a short period of time, $O(\ln L/L)$, it may be omitted. Figure 2 shows the typical structure of the solution, which consists of a "boundary layer", where $\|\dot{x}\| = O(L)$, and the slow motion interval, $O(\ln L/L) \leq t \leq T$, where $\|\dot{x}\| = O(1)$. The latter sometimes is referred to as a "quasisteady state." To insure smoothness of the quasisteady part of the trajectory, we need a step size, $\tau = o(1) \gg o(1/\|f_x\|)$, to compute it. But this smooth solution is surrounded by nonsmooth solutions. Figure 2 shows the structure of the directions field for successive magnifications. An attempt at an approximate solution through application of the simple scheme $x_{n+1} = x_n + \tau A x_n$ $\big($where x_n is an approximation of $x(n\tau)\big)$ produces

$$x_n = \sum_i C_i (1 + \tau\Lambda_i)^n \Phi_i + \sum_i c_j (1 + \tau\lambda_j)^n \varphi_j. \qquad (1.2)$$

The second term of the right-hand part of Equation (1.1) is correct, because $(1 + \tau\lambda_j)^n = e^{n\tau\lambda_j} + O(\tau\lambda)$; but the first term has the value of $(-\tau L)^n \to \pm\infty$. The implicit scheme, $x_{n+1} = x_n + \tau A x_{n+1}$, or $x_{n+1} = (E - \tau A)^{-1} x_n$, produces

$$x_n = \sum_i C_i (1 - \tau\Lambda_i)^{-n} \Phi_i + \sum_i c_j (1 - \tau\lambda_j)^{-n} \varphi_j. \qquad (1.3)$$

The slow part of solution (1.3) approximates the corresponding part of the accurate solution in the usual sense of the word, because $(1 - \tau\lambda)^{-n} = e^{n\tau\lambda} + O(\tau\lambda)$. The first term in the right-hand part of Equation (1.3) has the value $1/(\tau L)^n$. It tends toward zero as $\exp(\Lambda_i n\tau) \to 0$. The numerical boundary layer by no means approximates the accurate one. Its length, τ, which may be 2τ, or 3τ, or $\ldots$, is far greater than that of the real boundary layer, which is of $O(\ln L/L)$; but such an error

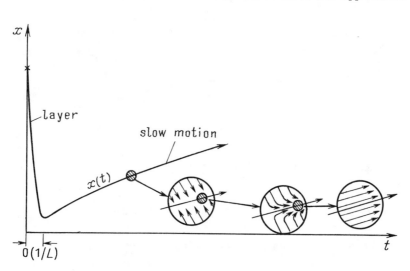

FIGURE 2. The trajectory of SS and the directions field near it.

is acceptable. Figure 3 shows the numerical solution denoted by an asterisk ($*$). It approximates the accurate solution $x(t)$ in the usual sense of the word at any t, except for the short interval $[0, k\tau]$, $k = \mathrm{O}(1)$ (which may be $1, 2, 3, \ldots$) If necessary, the boundary layer can be computed with required accuracy through application of an explicit scheme with a small step size, $\tau^* < 1/\|f_x\|$. This needs only $\mathrm{O}\left(\ln L/(L\tau^*)\right) \approx$

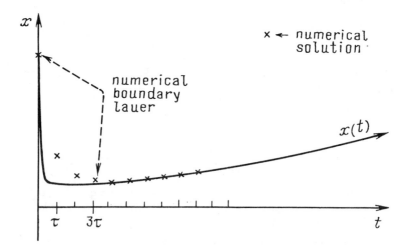

FIGURE 3. The numerical solution of SS obtained by any good scheme.

O(ln L) steps of numerical integration. This reasoning is based on an implicit supposition that all the stiff eigenvalues are O(L). If the stiff spectrum contains essentially different values (e.g., Re$\Lambda_1 = -L \gg$ ReΛ_2, etc.), accurate computation of the boundary layer's structure may be exceptionally difficult. It is thus evident that the implicit scheme does resolve the problem of numerical integration of a linear SS with a large $\tau \ll 1$, $\tau L \gg 1$. But every step of the algorithm requires solution of the linear equation system $(E - \tau A)x = x_n$, and we do not analyze related computational errors. This problem is not simple, because matrix $(E - \tau A)$ is ill-conditioned: its eigenvalues vary from $(1 - \tau\lambda_j) = $ O(1) to $(1 - \tau\Lambda_i) = $ O(τL) $\gg 1$. Estimation of numerical errors arising in this case poses a special problem which is examined, for example, by Godunov.[1]

Another technique used to solve a linear SS is based on the equation

$$x(t + \tau) = e^{A\tau}x(t), \quad \text{or } x_{n+1} = e^{A\tau}x_n. \tag{1.4}$$

Its application requires an efficacious method for computation of the matrix $\exp(A\tau)$. The approximation obtained by summing

$$e^{A\tau} \approx F + \tau A + \frac{\tau^2}{2!}A^2 + \cdots + \frac{\tau^k}{k!}A^k \tag{1.5}$$

is unusable for two reasons. The first is the number of operations: using a rough estimate $k! \approx (k/e)^k$, one can compute the value k required to obtain $\tau^k L^k/k! \approx 1$. This value k is nearly $e\tau L$. The second reason is the numerical instability; $e^{A\tau} = $ O(1) is computed by summing the terms of the value O($e^{||A||\cdot\tau}$), because $e^{-A\tau} = $ O($e^{L\tau}$) is the sum of just the same terms. The method of the "redoubling argument" has been developed for computing $\exp(A\tau)$: first, the matrix $B = \exp(A\tau/2^p)$ is computed. Selecting an integer, p, such that $||A\tau/2^p|| < 1$, one can compute B using Equation (1.5) with a small $k = 1, 2$. Then $e^{A\tau}$ can be computed approximately as B^{2^p} by p multiplications of the matrix $(B^2 = B \cdot B, B^4 = B^2 \cdot B^2$, etc.). Application of this method incurs the risk of an implicit error. Let α be one of the eigenvalues of the matrix A (it may be Λ_i or λ_j). Then $\beta = 1 + \alpha\tau/2^p + \cdots + 1/k!(\alpha\tau/2^p)^k$ is the eigenvalue of B. If $\alpha = \Lambda_i$, only $|\beta| < 1$ and $|\beta^{2^p}| \ll 1$ is required. If $\alpha = \lambda_j$, the computation of the eigenvalue, β, may be too crude at a certain correlation between λ_j, L, p, and the relative error, ε, of the number's representation in the computer. Indeed, instead of $1 + \lambda\tau/2^p$, the value $1 + (\tau\lambda/2^p) + \varepsilon = 1 + \tau/2^p(\lambda + 2^p\varepsilon/\tau)$ is actually computed. B thus carries information about λ with an error, $\delta = 2^p\varepsilon/\tau$, which is an optimistic estimate. The real error is greater than this: $2^p \approx \tau||A||$, $\delta \approx \varepsilon \cdot ||A||$, and sometimes B^{2^p} does not approximate $\exp(A\tau)$ in the "soft" subspace. This error is the more hazardous because it is not immediately obvious; though incorrect, a numerical solution can look good. Using the sum from Equation (1.5), $k > 1$ is especially doubtful, because the term $(\tau\lambda/2^p)^k/k!$ may be less than the error, ε.

A large value of $p \approx \ln_2 L$ and corresponding errors can be avoided by computing $\exp(A\tau)$ in another way, for example,

$$e^{A\tau} = \left(\left(e^{-A\tau/2^p}\right)^{-1}\right)^{2^p}.$$

The matrix $\exp(-A\tau/2^p)$ can be approximated by the sum

$$B = E - A\frac{\tau}{2^p} + \cdots + (-1)^k \left(\frac{\tau}{2^p}\right)^k \frac{A^k}{k!}.$$

The integers p and k are selected to provide the required accuracy of the approximation $\exp(\lambda\tau/2^p)$ only for the soft spectrum ($|\lambda\tau| < 1$). For the stiff eigenvalue Λ, the corresponding eigenvalue of B is $\beta = 1 - (\Lambda\tau/2^p) + \cdots + \cdots (-1)^k (\Lambda\tau/2^p)^k/k!$. Evidently $|\beta| \gg 1$ (because p is small), the stiff eigenvalue of B^{-1} is β^{-1}, $|\beta^{-1}| \ll 1$, and qualitative approximation of the value $\exp(\Lambda\tau) \approx \exp(-L\tau) \ll 1$ by a very small β^{-2^p} is sufficient for numerical purposes. Calculation of $\exp(A\tau)$ is of interest because it is used in certain algorithms for numerical integration of nonlinear stiff systems. An example of such an algorithm was proposed in reference 2. The standard steps of this process are:

1) let the point x_n be known. Compute the matrix $A = f_x(x_n)$;

2) introduce the new variable u: $x(t) = x_n + u(t)$. For this we have the equation

$$\frac{du}{dt} = f(x_n + u(t)) = Au(t) + \{f(x_n + u(t)) - Au(t)\}, \qquad u(t_n) = 0;$$

3) invert the operator $(d/dt - A)$ to obtain

$$u(t_n + \tau) = e^{A\tau} \int_0^\tau e^{-A\xi} \{f(x_n + u(t_n + \xi)) - Au(t_n + \xi)\} \, d\xi.$$

This is an accurate formula. For the next step, notice that the function $\exp(-A\xi)$ is at the maximum on the right-hand boundary of the interval $(0, \tau]$. The small neighborhood of the point $\xi = \tau$ thus makes the key contribution to the integral. Therefore, apply the approximation

$$u_{n+1} = e^{A\tau} \int_0^\tau e^{-A\xi} \{f(x_n + u_{n+1}) - Au_{n+1}\} \, d\xi,$$

where u_{n+1} is an approximation of $u(t_n + \tau)$. Now compute the integral

$$u_{n+1} = A^{-1}(e^{A\tau} - E)\{f(x_n + u_{n+1}) - Au_{n+1}\}.$$

Another proof of this operation is the fact that $f(x_n + u) - Au = f(x_n) + O(u^2)$; u must be small so that the expression $\{\ldots\}$ might be assumed to be constant. The nonlinear equation for u_{n+1} can be solved by any simple iteration process which requires computation of the matrix $B = A^{-1}(e^{A\tau} - E)$. The method described previously was proposed specifically for numerical integration of ODE in chemical kinetics, where one first integral is often known because components of the vector, x, are relative concentrations of reacting compounds, $(x(t), e) = 1$, where $e = \{1, 1, \ldots, 1\} \in R^N$, and A^{-1} does not exist. Indeed, $d/dt(x, e) = (\dot{x}, e) = (f(x), e) = 0$. Differentiation of this equation in relation to x produces $(f_x(x), e) = 0$. This means that $\lambda = 0$ is the point of the spectrum of the matrix $f_x(x)$ and e is the corresponding eigenvector. To compute $A^{-1}(e^{A\tau} - E)$ in this situation, do simple formal mathematics.

$$A^{-1}(e^{A\tau} - E) = A^{-1}(e^{1/2A\tau} - E)(e^{1/2A\tau} + E)$$

$$= A^{-1}(e^{1/2A\tau} - E)AA^{-1}(e^{1/2A\tau} - E + 2E).$$

Let us denote $B_\tau = A^{-1}(e^{A\tau} - E)$ to obtain the recurrent equation

$$B_\tau = B_{\tau/2} A(B_{\tau/2} + 2E)$$

which makes it possible to compute B_τ by $2p$ multiplications of the matrices starting from

$$B_{\tau/2^p} \approx A^{-1}\left(E + \frac{\tau}{2^p} A - E\right) = \frac{\tau}{2^p} E.$$

The definition of B_τ is correct, because $(e^{A\tau} - E)e = 0$.

2.   A-STABILITY OF DIFFERENCE SCHEMES

Theoretical analysis of difference schemes for numerical integration of SS with large step sizes is based on two notions. The first is the standard procedure of approximation used in the analysis of common difference schemes. This remains the same for SS.

The second is a special requirement that the difference scheme be A-stable. A-stability is a new specific feature. To clarify, let us examine a simple example of the A-stable scheme proposed by Rosenbrock:[9]

$$x_{n+2} - \frac{4}{3}\,x_{n+1} + \frac{1}{3}\,x_n - \frac{2}{3}\tau f(x_{n+2}) = 0, \qquad n = 0, 1, \ldots \quad (2.1)$$

It is an implicit scheme; every step of numerical integration requires solution of the nonlinear equations. To verify A-stability of any scheme, the scheme must be tested for linear stiff system $\dot{x} = Ax$, because the numerical solution can be effectively calculated and analyzed in this case. Applied to a linear equation, this scheme produces

$$x_{n+2} - \frac{4}{3}\,x_{n+1} + \frac{1}{3}\,x_n - \frac{2}{3}\tau A x_{n+2} = 0. \qquad (2.2)$$

Let the matrix P be such that the matrix PAP^{-1} is diagonal. Replacing x by $z = Px$ and multiplying Equation (2.2) by P, obtain scalar equations of the same type as Equation (2.2), where A must be substituted for its eigenvalues. This latter has a known accurate solution of the form

$$x_n = C_1 q_1^n + C_2 q_2^n, \qquad (2.3)$$

where q_1, q_2 are roots of the characteristic equation

$$q^2 - \frac{4}{3}\,q + \frac{1}{3} - \frac{2}{3}\lambda\tau q^2 = 0.$$

These roots depend on $\tau\lambda$,

$$q_1(\tau\lambda) = \frac{2 + \sqrt{1 + 2\tau\lambda}}{3 - 2\tau\lambda}, \qquad q_2(\tau\lambda) = \frac{2 - \sqrt{1 + 2\tau\lambda}}{3 - 2\tau\lambda}.$$

Using these expressions, analyze the solutions of Equation (2.3) at different values of $\tau\lambda$ which can be met if matrix A is stiff,

1) let λ be a soft eigenvalue; thus, $|\lambda\tau| \ll 1$. Evidently, $q_1 = e^{\tau\lambda} + O((\tau\lambda)^3)$, $q_2 \approx 1/3$. So $C_1 q_1^n$ approximates $e^{n\tau\lambda}$ at an accuracy of $O(\tau\lambda)^2$;

2) let $|\tau\lambda| \gg 1$. Then, because $q_1 \approx q_2 \approx 1/\sqrt{\tau\lambda}$ and Equation (2.3) approximates the exact solution qualitatively, x_n tends toward zero as $\exp(\lambda n\tau) \to 0$ (at SS $\mathrm{Re}\lambda < 0$).

The characteristic equation has one main root (such as q_1 in the previous example) and can have several extraneous ones (such as q_2). Extraneous roots in any case must be smaller by module than $q_0 < 1$. The main root, $q_1(\tau\lambda)$, must have certain properties; in the complex plane of the variable $\xi = \tau\lambda$, two regions are defined. The first is the region of accuracy. It is a small neighborhood of 0; $q_1(\tau\lambda)$ approximates $e^{\tau\lambda}$. This region may be divided into subregions where $\ln q_1(\tau\lambda)$ approximates $\tau\lambda$ at an accuracy of 0.1%, 1%, 2%, and so forth. The wider the subregions, the more accurate is the scheme.

The second region is the region of stability. Here $|q(\tau\lambda)| \leq q_0 < 1$. This region must include the half plane $\mathrm{Re}\xi < 0$ so that the scheme is A-stable. There are different versions of this notion. For example, the region of stability may be only the angle $\{\mathrm{Re}\xi < 0, |\mathrm{Im}\xi| \leq \beta|\mathrm{Re}\xi|\}$, or there may be imposed an extra constraint, $|q(\xi)| \to 0$ while $\mathrm{Re}\xi \to -\infty$, and so forth.

The difference scheme is so far believed to be usable for numerical integration of SS if it is A-stable and approximates the differential equation. Practical application of the scheme to some model problems used as tests avoids possible errors. But these two requirements (approximation and A-stability) are insufficient as the basis of a theoretical concept. For example, there were built explicit A-stable schemes of the type (for linear SS)

$$x_{n+1} = x_n + \beta(\tau, r_n)Ax_n, \qquad r_n = \frac{(Ax_n, x_n)}{(x_n, x_n)}$$

$$\beta(\tau, r) = \frac{(e^{r\tau} - 1)}{r}, \qquad \tau\|A\| \gg 1 \qquad (2.4)$$

if $\tau \to 0$, $\beta \to \tau$, and scheme (2.4) approximates the equation $\dot{x} = Ax$. For the stiff equation $\dot{x} = -Lx$, $L\tau \gg 1$, $\beta \approx 1/L$, and scheme (2.4) is A-stable. To be accurate one must integrate the time using the same formula, $t_{n+1} = t_n + \beta(\tau, r)$. This is nothing more than integration with an adaptive step size. The boundary layer (where $r \approx \|A\| = O(L)$) is thus integrated with a small step size, $\beta \approx 1/\|A\|$. If the point x_n consists mainly of soft eigenvectors and $r_n \approx O(1)$, step-size $\beta \approx \tau \gg 1/\|A\|$. But small stiff components of x_n increase as r_{n+1} increases in this case, whereas the step size decreases, and r_n does so also, β tends to τ, and so forth. It is difficult to estimate the efficacy of this algorithm.

V.I. Lebedev studied the applicability of explicit schemes in numerical integration of the SS (see his paper in this volume).[8] His methods are based on the advanced theory of numerical stability of difference schemes and Chebyshev-type iterative processes.

The well-known implicit scheme of the second order of approximation,

$$\frac{x_{n+1} - x_n}{\tau} = \frac{f(x_n) + f(x_{n+1})}{2}, \qquad (2.5)$$

is A-stable because $|(1 - \tau\lambda/r)/(1 + \tau\lambda/2)| \leq 1$ for all $\mathrm{Re}\lambda \leq 0$. Sometimes it is claimed to be usable for numerical integration of "stiff oscillating" systems with large step sizes. This claim appears to be wrong: the numerical solution of the equation $x = iLx$, obtained by using scheme (2.5) with a step $\tau L \gg 1$, oscillates as the exact solution does; they have nothing more in common, so it is inapplicable.

3. SINGULARLY-PERTURBATED SYSTEMS

The singularly-perturbated class of stiff systems is applicable for theoretical analysis of numerical methods because it is covered by an advanced asymptotic theory,[3] and the qualitative behavior of the trajectory is clear enough. This makes possible a statement of the requirements numerical integration methods must meet, and checking if one or another algorithm meets these requirements. Examine a system $\dot{x} = f$ of the form

$$\dot{y} = Y(y, z)$$

$$\dot{z} = LZ(y, z) \quad \left(\text{or } \varepsilon\dot{z} = Z(y, z), \quad \varepsilon = \frac{1}{L} \ll 1 \right). \tag{3.1}$$

Here Y, Z are smooth functions $\big(Y, Z,$ and their derivatives have values of $O(1)\big)$, the dimensions of the vectors y and z are, respectively, J and I, the vector $x = \{y, z\}$, and $f = \{Y, Z\}$. The spectrum of the variational matrix, f_x, is defined by the equation

$$\det \begin{pmatrix} Y_y - \lambda E_J & Y_z \\ LZ_y & LZ_z - \lambda E_I \end{pmatrix} = 0. \tag{3.2}$$

Evidently the stiff spectrum is defined by the spectrum of the matrix LZ_z (if it does not have eigenvalues $O(1/L)$, of course), and the corresponding eigenvectors, Φ_i, have essentially z components; their y components are $O(1/L)$. The qualitative structure of the trajectories is well known. It is defined by the surface Γ,

$$\Gamma = \big\{ x = \{y, z\} \; : \; Z(y, z) = 0 \big\}. \tag{3.3}$$

It divides the phase space into two parts, $Z(y, z) > 0$ and $Z(y, z) < 0$. System (3.1) is stiff only at points x, where the spectrum of the matrix, Z_z, is "stable" ($\text{Re}\Lambda_i < 0$). Figure 4 shows the typical trajectory of system (3.1) when y, z are scalars; the picture is more complicated, though basically of the same kind, in the common case. The phase velocity is nearly horizontal and very large $\big(\|\dot{x}\| = O(L), \|\dot{y}\| = O(1), \|\dot{z}\| = O(L)\big)$ outside the small $O(1/L)$ neighborhood of Γ. The point $x(t)$ travels rapidly to the right if $Z(y, z) > 0$, or to the left if $Z(y, z) < 0$. Within a short period of time, $O(1/L)$, the point $x(t)$ travels from the initial position, X_0, to the $O(1/L)$ neighborhood of Γ (we denote it as $\Gamma_{O(1/L)}$). Here, $\dot{x} = O(1)$ and the point $x(t)$ travels slowly (as in a quasisteady state) along Γ. The surface, Γ, is divided into two parts according to the sign of Z_z: the stable branch of Γ (if $Z_z < 0$ in the scalar case, or if the spectrum of matrix Z_z is stable, in the common case) and the unstable branch (if $Z_z > 0$ in the scalar case, or if Z_z has at least one eigenvalue, Λ, with $\mathrm{Re}\Lambda > 0$). In Figure 4, the stable branches of Γ are AB and CD, and the unstable branch is BD. System (3.1) is not "stiff" in the small neighborhood of the unstable branch of Γ, because f_x has large $O(L)$ eigenvalues with a positive real part.

The most interesting phenomenon occurs near the point B or D (see Figure 4), where Γ is losing stability and one of the eigenvalues of Z_z moves from the left side of the complex plane to the right side. Then the point $x(t)$ moves from one stable branch of Γ into the point C on another stable branch within a short period of time, $O(\ln L/L)$. This part of the trajectory forms the so-called "inner layer". Then $x(t)$ moves, according to the sign of Y, along Γ at a velocity $O(1)$ to the point D (or E), and so forth. The trajectory, $x(t)$, is similar to the curve in Figure 5; the slow-motion intervals are separated by short $O(1/L)$ boundary or inner layers of the quick motion $\big($at a rate of $O(L)\big)$. The trajectory appears to be discontinuous. Similar layers, separated by slow-motion intervals, can be seen, for example, on the trajectories of equations in chemical kinetics. There is a point of view that stiff systems may be regarded as systems derived from singularly-perturbated systems by any unknown change of variables. Stiff systems are thus systems with large implicit parameters because there is no explicit division of variables into quick and slow ones $\big(z$ and y for system $[3.1]\big)$, and we have no simple equation for Γ such as $Z(x,y) = 0$ for system (3.1). Therefore, there is no asymptotic theory similar to the theory of singularly-perturbated systems.[3] An attempt at such a theory is discussed in reference 4; also see Section 6 of this paper.

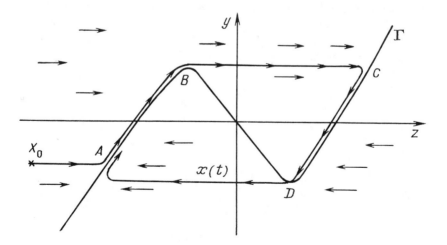

FIGURE 4. A phase diagram of SPS, $\dot{y} = z$, $\varepsilon\dot{z} = -y + z - z^3/3$.

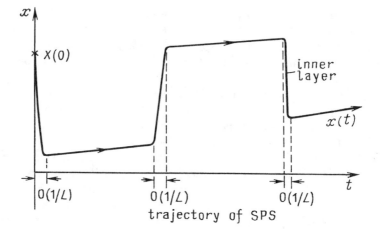

FIGURE 5. The trajectory of SS appears to be discontinuous.

4. NUMERICAL INTEGRATION OF A SINGULARLY-PERTURBATED SYSTEM

the stiff system (3.1) is carried out by any difference scheme with large step τ, that is, $\tau \ll 1$, but $\tau L \gg 1$. (In reality, $\tau = 0.01, 0.1, \tau L$ may have the value of $10^3, 10^6$, and more). Assume also that the numerical description of the layers (both boundary and inner) must be but qualitative; their time may be $O(\tau) \gg O(1/L)$, whereas the structure may not be described at all. But slow-motion intervals must be described at a good approximation in the ordinary sense. So we use the standard notion of accuracy of numerical solution in the full interval [0,T] except in the small $O(\tau)$ neighborhoods of the layers. Let us examine the results which may be produced by a simple implicit scheme

$$\frac{y_{n+1} - y_n}{\tau} = Y(y_{n+1}, z_{n+1}), \qquad \frac{z_{n+1} - z_n}{\tau} = LZ(y_{n+1}, z_{n+1}),$$

$$n = 0, 1, \ldots \quad (4.1)$$

Typical situations are:

1) the first step of the numerical integration ($n = 0$); the assigned initial point, $\{y_0, z_0\}$, is placed far from Γ; the next point, $\{y_1, z_1\}$, must be found by solving the nonlinear equations in y and z,

$$\begin{cases} y - y_0 - \tau Y(y, z) = 0 \\ Z(y, z) - \dfrac{(z - z_0)}{\tau L} = 0. \end{cases} \quad (4.2)$$

The first equation defines the J dim surface. Because $\tau \ll 1$, it is close to the plane, $y = y_0$. The second equation defines the I dim surface, Γ^*. Because $\tau L \gg 1$, it is close to Γ; therefore, Γ^* is located in $\Gamma_{O(1/\tau L)}$ (we are examining a common, nondegenerate case, of course). According to the asymptotic theory,[3] the first approximation to the exact solution is defined by the equations of the boundary layer

$$\dot{z} = LZ(y, z), \qquad \dot{y} = 0, \qquad z(0) = z_0, \qquad 0 \le t \le O\left(\frac{\ln L}{L}\right).$$

$$(4.3)$$

The right boundary of the layer, with an accuracy of $O(1/L)$, is the solution of the equation $Z(z, y_0) = 0$.

The solution of system (4.2) coincides with that of system (4.3) at an accuracy of $O(\tau) + O(1/\tau L)$ (in practice, $O(\tau)$ is the main term of the error). This error is reasonable from the generally adopted point of view, although the length of the boundary layer is $O(\tau) \gg O(\ln L/L)$, and the layer has no structure. Unfortunately, though system (4.2) may have several solutions, only one of them happens to be the right one. If system (4.2) is solved by any effective iterative method (say, Newton's method, starting at the point z_0, y_0), it will produce not the right solution, but the "nearest" to the starting point. Thus, the point $\{y_1, z_1\}$ may be wrong. For example, $\{y_1, z_1\}$ may be found in the $O(\tau)$ neighborhood of the unstable branch of Γ. To sum up, one can say that an attempt to pass through the layer in only one step of numerical integration may succeed but may produce a totally incorrect result. It is thus necessary to use a very small step size, $\tau^* \ll 1/L$ (say, $\tau^* \approx 0.1 \div 0.01/L$), for numerical computation of the layer. Then the numerical solution $\{y_n, z_n\}$ approximates the exact solution $\{y(n\tau^*), z(n\tau^*)\}$. It requires $n = O(\ln L)$ steps, τ^*, of $O(1/L)$, which is quite acceptable. Moreover, this integration can be carried out by a simpler explicit scheme.

2) the slow-motion stage; according to the theory proposed by Vasil-jeva and Butuzov,[3] slow motion may be approximately described by the solution of the equations

$$\frac{dy}{dt} = Y(y, z)$$

$$Z(y, z) = 0 \qquad (\text{or } \{y, z\} \in \Gamma). \qquad (4.4)$$

The accuracy of this approximation is $O(1/L)$. Difference scheme (4.1) integrates almost the same equation, but Γ is substituted for Γ^*. The accuracy of the numerical solution is thus $O(\tau) + O(1/\tau L)$. It is $O(\tau^p) + O(1/\tau L)$ if a more accurate scheme is used (p is the degree of approximation). In reality, the stiffness of the system often is so great that just $O(\tau)$ is the main term of the error. At this stage of the numerical integration $\{y_n, z_n\} \in \Gamma^*$ there exists a solution of system (4.1), $x_{n+1} \in \Gamma^*$ in the $O(\tau)$ neighborhood of the point x_n. The iterative process starting at the point x_n is most likely to converge with the right solution, x_{n+1}. It may rarely, if ever, produce another, incorrect solution.

3) the inner layer; examine the most exciting and complicated event in "the life of trajectory", restricting consideration to a simple case where y and z are scalars. Figure 6 shows the structure of the trajectory in the neighborhood of the critical point B (the beginning of the inner layer). Let point x_n be the point B or very close to B.

It will become clear from subsequent discussion how point x_n can differ from B to retain the main fact: system (4.1) has no solution in the small neighborhood of x_n, and the point x_{n+1} must make a great leap on another branch of Γ^*. It may be either right or wrong, stable or unstable.

Using $y = y_n + \xi$, $z = z_n + \eta$, rewrite Equations (4.1) as

$$\xi = \tau Y(y_n, z_n) + \tau Y_y \xi + \tau Y_z \eta + \dots \qquad (4.5a)$$

$$\xi = O\left(\frac{1}{\tau L}\right) - b\eta^2 + \dots, \qquad b > 0. \qquad (4.5b)$$

Because the point $x(t)$ comes from A to B, $Y(x_n) = a > 0$. The solution of system (4.5) is at the intersection of two curves. The first is determined by Equation (4.5a). It is located in a narrow cone in the neighborhood of B. Its pole is the point $\xi = a$, $\eta = 0$, and the angle of the cone is $O(\tau)$. The second curve determined by Equation (4.5b) is in the half plane $\xi < O(1/\tau l)$. Because $O(1/\tau L) \ll a\tau$ (there may certainly be assumed any relation between L and τ) Equations (4.5) have no solution for a small $\xi, \eta = O(\tau)$ or $O(\sqrt{\tau})$. In the common case where y, z are vectors, equations of type (4.5b) must be written for each of the $\dot{I}$ scalar components of z. At least one will have as the main term form (4.5b) and another the general linear form. To conclude, it is sufficient to say that system (4.5) has no solution $\xi, \eta = O(\tau)$.

An attempt to pass through the inner layer in only one step of numerical integration may succeed, but this could also generate a considerable error. For example, the point x_{n+1} may be found on the unstable branch of Γ.

4) computations on the unstable branch of Γ; let us examine a situation where the point (y_n, z_n) sits on the unstable branch of Γ. In real practice it may be the result of a mistake in previous computations. System (3.1) is not stiff in this case, because the matrix Z_z has at least one eigenvalue with a positive real part. This point, x_n, is unstable for a differential equation. The trajectory starting at x_n passes into another branch of Γ (or tends to infinity)

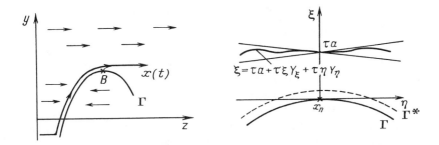

FIGURE 6. The structure of the curves defined by Equations (4.5).

in a short period of time, $O(\ln L/L)$. But for difference Equation (4.1), with a large $L\tau \gg 1$, this branch of Γ is as stable as a really stable one. To be precise, it is stable if any good iterative process for solving system (4.1) is used (e.g., Newton's method). There are certainly bad iterative processes in which this point is unstable. But such iterative algorithms are similar to integration of a difference equation with a very small step size, $\tau^* < 1/\|f_x\|$. They are not applied practically.

It is clear from this discussion that the layers must be passed at very small step size, $\tau^* \ll 1/L$. But there arises the question of how to recognize the beginning and the end of the layer. Prematurely replacing a large step size, τ, with a small τ^* at the beginning of the layer and a delayed switch back from τ^* to τ at the end of the layer leads to an undesirable (and often unacceptable) waste of computer time. The beginning of the layer may be recognized by the large change in the value $\|x_{n+1}-x_n\|$. Immediate adoption of a small τ^* and repetition of the calculations, starting from x_n, is not a good solution to the problem: the beginning of the layer may be found at the end of the interval $[t_n, t_n + \tau]$, so that this interval will be passed through in a very great number of steps τ/τ^* (the value may be $10^6, 10^9$, etc.). Restarting the numerical integration at the point x_n with, say, $\tau := \tau/3$ is preferable. Therefore, methods of changing τ for τ^* and vice versa constitute a very important part of numerical integration algorithms of SS, such as Gear's well-known algorithm.

5. ON REGULAR STIFF SYSTEMS

Stiff systems are numerically integrated under conditions which invalidate well-known theorems on the convergence of difference schemes. A simple analysis of such theorems shows that the deduction "approximation + stability = convergence" works only if $\tau\|f_x\| < 1$. It fails in the case we are examining. Moreover, the standard estimate of the value $||x(n\tau) - x_n^\tau||$ contains the term $\tau^k \exp(Cn\tau)$. Here $x(t)$ is the exact solution of the SS, x_n^τ is the numerical solution obtained by any method of the kth order of approximation, and τ is the step size. We refer to the theorems using only information on Lipschitz's constant for the function f. This constant, C, is a value close to $||f_x||$. In the case we are examining, the value $\exp(\tau||f_x||)$ is so large that such estimates are useless. There are other convergence theorems using special properties of the matrix f_x. For example, if $\mathrm{Re} f_x = 0.5(f_x + f_x^*) \leq \alpha < 0$, there is no exponential factor in the estimate of the numerical error. This theorem is well known, but its proof does use the condition $\tau||f_x|| < 1$. Moreover, references to the theorem on $\lim_{\tau \to 0} x_n^\tau$, $n = T/\tau$ make no sense for SS, because they are numerically integrated in a situation with two small parameters. These parameters are τ and $1/\tau L$. An interesting and useful theory is one which would give an estimate of the numerical errors from these two parameters. We have every reason to believe that an estimate of this kind cannot be obtained without an asymptotic theory for SS, similar to the theory of singularly-perturbated systems. An attempt to work out such a theory for a special class of SS, called regular SS, is discussed in reference 4.

The essentials of this theory and its application to analysis of some numerical methods will be discussed later. This theory proceeds from an analogy between stiff systems and a thoroughly studied class of singularly-perturbated systems (SPS). Many stiff systems are believed to result from any SPS by an unknown change of variables. Therefore, it may be reduced back to a SPS by an inverse change. Subsequently we will build this inverse change, ineffective to a certain extent practically, but useful in the theoretical analysis. The theory subsequently proposed is an analogy to the part of the SPS theory which describes the slow motion of the phase point, $x(t)$, in the proximity of the stable branch of Γ. This part of the trajectory is the most difficult to integrate numerically, and it is only this part that requires special difference schemes to be used and studied. Computation of layers may employ standard schemes with small $\tau^* < 1/||f_x||$. Particular difficulties in the SS asymptotic theory, as compared with the SPS theory, arise from:

1) having no explicit division of the components of the vector x into the slow and quick components (such as the separation of x into y and z for SPS Equation [3.1]);

2) having no explicit equation describing Γ (such as $Z(y, z) = 0$ for Equation [3.1]).

We will use groups of values which differ essentially from one another in the subsequent analysis and use the scale of values of different orders,

$$\mathrm{O}(L) \gg \mathrm{O}(1) \gg \mathrm{o}(1) \gg \mathrm{O}(1/L)\ldots \qquad (5.1)$$

It is assumed that several additions of values of the same order does not change the position of the result in the scale given in Equation (5.1). We use the obvious rules $O(\alpha) + O(\alpha) = O(\alpha)$, $O(\alpha) * O(1) = O(\alpha)$, and so on.

Let us define the manifold Γ:

Definition 2. $x \in \Gamma$ if $\big(f(x),\, \Phi_i^*(x)\big) = 0$, $i = 1, 2, \ldots, I$.
The dimension of Γ is J.

Corollary. If $x \in \Gamma$ then $f(x) = \sum_{j=1}^{J} c_j(x)\varphi_j(x)$, where $c_j(x) = \big(f(x),\, \varphi_j^*(x)\big)$.

This means that the vector of the phase velocity of the trajectory, $x(t)$, lies on Γ in the J dim hyperplane formed by "soft" eigenvectors of the matrix $f_x(x)$.

We make the additional assumptions:

Assumption 1. $f(x) = O(1)$ if $x \in \Gamma$.

Assumption 2. $f_x(x) = O(L)$ if $x \in \Gamma_{o(1)}$.

Assumption 3. $f_{xx}(x) = O(L)$ if $x \in \Gamma_{o(1)}$.

Remember that Γ_ε is the ε neighborhood of Γ.

Let us discuss these assumptions. Properties of any function $f(x)$: $f(x) = O(1)$, $f_x(x) = O(L) \gg 1$ are usually associated with the functions of the type $\sin(Lx)$ for which $f_{xx} = O(L^2)$. In the case we are examining there is another mechanism: $f(x) = O(1)$ on Γ because large items are canceled. These assumptions define a special class of SS we refer to as "regular". It is not empty: a system derived from any SPS by a smooth change of variables is a regular SS.

Assumption 4. all the functions we shall use later are assumed to be smooth.

This means that the function and all its derivatives which we are examining have values of $O(1)$. The only exception is the function $f(x)$. But its restriction on Γ is assumed to be smooth.

Assumption 5. the basis formed by the eigenvectors $\Phi_i(x)$, $i = 1, \ldots, I$, $\varphi_j(x)$, $j = 1, \ldots, J$ is assumed to be well conditioned.

The sense of this assumption is this: let $B(x)$ be a matrix which transforms the coordinates of a point in the initial basis in R^N into coordinates in the (Φ, φ) basis. It is assumed that $\|B(x)\|$ and $\|B^{-1}(x)\|$ are values of $O(1)$, uniform with regard to all the points x to be examined. Let us prove that Γ defined in this way is similar to the stable branch of Γ in the SPS theory.

Statement 1. If $x \in \Gamma$, $f(x)$ is "almost tangent" to Γ (tangent at an accuracy of $O(1/L)$).

Proof (in brief). Let $x \in \Gamma$ and l be a vector tangent to Γ in point x ($\|l\| = 1$). Upon differentiation of equations $\left(f(x), \Phi_i^*(x)\right) = 0$ with regard to direction l, obtain

$$0 = (f_x l, \Phi_i^*) + (f, \frac{\partial \Phi_i^*}{\partial x} l) = (l, \Lambda_i^* \Phi_i^*) + \mathrm{O}(1)$$

and so $(\Phi_i^*, l) = \mathrm{O}(1/L)$. Consequently, the vector l in the basis (Φ, φ) has the form

$$l = \sum_i \mathrm{O}\left(\frac{1}{L}\right) \Phi_i(x) + \sum_j \varepsilon_j \cdot \varphi_j(x). \tag{5.2}$$

Let P be a J dim hyperplane tangent to Γ at the point x. It is spread over J vectors l_k; each can be represented in the form of Equation (5.2). Let P_1 be a hyperplane spread over the vectors $\varphi_j(x)$, $j = 1, \ldots, J$. The angle between P and P_1 is $\mathrm{O}(1/L)$. Because $f(x) \in P_1$ and $f(x) = \mathrm{O}(1)$, there exists a small vector $\delta = \mathrm{O}(1/L)$ such that $f(x) + \delta$ is tangent to Γ at point x. Statement 1 is proven.

According to Assumption 4, Γ is a smooth manifold and it can be represented by a smooth mapping,

$$x = X(\alpha), \qquad \alpha \in R^J; \qquad X : R^J \to R^N \qquad (N = I + J). \tag{5.3}$$

Components of α are thus the coordinates on Γ. Let us introduce in a small neighborhood of Γ a special change of variables which will reduce the SS $\dot{x} = f(x)$ to the form of a SPS. This change is carried out by mapping $(\alpha, s) \to x$, where $s = \{s_1, s_2, \ldots, s_I\} \in R^I$,

$$x = X(\alpha) + \sum_i s_i * \Phi_i X(\alpha). \tag{5.4}$$

Henceforth we will use special designations for superpositions of the functions: $\Phi_i X(\alpha)$ will mean $\Phi_i\left(X(\alpha)\right)$, and so forth. Omitting the brackets, we may not omit the sign of multiplication. To avoid confusion, use fixed letters to denote points of different spaces: x for R^N, α for R^J, s for R^I. Mapping Equation (5.4) is fairly simple; Figure 7 shows its structure. By virtue of Assumption 5, the inverse mapping $x \to \{\alpha, s\}$ exists (at least in a neighborhood of Γ). The inverse mapping may be described by two smooth functions of x,

$$\alpha = A(x), \qquad A : R^N \to R^J;$$
$$s = S(x), \qquad S : R^N \to R^I. \tag{5.5}$$

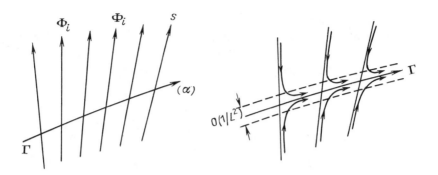

FIGURE 7. The structure of the mapping of Equation (5.4) and the trajectories of SS.

Statement 2. A regular stiff system, $\dot{x} = f(x)$, can be reduced to a singularly-perturbated system in a small $o(1)$ neighborhood of Γ.

Statement 2 is the union of Statements 3, 4, and 5, which have independent significance.

Proof. Let us denote $\alpha[t] = Ax(t)$, $s[t] = Sx(t)$. Assume that $x(t) \in \Gamma_{o(1)}$; it is sufficient to assume that $x(0) \in \Gamma_{o(1)}$. Differential equations can be derived for these functions, which are the slow and quick components of $x(t)$, respectively. Differentiation with regard to time produces

$$\dot{s}[t] = S_x\,x(t) * fx(t).$$

Here
$$x(t) = X\alpha[t] + \sum_i s_i[t] * \Phi_i X\alpha[t]. \qquad (5.6)$$

Because $s_i[t] = o(1)$, the function $fx(t)$ can be expanded, for example

$$fx(t) = f\left(X\alpha[t] + \sum_i s_i[t] * \Phi_i X\alpha[t]\right)$$

$$= fX\alpha[t] + f_x X\alpha[t] * \sum_i s_i[t] * \Phi_i X\alpha[t] + \mathrm{O}\!\left(Ls^2\right).$$

(we use s^2 instead of $||s||^2$). Use

$$\frac{\partial S_k(x)}{\partial x}\,\Phi_i X A(x) = \delta_{i,k}, \qquad i,k \in [1:I]. \qquad (5.7)$$

Here S_k is the kth component of the vector function $S(x)$. Equation (5.7) is almost evident. Compute the derivative of $S_k(x)$ with regard to the direction, $\Phi_i XA(x)$, at the point $x \in \Gamma_{o(1)}$. Computing this derivative directly, evaluate

$$\lim_{h \to 0} \{S_k(x + h * \Phi_i XA(x)) - S_k(x)\}/h.$$

Considering that

$$x = XA(x) + \sum_l S_l(x) * \Phi_l XA(x)$$

and $\quad S_k\left(XA(x) + \sum_i s_i * \Phi_i XA(x)\right) = s_k \quad$ for all $x, s_i, k,$

which is evident from Equation (5.4), Equation (5.7) is easily obtained. Continuing the analysis of $\dot{s}[t]$,

$$\dot{s}[t] = S_x(X\alpha[t] + O(s))$$
$$* \left\{ fX\alpha[t] + \sum_i s_i[t] * \Lambda_i X\alpha[t] * \Phi_i X\alpha[t] + O(Ls^2) \right\}$$

$$= \{S_x X\alpha[t] + O(s)\}$$
$$* \left\{ fX\alpha[t] + \sum_i s_i[t] * \Lambda_i X\alpha[t] * \Phi_i X\alpha[t] + O(Ls^2) \right\}$$

$$= S_x X\alpha[t] * fX\alpha[t] + O(s) * fX\alpha[t]$$
$$+ \sum_i s_i[t] * \Lambda_i X\alpha[t] * S_x X\alpha[t] * \Phi_i X\alpha[t]$$

$$+ O(s) * \sum_i s_i[t] * \Lambda_i X\alpha[t] * \Phi_i X\alpha[t]$$
$$+ O(Ls^2) * S_x X\alpha[t] * fX\alpha[t] + O(Ls^3).$$

When we evaluate each term of the result, then

1) $fX\alpha[t] = f(X(\alpha[t])) = l + O(1/L)$, where l is a vector tangent to Γ at the point $X(\alpha[t]) \in \Gamma$. Evidently $S_x X\alpha[t] l = S_x(X(\alpha[t])) l = 0$, because it is a derivative of S with regard to the direction, l, tangent to Γ, computed at the same point $X\alpha[t] \in \Gamma$. The first term has a value of $O(1/L)$;

2) the second term has a value of $O(s)$, because $X\alpha[t] \in \Gamma$; consequently, $fX\alpha[t] = O(1)$;

3) by virtue of Equation (5.7), the ith component of the third term is $s_i[t] * \Lambda_i X\alpha[t]$;

4) the fourth term is $\mathrm{O}(Ls^2)$;

5) the fifth term is $\mathrm{O}(Ls^2)$ because $fX\alpha[t] = \mathrm{O}(1)$.

Summing up, we arrive at

Statement 3. The functions $s_i[t]$ satisfy the equations

$$\dot{s}_i[t] = \Lambda_i X\alpha[t] * s_i[t] + \mathrm{O}(Ls^2) + \mathrm{O}\left(\frac{1}{L}\right),$$

$$i = 1, \ldots, \dot{I}. \qquad (5.8)$$

Simple analysis of the solutions leads to

Statement 4. Γ is an attractive, stable, "near-integral" manifold.

In a refined mathematical sense, this statement says that every trajectory starting at the point $x(0) \in \Gamma_{o(1)}$ within a short period of time, $\mathrm{O}(\ln L/L)$, attains the $\mathrm{O}(L^{-2})$ neighborhood of Γ and remains in $\Gamma_{\mathrm{O}(1/L^2)}$ while the situation described in Definition 1 holds.

> **Proof.** The proof is fairly simple. Assuming that $\mathrm{Re}(\Lambda_i X\alpha[t])$ $\leq -L$, the exponential decrease, $s_i[t]$, continues until the first term of the right-hand part of Equation (5.8) remains the main term. The decrease ceases if $\Lambda s \approx \mathrm{O}(1/L)$, which means $s[t] = \mathrm{O}(1/L^2)$.

Let us derive an equation for $\alpha[t]$. Differentiation with regard to time produces

$$\dot{\alpha}[t] = Ax(t) * \dot{x}(t) = A_x x(t) * fx(t).$$

Using Equation (5.6) and assuming that $s = \mathrm{O}(L^{-2})$ yields

$$x(t) = XAx(t) + \mathrm{O}(L^{-2}) = X\alpha[t] + \mathrm{O}(L^{-2}).$$

Moreover,

$$fx(t) = f\left(XAx(t) + \mathrm{O}(L^{-2})\right)$$
$$= fXAx(t) + \mathrm{O}(L^{-1}) = fX\alpha[t] + \mathrm{O}(L^{-1}).$$

We arrive thus at

Statement 5. The function $\alpha[t]$ satisfies the equation

$$\dot{\alpha}[t] = D(\alpha[t]) + \mathrm{O}(L^{-1})$$

where $\qquad D(\alpha) = A_x X(\alpha) * fX(\alpha). \qquad (5.9)$

This equation is not stiff; it holds only in $\Gamma_{O(L^{-2})}$. A similar equation can be obtained for the "rest of the layer". It holds in $\Gamma_{o(1)}$ only within a short period of time, $O(L^{-1})$, and has little significance. More interesting is another object linked with the trajectory: $\tilde{x}(t) \equiv X\alpha[t] \equiv XAx(t)$. An equation effective in $\Gamma_{O(L^{-2})}$ can be derived,

$$\dot{\tilde{x}}(t) = X_\alpha \alpha[t] * D\alpha[t] + O(L^{-1}),$$

and can be simplified to

$$X_\alpha(\alpha) * D(\alpha) = X_\alpha(\alpha) * A_x X(\alpha) * fX(\alpha) = fX(\alpha) + O(L^{-1}).$$

This is evident from the fact that $XA(x) \equiv x$ for all $x \in \Gamma$. Upon differentiating this equation with regard to a direction, l, tangent to Γ at the point x, one can obtain $X_\alpha(A(x))A_x(x)l = l$. Because $f(x) = l + O(L^{-1})$, the equation for $\tilde{x}(t)$ can be written as

$$\frac{d\tilde{x}}{dt} = f(\tilde{x}) + O(L^{-1}), \qquad \tilde{x}(t) \in \Gamma. \tag{5.10}$$

This equation is not stiff. We refer to the function $\tilde{x}(t)$ as "the shadow of the trajectory" on Γ. Equations (5.8) and (5.9) form the SPS mentioned in Statement 2. Equation (5.10) is similar to Equation (4.4).

Note that by using the results obtained we can imagine the behavior of the trajectories of SS in a small $o(1)$ neighborhood of Γ. Figure 7 shows a qualitative picture of the trajectories. The manifold Γ as defined by Definition 2 is more precise from the asymptotic point of view than Γ defined by an equation for a SPS. The first has an accuracy of $O(L^{-2})$, whereas the second is only accurate to $O(L^{-1})$. To estimate the accuracy of the numerical solution $x_n, n = 0, 1, \ldots$, introduce these objects: $x(t_n)$ as the exact solution of a SS, $\alpha^n = Ax(t_n)$, $s^n = Sx(t_n)$, $\alpha_n = A(x_n)$, and $s_n = S(x_n)$. The numerical solution x_n can be considered good if:

1) the quick components are small: $\|s_n\| \ll O(1)$;

2) the slow components of the numerical solution are close to the exact ones: $\|\alpha^n - \alpha_n\| = O(\tau^p)$, where p is the order of accuracy.

6. SOME APPLICATIONS OF THE THEORY

The asymptotic theory of regular stiff systems discussed previously is not convenient for immediate application in practical work. It uses objects like eigenvectors $\Phi_i(x)$ and $\varphi_j(x)$; their computation at many points x

is too time consuming. This theory, however, makes possible analysis of some numerical methods. Let us discuss examples of this analysis.

6.1 DEMIRCHYAN–RAKITSKY SCHEME

A special method of numerical integration of SS described in reference 5 was referred to by the authors as "the method of quick motions filtering" (MQMF). The authors referred to their scheme as "explicit", a rather arbitrary term. We first describe and analyze another scheme. It is built on the main ideas of the original MQMF and the asymptotic theory as discussed previously. Let us examine one standard step of the numerical integration.

1) Let the point x_n be known and let $x_n \in \Gamma_{O(1)}$. This means that some other method, for example an explicit method with a small step size, $\tau^* \ll L^{-1}$, was used to pass through the boundary or inner layer.

2) Denote $\alpha^n = A(x_n)$, $s^n = S(x_n)$, and compute the eigenvectors $\Phi_i^n = \Phi X(\alpha^n)$, $i = 1, \ldots, \dot{I}$. A linear subspace is formed,

$$x(\xi) = x_n + \sum_i \xi_i \Phi_i^n. \tag{6.1}$$

3) Then solve the problem

$$\min_\xi \|fx(\xi)\|, \qquad s^a = \arg\min_\xi \|fx(\xi)\|. \tag{6.2}$$

Refer to the point $x_n^a = x_n + \sum_i s_i^a * \Phi_i^n$ as "asymptotic".[5]

4) Finally, make an explicit step of numerical integration,

$$x_{n+1} = x_n^a + \tau f(x_n^a). \tag{6.3}$$

Let us analyze this algorithm. First, is the point x_n^a obtained by operation (6.2)?

Statement 6.

1) Operation (6.2) does not alter the values of slow variables: $A(x_n^a) = A(x_n) = \alpha^n$;

2) quick variables are $s_n^a = S(x_n^a) = O(1/L)$.

The first part of Statement 6 is obvious, because the hyperplane $x(\xi)$ (see Equation [6.1]) was built as the $\alpha = \text{const}$.

The second part follows from the simple facts that

1) $XA(x_n) \in \Gamma$, so $\|fXA(x_n)\| = O(1)$;

2) $XA(x_n)$ is a point on the hyperplane $x(\xi)$, defined by the values $\xi = S(x_n)$; therefore, $\min_\xi \|fx(\xi)\| = O(1)$;

3) since the function f varies at a rate of $O(L)$ while s moves away from O, $\min \|fx(\xi)\|$ can be found only at a distance of $O(1/L)$ from Γ (implying, of course, the local minimum nearest to the point x_n). As a result,

$$x_n^a = X(\alpha^n) + \sum_i s_i^a * \Phi_i^n,$$

$$\alpha_n = A(x_n), \quad s^a = O\left(\frac{1}{L}\right). \qquad (6.4)$$

Computing $f(x_n^a)$,

$$f(x_n^a) = f(X\alpha^n + \sum s_i^a \Phi_i^n)$$

$$= fX(\alpha^n) + \sum s_i^a * \Lambda_i^n * \Phi_i^n) + O\left(\frac{1}{L}\right). \qquad (6.5)$$

The function $\tilde{x}(t) = XAx(t)$, where $x(t)$ is an exact solution of SS, can be used to estimate the result of the "explicit" step, Equation (6.3). It satisfies Equation (5.10) introduced earlier. According to the definition of $\tilde{x}$ we have $X(\alpha^n) = \tilde{x}_n$ and $fX(\alpha^n) = f(\tilde{x}_n)$. Substitute Equations (6.4) and (6.5) into Equation (6.3) to obtain

$$x_{n+1} = X(\alpha^n) + \sum_i s_i^a * \Phi_i^n + \tau fX(\alpha^n) + \tau \sum_i s_i^a * \Lambda_i^n * \Phi_i^n + O\left(\frac{\tau}{L}\right)$$

$$x_{n+1} = \tilde{x}_n + \tau f(\tilde{x}_n) + \tau \sum_i s_i^a * \Lambda_i^n * \Phi_i^n + O\left(\frac{1}{L}\right) + O\left(\frac{\tau}{L}\right). \qquad (6.6)$$

The first and second terms of the result may be considered as the simplest approximation of nonstiff Equations (5.10), with a step size τ small enough to produce a numerical solution with required accuracy. The term $O(1/L)$ in Equation (5.10) cannot be approximated explicitly: its only function is to make an exact tangent from the vector $f(\tilde{x})$ to Γ at the point $\tilde{x} \in \Gamma$ and to be the value of $O(1/L)$. Otherwise, it may be arbitrary. The two last terms of the result obtained previously can be regarded as an acceptable error of approximation. This means that the stiffness of the system L is so great that $\tau^2 L \gg O(1)$. This assumption is acceptable from a practical point of view. A good numerical solution can be produced by a difference scheme of the type

$$\tilde{x}_{n+1} = \tilde{x}_n + \tau f(\tilde{x}_n) + O(\tau^2),$$

where the term $O(\tau^2)$ may be an arbitrary vector which transfers the point $\tilde{x}_n + \tau f(\tilde{x}_n)$ to Γ or to a small neighborhood of Γ. This neighborhood may be not only $L_{O(L^{-2})}$, but also $\Gamma_{O(\tau/L)}$. To prove this, assume that $x_{n+1} \in \Gamma_{O(\tau/L)}$. Then

$$x_{n+1} = XA(x_{n+1}) + \sum s_i \Phi_i A(x_{n+1}), \quad \text{where } s_i = O\left(\frac{\tau}{L}\right)$$

and $\tilde{x}_{n+1} = XA(x_{n+1}) \in \Gamma$. Computation of the function f at the point x_{n+1}, in contrast with its computation in the "right" point $\tilde{x}_{n+1}$, produces

$$f(x_{n+1}) = f(\tilde{x}_{n+1} + \sum s_i \Phi_i) = f(\tilde{x}_{n+1}) + \sum_i s_i \Lambda_i * \Phi_i + O(Ls^2).$$

The main term of the errors is $O(\tau)$, since $s = O(\tau/L)$. Such errors are acceptable in difference schemes of the first order of approximation.

The third term of the right-hand part of Equation (6.6) requires particular attention. Because $s_i^a = O(1/L)$, its value is $O(\tau)$. Such an error is unacceptable in the common case. But a special structure of this term and operation (6.2) of the algorithm we are examining change the situation. To prove this, examine the expression derived from Equation (6.6),

$$x_{n+1} = \tilde{x}_{n+1} + \tau \sum s_i^a \Lambda_i^n * \Phi_i^n. \tag{6.7}$$

Here, $\tilde{x}_{n+1} \in \Gamma$ is the right value from the point of view of numerical integration of nonstiff Equation (5.10). Examine the next step of the MQMF starting at the point $x_{n+1} \in \Gamma_{O(\tau)}$.

First form a new subspace

$$x(\xi) = x_{n+1} + \sum_i \xi_i \Phi_i^{n+1}, \quad \text{where } \Phi^{n+1} = \Phi XA(x_{n+1}). \tag{6.8}$$

Evidently $\|\Phi_i^{n+1} - \Phi^n\| = O(\tau)$, because $\|x_{n+1} - x_n\| = O(\tau)$. Therefore x_{n+1} (Equation [6.7]) can be represented as

$$x_{n+1} = \tilde{x}_{n+1} + \tau \sum_i s_i^a \Lambda_i^n * \Phi_i^{n+1} + O(\tau^2). \tag{6.9}$$

Notice that $\tilde{x}_{n+1}$ is not the "shadow" of the point x_{n+1} here, that is, $\tilde{x}_{n+1} \neq XA(x_{n+1})$. This point may be any point on Γ which can be considered as the point obtained by any algorithm of numerical integration of differential Equation (5.10). These points are defined at an accuracy of $O(\tau^2)$. The term $O(\tau^2)$ in Equation (6.8) can thus be omitted by a corresponding correction of $\tilde{x}_{n+1}$. Assume that this correction is made and use Equation (6.9) without this term. The point $\tilde{x}_{n+1}$ is thus one of the points $x(\xi)$ of Equation (6.8) used in the next step of the MQMF algorithm. Let x_{n+1}^a be a point obtained after operation (6.2).

It is clear that $||x_{n+1}^a - \tilde{x}_{n+1}|| = O(1/L)$ and $A(x_{n+1}^a) = A(\tilde{x}_{n+1})$. The term $O(\tau)$ in Equation (6.9) thus produces a considerable error only in relation to quick components of the vector x, but there are no errors in the slow components except the acceptable errors $O(\tau^2)$. At the next step of the algorithm, the error in the quick components will be eliminated by operation (6.2). Therefore, the point x_n^a, $n = 0, 1, \ldots$, obtained by MQMF, can be considered the result of the numerical solution of Equation (5.10) by the explicit scheme of the first order of approximation. To be more precise, we can say that if $x(t)$ is the exact solution, $||Ax(t_n) - A(x_n^a)|| = O(\tau)$ and $S(x_n^a) = O(1/L)$ (instead of $Sx(t_n) = O(1/L^2)$).

Figure 8 represents the qualitative picture which illustrates the process of numerical integration by the MQMF. The algorithm as discussed previously is too difficult for any practical use. Some possible ways to make it simpler are:

1) Instead of eigenvectors $\Phi_i^n = \Phi_i X A(x_n)$ the eigenvectors $\Phi_i(x_n)$ computed immediately at the point x_n can be used in Equation (6.1). It is clear that $||\Phi_i^n - \Phi_i(x_n)|| = O(\tau)$, and such an error is acceptable.

2) Computing the vectors $\Phi_i(x)$ is too time consuming if done at every step of the algorithm; sometimes it can be avoided. The stiffness of the matrix f_x just makes it possible to obtain an approximation, $x(\xi)$, of the hyperplane of Equation (6.1). Obviously, it is this hyperplane which is needed rather than its representation through eigenvectors Φ_i. Let l_m, $m = 1, \ldots, M$ be vectors

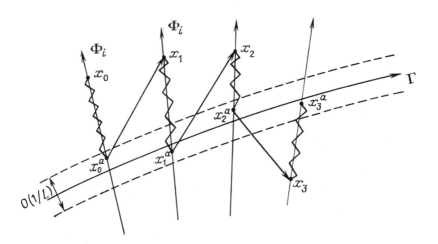

FIGURE 8. Points x_n, x_n^a produced by MQMF.

$(||l_m|| = 1)$. Each can be represented as

$$l_m = \sum_i C_m^i \Phi_i(x) + \sum c_m^j \varphi_j(x)$$

Examine the differences

$$r_m = f(x + \varepsilon l_m) - f(x), \qquad m = 1, \ldots, M,$$

where $\varepsilon = O(1/L)$. A simple mathematic calculation produces the representation

$$r_m = \varepsilon \sum_i C_m^i \Lambda_i(x) \Phi_i(x) + \varepsilon \sum c_m^j \lambda_j \varphi_j + O\left(\frac{1}{L}\right).$$

The two last terms in the right-hand part are the value of $O(1/L)$, and the first is the value of $O(1)$. Thus the hyperplane

$$x(\xi) = \sum_{m=1}^{M} \xi_m r_m + x_n \qquad (6.10)$$

can be considered as a good approximation of the hyperplane of Equation (6.1); the angle between these two hyperplanes is the value $O(1/L)$. This approach is not as simple as it appears. There arise at least two questions. First, how should the vectors l_m and the number M be selected? The obvious choice, $M = I$, is not very good, even though the exact value of I is known (whereas it is often unknown). There are pitfalls to be avoided. The first is underestimation of the dimension of the hyperplane of Equation (6.1). This can result from an erroneous choice of the vectors l_m, $m = 1, \ldots, M = I$. This error manifests itself by a very large value of $||f(x(\xi))||$ resulting from operation (6.2). The second pitfall is overestimation of the dimension of the hyperplane of Equation (6.1) arising from an incorrect value of $M > I$. Operation (6.2) produces a good result, $\min ||f(x(\xi))|| = O(1)$, but it can alter the slow variables $\alpha[t] = Ax(t)$ to the unacceptable value $O(\tau)$. This means that the speed, $\dot{\alpha}[t]$, is computed with an error of $O(1)$. The value $M > I$, even $M > \dim x$, is, however, usable. Indeed, points from Equation (6.10) are a linear combination of the vectors r_m. If $M = \dim x$ and the vectors r_m are linearly independent, every point in the $\dim x$ space can be represented as in Equation (6.10). But this representation requires very large values of ξ_i if this point happens to be far from the hyperplane of Equation (6.1). Thus, Equation (6.10) with $M > I$ is acceptably usable if the change of parameters, ξ_i, is limited to the value $O(1)$.

(For our purposes, we only need $\xi_i = O(\tau)$). Equation (6.2), for example, may be substituted for

$$\min_{\xi} \left\{ \|f(x(\xi))\| + \frac{1}{\sqrt{\tau}} \|\xi\| \right\}. \qquad (6.2^*)$$

3) The efficacy of the MQMF depends on the algorithm used for the solution of Equation (6.2) $\big($or Equation $[6.2^*]\big)$. The "method of repeated experiments", described on page 169 in reference 6, is recommended for use by Demirchyan and Rakitsky.[5] Notice that Equation (6.2) needs no exact solution. It is clear from the previous discussion that only $\|f(x(\xi))\| = O(1)$ is required.

4) The original MQMF as described in reference 5 differs from the method discussed here. The main distinction is the construction of the vectors r_m used in Equation (6.10). In reference 5

$$r_1 = f(x_n), \qquad r_2 = \frac{[f(x_n + \xi r_1) - f(x_n - \varepsilon r_1)]}{2\varepsilon},$$

$$r_3 = \frac{[f(x_n + \varepsilon r_1) - 2f(x_n) + f(x_n - \varepsilon r_1)]}{\varepsilon^2}$$

and so on. ξ_m are defined by minimizing a function which has the sense of $\|\ddot{x}(\xi)\|$.[5] We find these recommendations unjustified. But the two main ideas of the original MQMF[5] were used in our version of the method. The first is the replacement of the very difficult equation in Γ in Definition 2 by the relatively simple problem in Equation (6.2). The second idea is the use of the hyperplane of Equation (6.1), on which $\|f(x)\|$ must be minimized.

7. SLOW PROCESSES IN A NUCLEAR REACTOR

We will discuss a method of computing the problems for a special class of slow processes in nuclear reactors[7] as yet another application of the asymptotic theory of SS. Processes of this type are described by a system of partial differential equations (with homogeneous boundary conditions which we will not discuss here),

$$\frac{1}{V_1} \frac{\partial \psi^1}{\partial t} = \operatorname{div} D_1 \operatorname{grad} \psi^1 - A_{11}\psi^1 + A_{12}\psi^2$$

$$\frac{1}{V_2} \frac{\partial \psi^2}{\partial t} = \operatorname{div} D_2 \operatorname{grad} \psi^2 - A_{22}\psi^2 + A_{21}\psi^1 \qquad (7.1)$$

$$\frac{\partial \alpha}{\partial t} = A(\alpha, \psi), \qquad \psi = \{\psi^1, \psi^2\}. \qquad (7.2)$$

Here $\psi^1, \psi^2(t, r)$ are the functions describing the distribution of two groups of neutrons, r a space variable and $\alpha(t, r)$, which is the vector of variables we refer to as "slow" (the temperature, xenon's concentration, and some others). The coefficients of these equations (D_i, A_{ij}) depend on r and $\alpha(t, r)$. The values V_1 and V_2 are large parameters, therefore systems (7.1) and (7.2) describe the interaction of two different types of processes. The first is a very quick process of neutron relaxation; the second is a slow evolution of the variables α. Rewrite Equations (7.1) and (7.2) as

$$\frac{\partial \psi}{\partial t} = L\psi, \qquad (7.1^*)$$

$$\frac{\partial \alpha}{\partial t} = A. \qquad (7.2^*)$$

Here, L is an operator (in a Banach space) containing a large parameter. Equations (7.1*) and (7.2*) thus look like a singularly-perturbated system. The SPS theory is immediately applied to the system to yield an equation for slow motion (similar to Equation [4.4]) which produces

$$\frac{d\alpha}{dt} = A(\alpha, \psi), \qquad L\psi = 0.$$

This is evidently wrong, because the equation $L\psi$ has in the common case only the trivial solution, $\psi = 0$.

The error arises because of an implicit small parameter, the first eigenvalue of the spectral problem

$$L\psi = \lambda\psi. \qquad (7.3)$$

In order to apply the standard SPS theory and the asymptotic theory of the slow motion of SS as previously discussed, denote the first eigenvalue and eigenvector of the operator L as λ_1, φ_1 and all others as Φ_i, Λ_i, $i = 1, 2, \ldots$ It is more precise to denote them as $\Phi_i(r, \alpha(t, \cdot)), \Lambda_i(\alpha(t, \cdot))$, because the functions Φ_i, Λ_i depend on $\alpha(t, r)$ as functionals. All the spectrum of the operator L except λ_1 will be assumed to be stiff. This means that for the time Δt, which is small as concerns the variables α,

$$\exp(\lambda_1 \Delta t) = O(1), \qquad \exp\left(\Delta t(\Lambda_i - \lambda_1)\right) \ll 1, \ i = 2, 3, \ldots \qquad (7.4)$$

Also, the change of variables α within the time Δt is rather small. The manifold Γ is thus determined within the full functional space of the functions α, ψ by

$$\alpha, \psi \in \Gamma \equiv \{(L\psi, \Phi_i^*) = 0 \quad \text{or} \ (\psi, \Phi_i^*) = 0, \ i = 2, 3, \ldots\}. \qquad (7.5)$$

This is a special case of Definition 2. Therefore, Γ is any manifold within the functional space. All the functions in Equation (7.5) are regarded as functions only of the space variable r; t is regarded as a parameter. Equations (7.5) can be solved. Let the function α be arbitrary. Let $\varphi_i(\alpha)$ be the first eigenvector of Equation (7.3), where the operator L depends on α, and c is any scalar. Thus,

$$\alpha, \psi \in \Gamma \quad \text{is equivalent to} \quad \psi = c \cdot \varphi_1(\alpha). \tag{7.6}$$

$\alpha(t), c_1(t)\varphi_1\big(\alpha(t)\big)$ is the approximate solution of Equations (7.1) and (7.2). The equation in $c_1(t)$ is easily obtained by substituting ψ into Equations (7.1*) and (7.2*) to yield

$$\frac{dc_1}{dt} \cdot \varphi_1 + c_1 \frac{d\varphi_1}{dt} = \lambda c_1 \varphi_1 + o(1). \tag{7.7}$$

Equation (7.7) is thus similar to Equation (5.10). The term $o(1)$ is added to get an "equation on Γ". This term can be easily computed, and the full system describing the slow motion "on Γ" is

$$\frac{dc_1}{dt} = \lambda_1 c_1 - c_1 \left(\frac{d\varphi_1}{dt}, \varphi_1^* \right)$$

$$\frac{d\alpha}{dt} = A\big(\alpha, c_1(t)\varphi_1(t)\big). \tag{7.8}$$

Here φ_1 is the first eigenvector of the operator L, determined for a fixed time, t, by the function $\alpha(t, r)$; φ_1^* is the eigenvector of L_1^* and λ_1 is a corresponding eigenvalue. Equation (7.8) was used in a special case, where α has only one component, the temperature. Computations showed that $d\varphi_1/dt$ was fairly small and the corresponding term in Equation (7.8) might be omitted. It is true, of course, only for problems discussed in reference 7. The equation for c_1 can be integrated to yield

$$c_1(t) = c_1(0) \exp \left(\int_0^t \lambda_1(\tau) d\tau \right).$$

The equation in α $\big($Equation [7.2]$\big)$ was integrated by a simple explicit scheme. Every step of the numerical integration requires solution of the main eigenvalue problem $\big($Equation [7.3]$\big)$. For more details see reference 7.

8.   ON THE ACCURACY OF EULER'S IMPLICIT SCHEME

Let us analyze the simplest numerical method often used for an approximate solution of SS, Euler's well-known scheme. We examine it only in a small neighborhood of Γ. Let us see how this method works at the stage of slow motion in the proximity of Γ. Examine only one standard step of the algorithm, starting at the point $x^0 \in \Gamma_{O(1)}$. The next point, x^1, must be found by solving the nonlinear equation

$$x - \tau f(x) = x^0. \tag{8.1}$$

It is assumed that the step size, τ, is large with respect to $\|f_x\| = O(L)$ and small with respect to the slow motion. This means that we are dealing with two small parameters, $\tau \ll 1$ and $1/\tau L \ll 1$. Denote

$$x^0 = X\alpha^0 + \sum s_i^0 * \Phi_i^0, \quad \text{where } \Phi_i^0 = \Phi_i X A\left(x^0\right), \ \alpha^0 = A\left(x^0\right).$$

The unknown x may be similarly represented as

$$x = X\alpha + \sum s_i \Phi_i, \quad \text{where } \Phi_i = \Phi_i X A(x), \ \alpha = A(x).$$

The values α and s_i must be found from Equation (8.1). Local errors of the step Equation (8.1) can be estimated by the two small parameters indicated previously.

With the new unknown α and s, rewrite Equation (7.8) as

$$X\alpha + \sum s_i * \Phi_i + \tau f(X\alpha + \sum s_i * \Phi_i) = X\alpha^0 + \sum s_i^0 * \Phi_i^0.$$

Applying simple mathematics, obtain

Statement 7. Equation (8.1) can be represented as

$$R(\alpha) + \sum_i \left[(1 - \tau\Lambda_i) * s_i - s_i^0 \right] * \Phi_i$$

$$+ R_1(\alpha, s) + R_2(\alpha) = 0, \tag{8.2}$$

where

$$R(\alpha) = X\alpha - \tau f X\alpha - X\alpha^0, \tag{8.3a}$$

$$R_1(\alpha, s) = \tau * [f(X\alpha + \sum s_i * \Phi_i)$$

$$- fX\alpha - \sum s_i * \Lambda_i * \Phi_i], \tag{8.3b}$$

$$R_2(\alpha) = \sum s_i^0 * (\Phi_i - \Phi_i^0), \tag{8.3c}$$

$$\Lambda_i = \Lambda_i X\alpha. \tag{8.3d}$$

The system of $I + J$ equations $\left(\text{Equation } [8.2]\right)$ must be divided into two groups, J equations in α and I equations in s. These groups of equations are separated in main terms. We introduce an unknown, $\beta = \alpha - \alpha^1$, where α^1 is the solution of the equation

$$\alpha^1 - \tau D(\alpha^1) = \alpha^0, \quad D(\alpha) \equiv A_x X\alpha * fX\alpha. \tag{8.4}$$

The value α^1 may be regarded as the correct value of α because it is obtained by one step of numerical integration of a nonstiff equation in slow variables, Equation (5.9). Therefore β is an error to be estimated. Evidently $\alpha^1 - \alpha^0 = O(\tau)$.

Statement 8. For functions R_1 and R_2 $\big($Equation [8.1]$\big)$ we have almost evident estimates:

$$R_1(\alpha, s) = O\big(\tau L s^2\big), \qquad (8.5a)$$

$$R_2(\alpha) = O(s^0 \tau) + O(s^0 \beta). \qquad (8.5b)$$

Obtain the second estimate, Equation (8.5b), by simple computation:

$$\begin{aligned}
R_2(\alpha) &= \sum s_i^0 * \Phi_i X(\alpha^1 + \beta) - \sum s_i^0 * \Phi_i X(\alpha^0) \\
&= \sum s_i^0 * \Phi_i X \alpha^1 + O(s\beta) - \sum s_i^0 * \Phi_i X \alpha^0 \\
&= O(s\beta) + O(s\tau).
\end{aligned}$$

The estimates in Equations (8.5) will subsequently be used only in a small neighborhood of the point α^1.

The most difficult task is to prove that $R(\alpha)$ in Equation (8.1) is a vector formed basically by the "soft" eigenvectors $\varphi_j(X\alpha^1)$. Multiply Equation (8.2) by the eigenvectors Φ_i^* and φ_j^* to obtain simple equations in s_i and α, separately. These equations will certainly include small terms depending on both variables, s and α.

Statement 9. The function $R(\alpha^1 + \beta)$ can be represented as

$$R(\alpha^1 + \beta) = \sum_j (1 - \tau \lambda_j^1)\gamma_j \varphi_j^1 + O\big(\tau^2\big) + O\big(\beta^2\big)$$
$$+ O\left(\frac{\tau}{L}\right) + O\left(\frac{\beta}{L}\right) + O(\tau\beta), \quad (8.6)$$

where $\lambda_j^1 = \lambda_j X\alpha_j^1$, $\varphi_j^1 = \varphi X\alpha_j^1$ and γ_j are new scalar variables determined by β, and $\gamma = O(\beta)$. Evidently,

$$R(\alpha^1 + \beta) = X(\alpha^1 + \beta) - \tau f X(\alpha^1 + \beta) - X\alpha^0. \qquad (8.7)$$

Examine the first term in the right-hand part,

$$X(\alpha^1 + \beta) = X\alpha^1 + X_\alpha \alpha^1 * \beta + O\big(\beta^2\big). \qquad (8.8)$$

Similarly, $\quad f X(\alpha^1 + \beta) = f X\alpha^1 + f_x X_\alpha \alpha^1 * \beta + O\big(\beta^2\big). \qquad (8.9)$

Remember that the function $fX(\alpha)$ (according to Assumption 4) is supposed to be smooth. Therefore,

$$R = \{X\alpha^1 - \tau * fX\alpha^1 - X\alpha^0\} + (E - \tau * f_x X\alpha^1) * X_\alpha \alpha^1 * \beta$$
$$+ O(\beta^2) + O(\tau\beta^2). \qquad (8.10)$$

To estimate the first term, use a simple expansion

$$X\alpha^0 = X(\alpha^0) = X\left[\alpha^1 - (\alpha^1 - \alpha^0)\right]$$
$$= X\alpha^1 - X_\alpha\alpha^1 * (\alpha^1 - \alpha^0) + O\left(\tau^2\right)$$
$$= X\alpha^1 - \tau * X_\alpha\alpha^1 * A_x X\alpha^1 * fX\alpha^1 + O\left(\tau^2\right).$$

We used two relationships, $\alpha^1 - \alpha^0 = O(\tau)$ and $\alpha^1 - \alpha^0 = \tau D(\alpha^1)$, and the formula for D (Equation [8.4]). Notice that $X_\alpha\alpha * A_x X\alpha * l = l$ for every vector l tangent to Γ at the point $X\alpha$. The vector $fX\alpha^1$ is tangent to Γ at the point $X\alpha^1$ at an accuracy of $O(1/L)$ (Statement 1). Therefore,

$$X\alpha^0 = X\alpha^1 - \tau * fX\alpha^1 + O\left(\frac{\tau}{L}\right) + O(\tau^2),$$

and the expression for R (Equation [8.10]) can be rewritten as

$$R(\alpha) = (E - \tau * f_x X\alpha^1) * X_\alpha\alpha^1 * \beta + O(\beta^2)$$
$$+ O(\tau\beta^2) + O\left(\frac{\tau}{L}\right) + O(\tau^2).$$

The vector $X_\alpha\alpha^1 * \beta$ is tangent to Γ at the point $X\alpha^1$. Therefore, according to Equation (5.2) it can be represented as

$$X_\alpha\alpha^1 * \beta = \sum_j \gamma_j * \varphi_j^1 + O\left(\frac{\beta}{L}\right), \qquad \gamma = O(\beta).$$

In this case we have

$$(E - \tau * f_x X\alpha^1) * (X_\alpha\alpha^1 * \beta)$$
$$= (E - \tau f_x X\alpha^1)\left(\sum \gamma_j * \varphi_j^1 + O\left(\frac{\beta}{L}\right)\right)$$
$$= \sum_j \gamma_j(1 - \tau\lambda_j^1) * \varphi_j^1 + O\left(\frac{\beta}{L}\right) + O(\tau\beta).$$

This result in Equation (8.10) takes us to Statement 9. From Statements 7, 8, and 9 follows Statement 10:

Statement 10. Equation (8.1) can be represented as

$$\sum_j (1 - \tau\lambda_j^1) * \gamma_j * \varphi_j^1 \left[\left(1 - \tau\Lambda_i^1 \right) s_i - s_i^0 \right] * \Phi_i^1$$

$$+ \, O(\tau L s^2) + O(s^0 \beta)$$

$$+ \, O(\tau^2) + O(\beta^2) + O(\tau\beta). \tag{8.11}$$

Immediate application of estimates (8.2), (8.5), and (8.6) produces the sum of the terms of the $O(\ldots)$ type,

$$O(\tau L s^2) + O(s^0 \tau) + O(s^0 \beta) + O(\tau^2) + O(\beta^2)$$

$$+ \, O\left(\frac{\tau}{L}\right) + O\left(\frac{\beta}{L}\right) + O(\tau\beta).$$

Notice that $1/L \ll \tau$, $s^0 = O(\tau)$ (though we will see that, more typically, $s^0 = O(\tau/L) \ll \tau$). In Equation (8.11) we retain only the main small terms. Multiply Equation (8.11) by vectors $\Phi_i^* X \alpha^1$ and $\varphi_j^* X \alpha_j^1$, remembering that $(1 - \tau\lambda_j) \approx 1$ and $1 - \tau\Lambda_i = O(\tau L)$, to obtain

Statement 11. Equation (7.8) is reduced to a system,

$$\gamma = g(\gamma, s), \qquad s = z(\gamma, s), \tag{8.12}$$

where g and z are mappings: $g = R^J \times R^I \to R^J$, and $z = R^J \times R^I$.

Estimates for these small functions are

$$\|g(\gamma, s)\| \leq C_1 \tau L s^2 + C_2 s^0 \tau + C_3 s^0 \beta + C_4 \tau^2 + C_5 \tau\beta, \tag{8.13}$$

$$\|z(\gamma, s)\| \leq B_1 s^2 + \frac{B_2 s^0}{L} + \frac{B_3 s^0 \beta}{L} + \frac{B_4 \tau^2}{L} + \frac{B_5 \beta}{L} + \frac{s^0}{\tau L}. \tag{8.14}$$

Remember that we use the symbol O in the sense that $O(\alpha)$ means $\|O(\alpha)\| \geq a\alpha$ and $\|O(\alpha)\| \leq A\alpha$, $A/a = O(1)$.

In order to use the well-known fix-point theorem, let us construct a convex domain, Ω, such that the mapping $\{\gamma, s\} \to \{g(\gamma, s), z(\gamma, s)\}$ maps Ω into itself. Determine this domain by the inequalities

$$\|\gamma\| \leq p\tau^2, \qquad \|s\| < \frac{q}{L}. \tag{8.15}$$

The parameters p and q must be chosen to assure that every term in the right-hand part of Equation (8.13) is estimated by the value of $p\tau^2/5$

and every term of Equation (8.14) by $q/6L$ (for $\{\gamma, s\} \in \Omega$, of course). Let us prove that the values $p = O(1)$ and $q = O(1)$ do really exist.

All the computations will be made in a standard procedure used to obtain a required inequality from an evident relation, for example, $C < 1/\tau$. Use $s^0 < C\tau$ for the value s^0. Omit the factors $1/5$, $1/6$, and certain other factors which are of $O(1)$, because all the computations are to be made at an accuracy of $O(1)$.

1) $C_1 < p\tau L/q^2 \Rightarrow C_1 q^2 \tau/L < p\tau^2 \Rightarrow C_1 \tau L(q^2/L^2) < p\tau^2 \Rightarrow C_1 \tau L s^2 < p\tau^2$;

2) $CC_2 < p \Rightarrow CC_2 \tau^2 < p\tau^2 \Rightarrow C_2 \tau(C\tau) < p\tau^2 \Rightarrow C_2 \tau s^0 < p\tau^2$;

3) $CC_3 < 1/\tau \Rightarrow C_3(C\tau) < p\tau^2 \Rightarrow C_3 s^0 \beta < p\tau^2$;

4) $C_4 < p \Rightarrow C_4 \tau^2 < p\tau^2$;

5) $C_5 < 1/\tau \Rightarrow C_5 \tau(p\tau^2) < p\tau^2 \Rightarrow C_5 \tau \beta < p\tau^2$.

Let us make similar calculations for the terms of Equation (8.14):

1) $B_1 \leq L/q \Rightarrow B_1 q^2/L^2 < q/L \Rightarrow B_1 s^2 < q/L$;

2) $B_2 < q/C\tau \Rightarrow B_2 c\tau/L < q/L \Rightarrow B_2 s^0/L < q/L$;

3) $B_3 < q/Cp\tau^3 \Rightarrow B_3(p\tau^2)c\tau < q \Rightarrow B_3 s^0 \beta/L < q/L$;

4) $B_4 < q/\tau^2 \Rightarrow B_4 \tau^2/L < q/L$;

5) $B_5 < q/p\tau^2 \Rightarrow B_5(p\tau^2)/L < q/L \Rightarrow B_5 \beta/L < q/L$;

6) $C < q \Rightarrow C\tau/L\tau < q/L \Rightarrow s^0/L\tau < q/L$.

It is almost evident that all the initial inequalities can be satisfied by values $p = O(1)$ (say, $p = \max\{CC_2, C_4\}$), and $q = O(1)$ (e.g., $q = C$). The convexity of the domain Ω, determined by Equation (8.15), is evident. Hence

Statement 12. System (8.12) has at least one solution in Ω.

Statement 13. System (7.8) has at least one solution, x^1. This point x^1 may be represented by values α^1 and s^1 as the form

$$x^1 = X(\alpha^1 + \beta) + \sum_i s_i^1 * \Phi_i X\alpha^1.$$

where $\beta = O(\tau^2)$, $s_i^1 = O(1/L)$ and α^1 is a value of the slow variable obtained by one step of numerical integration of nonstiff Equation (5.9). Simple scheme (8.4) is used for computation of α^1. Therefore, β may be regarded as an acceptable error of the numerical method. The existence of any suitable solution of system (8.1) is thus proved.

Now prove the uniqueness of such solutions in a special sense. It will be the uniqueness at an accuracy of $O(\tau^2)$ as concerns the slow variables

α, and $O(1/L)$ as concerns the variables s_i. Using the symbol γ in the sense of $||\gamma||$, obtain from Equations (8.12) and (8.13) the inequality

$$\gamma(1 - C_3 s^0 - C_5 \tau) - C_1 \tau L s^2 \leq C_2 s^0 \tau + C_4 \tau^2. \qquad (8.16)$$

We may use γ instead of β, possibly changing C_3 and C_5 by a factor of $O(1)$. Because $s^0 < C\tau$, Equation (8.15) can be rewritten as

$$\gamma - a\tau L s^2 \leq b\tau^2, \quad \text{where } a = O(1), \ b = O(1). \qquad (8.17)$$

Similarly, obtain from Equations (8.12) and (8.13)

$$s(1 - B_1 s) - \frac{B_5 \gamma}{L} - \frac{B_3 s^0 \gamma}{L} \leq \frac{B_2 s^0}{L} + \frac{B_4 \tau^2}{L} + \frac{s^0}{\tau L}.$$

Take into account only the main terms and obtain

$$s(1 - B_1 s) - \gamma \frac{d}{L} \leq \frac{l}{L}, \quad \text{where } d = O(1), \ l = O(1). \qquad (8.18)$$

We restrict our consideration to small values of s, for example, $s < 0.5/B_1$. The factor $(1 - B_1 s)$ in Equation (8.16) may thus be omitted; instead of Equation (8.16) use the inequality

$$s - \gamma \frac{d}{L} \leq \frac{l}{L}. \qquad (8.19)$$

Here d, l differ from d, l in Equation (8.16) by a factor of 2. We obtain, therefore, inequalities in Equations (8.15) and (8.17), which describe the region of possible values γ, s, as Figure 9 shows. A simple analysis produces

> **Statement 14.** All the possible solutions of Equation (7.8) have the properties of the point x^1 described in Statement 11. Each of these solutions suits the purposes of numerical integration of the stiff system $\dot{x} = f$.

To formulate the main result, let us use these objects:

$x(t)$ is the trajectory of a SS, starting at the point $x^0 \in \Gamma_{O(1)}$. $x_n, n = 0, 1, \ldots$ are points obtained by implicit scheme (7.8) with a large step size τ; $x_0 = x^0$;

$x^n \equiv x(n\tau), \ n = 0, 1, \ldots$ is the "exact" solution on the grid we are using;

$\alpha_n = A(x_n), \ \alpha^n = A(x^n), \ s_n = S(x_n), \text{ and } s^n = S(x^n), \ n = 0, 1, \ldots$

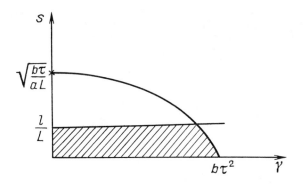

FIGURE 9. A region of possible solution of Equation (8.12).

Theorem. Euler's implicit scheme (7.8) used with a large step size, $\tau \ll 1$, $\tau\|f_x\| \gg 1$, is a scheme of the first degree of accuracy. This means that

1) $\|\alpha_n - \alpha^n\| = O(\tau)$;

2) $\|s_n\| = O(1/L)$, while $\|s^n\| = O(1/L^2)$.

Notice that the estimate for $\|s_n\|$ can be refined: $\|s_n\| = O(\tau/L)$. All the initial inequalities for B_k allow one to assume $q = O(\tau)$. The only obstacle to this is the inequality $C < q$. But the value C is connected with the assumption that $s^0 < C\tau$. The next step of numerical integration, however, starts at the point x_1, where $s_1 = O(1/L)$. Therefore, $C\tau$ can be used instead of C in the next and all subsequent steps.

REFERENCES

1. Godunov, S.K., *Solution of Systems of Linear Equations*, Nauka, Novosibirsk, 1980.

2. Pavlov, V.B., and Povzner, A.Ja., On one method of the numerical integration of ordinary differential equations, *Jour. Appl. Math. and Math. Phys.*, v. 13, 1056, 1973.

3. Vasiljeva, A.B., and Butuzov, V.F., *Asymptotic Expansions of Solutions of Singularly-perturbuted Equations*, Nauka, Moscow, 1977.

4. Fedorenko, R.P., On regular stiff systems of ordinary differential equations, *Dokl. Acad. Nauk. SSSR*, v. 273, No. 6, 1318, 1983.

5. Demirchyan, K.S., and Rakitsky, Jv.V., On filtration of components with large derivatives in differential systems, *DAN SSSR*, v. 279, No. 3, 525, 1984.

6. Rakitsky, Jv.V., Ustinov S.M., and Chernorutsky, I.G., *Numerical Methods of Solution of Stiff Systems*, Nauka, Moscow, 1979.

7. Strakhovskaya, L.G., and Fedorenko, R.P., On the numerical method for some quasi-stationary processes in nuclear reactor, *Jour. Appl. Math. and Math. Phys.*, v. 19, No. 5, 1237–1252, 1979.

8. Lebedev, V.I., *this volume*, 45, 1994.

9. Curtiss, C.F., and Hirschfelder, T.O., Integration of stiff equations, *Proc. National Academy of Science*, USA, 1952.

Convergence Rate Estimates of Finite-Element Methods for Second-Order Hyperbolic Equations

A.A. Zlotnik

In finite-element methods theory, convergence rate estimates for approximate solutions is one of the central questions. It is most fully studied for the case of elliptic and parabolic equations.

There are a lot of papers devoted to convergence rate estimates for the case of initial-boundary value problems for second-order hyperbolic equations.[5-10,13-15,17-20,22-24,32-34,39] For various methods, estimates of maximal order $p \geq 1$ are established under some smoothness conditions on the problem solution. In a few papers these estimates have been expressed in terms of conditions on the initial data;[6,32,33] the right-hand part of the equation is assumed equal to 0.

At the same time, the error analysis is of great interest for initial data and for the right-hand part not having sufficiently high smoothness, in particular discontinuous or concentrated data. This paper is devoted to this case, with results of a previous study [41] widely developed in it. Only second-order approximation methods ($p = 2$) are considered. This choice is motivated, to a certain extent, by the fact that this kind of method is widespread in practice. Namely, a three-level method with weight, a two-level method, and a method with a splitting operator (in other words, the alternating-direction Galerkin method) are studied.

If the initial data possess smoothness of order α, and the right-hand part possesses the mixed smoothness of order α_1 in x and of order α_2

0-8493-8947-X/94 /$0.00 + $.50

155

in $t(\alpha_1 + \alpha_2 = \alpha - 1)$, then the convergence rate estimates of order $O\big((h^2 + \tau^2)^{(\alpha-k)/3}\big)$ for $k \leq \alpha \leq k+3$ are obtained in the kth energetic norm, $(k = 0, 1)$. First, the estimates for $\alpha = k$ (the error boundedness) and $\alpha = k + 3$ (the maximal-order estimate) are derived; the estimates for $k < \alpha < k+3$ are derived with the help of interpolation of the error operator.[12] Also, the estimates in the $W^1_{2,h}$ norm and C_h norm are obtained for the one-dimensional problem with discontinuous coefficients.

The exactness of the indicated estimates is proved in the simplest case of the one-dimensional wave equation with constant coefficients (see reference 41, Theorem 2), for the three-level method with weight using piecewise-linear finite elements.[†] Recently the exactness was also proved in the case of the multidimensional wave equation.[48] In comparison with the corresponding estimates in terms of data, following from a priori error estimates in references 5, 13, 18, 22, and 23, the obtained error estimates are of better order (for $\alpha < k + 3$) or impose weaker requirements on data (for $\alpha = k + 3$).

1.   INITIAL-BOUNDARY VALUE PROBLEM FOR SECOND-ORDER MULTIDIMENSIONAL HYPERBOLIC EQUATIONS

1.1.   NOTATION

Let Ω be a bounded domain in n-dimensional Euclidean space $\mathbf{R}^n (n \geq 1)$ and $\partial\Omega$ be its boundary (for simplicity piecewise-smooth). Let $L_q(\Omega)$ be the Lebesgue spaces $(1 \leq q \leq \infty)$ and

$$H^{<1>} = \overset{\circ}{W}{}^1_2(\Omega)$$

be the Sobolev space with the norm $\|w\|_{H^{<1>}} = \| \, |Dw| \, \|_{L_2(\Omega)}$ (the condition $w|_{\partial\Omega} = 0$ is assumed), where $Dw = (D_1 w, \ldots, D_n w)$ is the gradient of function w, $| \cdot | = \| \cdot \|_{R^n}$, and D_i are generalized partial derivatives. The inequalities

$$\|w\|_{L_{2p_n/(p_n-2)}(\Omega)} \leq c^{(0)} \, \|w\|_{H^{<1>}},$$

$$\|w\|_\Omega \equiv \|w\|_{L_2(\Omega)} \leq c^{(1)} \, \|w\|_{H^{<1>}} \qquad (1.1)$$

are valid, where $p_1 = 2$, $p_2 > 2$, $p_n = n$ for $n \geq 3$ (see, e.g., reference 27).

[†]See the special corollary of this theorem in reference 34.

Let functions ρ, a_{ij}, a satisfy the conditions

$$\|\rho\|_{L_\infty(\Omega)} + \|a_{ij}\|_{L_\infty(\Omega)} + \|a\|_{L_{p_n/2}(\Omega)} \le N,$$

$$a_{ij} = a_{ji}, \quad 1 \le i \le n, \quad 1 \le j \le n,$$

where $N \ge 1$ is a parameter, and also the conditions

$$N^{-1} \le \rho(x), \quad N^{-1}|\xi|^2 \le a_{ij}(x)\xi_j\xi_i$$

$$\forall x \in \Omega, \quad \forall \xi = (\xi_1, \dots, \xi_n) \in \mathbf{R}^n.$$

By the convention in this paper, the summation over repeated subscripts i, j is performed from 1 to n.

Let us introduce the Hilbert spaces $H^{<0>} = L_2(\Omega)$, and also H^0, which is the space $L_2(\Omega)$ with weight ρ. Let us also introduce the scalar products,

$$(w, \varphi)_\Omega = \int_\Omega w\varphi\, dx, \qquad (w, \varphi)_{0,\Omega} = \int_\Omega \rho w\varphi\, dx,$$

and the symmetric bilinear form, $\mathcal{L}_\Omega(w, \varphi) = (a_{ij}D_j w, D_i\varphi)_\Omega + (aw, \varphi)_\Omega$ on $H^{<1>} \times H^{<1>}$. Let this bilinear form be $H^{<1>}$-elliptic, that is,

$$\nu_0\|w\|^2_{H^{<1>}} \le \|w\|^2_{H^1} \equiv \mathcal{L}_\Omega(w, w) \qquad \forall w \in H^{<1>}. \tag{1.2}$$

(The indicated condition is trivially satisfied if $a(x) \ge 0$ for $x \in \Omega$.) Thus, the Hilbert space H^1, coinciding with $H^{<1>}$ to within the norm equivalence, is defined.

Introduce the spaces $H^{<-1>}$ and H^{-1}, which are in slightly different senses dual to H^1. To define $H^{<-1>}$ let us consider $H^{<0>}$ as the main space (i.e., identical to its dual one), and H^1 as continuously and densely imbedded into the space $H^{<0>}$. Introduce the duality relation on $H^{<-1>} \times H^1$ as the continuation of the scalar product in $H^{<0>}$ and then the norm in $H^{<-1>}$ such that

$$\|w\|_{H^{<-1>}} = \sup_{\varphi \in H^1} \frac{(w, \varphi)_\Omega}{\|\varphi\|_{H^1}}$$

(see reference 3 for more details). The space H^{-1} is defined in a similar way but with H^0 as the main space. For $w \in H^{-1}$ we introduce $w_\rho \in H^{<-1>}$ such that $(w_\rho, \varphi)_\Omega = (w, \varphi)_{0,\Omega}$ for all $\varphi \in H^1$. It is obvious that $\|w_\rho\|_{H^{<-1>}} = \|w\|_{H^{-1}}$, and for $w \in H^0$ we have $w_\rho = \rho w$.

Also introduce the linear bounded operators $L : H^1 \to H^{<-1>}$ and $A : H^1 \to H^{-1}$ with the help of identities

$$(Lw, \varphi)_\Omega = \mathcal{L}_\Omega(w, \varphi), \quad (Aw, \varphi)_{0,\Omega} = \mathcal{L}_\Omega(w, \varphi) \qquad \forall w, \varphi \in H^1.$$

It is clear that $Lw = -D_i(a_{ij}D_j w) + aw$ and $Lw = (Aw)_\rho$.

Consider the eigenvalue problem

$$Aw = \lambda w, \qquad w \in H^1, \ w \not\equiv 0.$$

By virtue of the previous assumptions, this problem possesses a system of its eigenfunctions $\{\mu_k\}_{k \geq 1}$ such that $A\mu_k = \lambda_A^{(k)}\mu_k$ with $\lambda_A^{(k)} > 0$ (the sequence $\{\lambda_A^{(k)}\}_{k \geq 1}$ is nondecreasing), which is orthonormal complete in H^0 and orthogonal complete in H^1. Further introduce the Hilbert spaces ($\alpha \geq 0$)

$$H^\alpha = \{w \in H^0 | \ \|w\|^2_{H^\alpha} = \sum_{k \geq 1} \left(\lambda_A^{(k)}\right)^\alpha \tilde{w}_k^2\},$$

where $\tilde{w}_k = (w, \mu_k)_{0,\Omega}$, are the Fourier-series coefficients of the function w expansion according to the system $\{\mu_k\}_{k \geq 1}$. If we introduce the operator A powers with the help of the formula

$$A^{\alpha/2}\,w = \sum_{k \geq 1} \left(\lambda_A^{(k)}\right)^{\alpha/2} \tilde{w}_k\,\mu_k,$$

it is clear that $\|w\|_{H^\alpha} = \|A^{\alpha/2}w\|_{H^0}$. For $\alpha = 0, 1$ the spaces H^α are identical to those introduced previously.

Let B be a Banach space. Introduce the spaces $C(B)$, $L_r(B)$ ($1 \leq r \leq \infty$), and $V(B)$ of functions on $[0, T]$ with values in B respectively continuous, strongly measurable, and having bounded variation[19,21] with the norms

$$\|w\|_{C(B)} = \max_{0 \leq t \leq T} \|w(t)\|_B,$$

$$\|w\|_{L_r(B)} = \|\,\|w(t)\|_B\|_{L_r(0,T)},$$

$$\|w\|_{V(B)} = \sup_{0 \leq t \leq T} \|w(t)\|_B + \mathrm{var}_{[0,T]}w.$$

Written in more detail, this is

$$C(B) = C([0,T]; B),$$

$$L_r(B) = L_r((0,T); B),$$

and
$$V(B) = V([0,T]; B).$$

Let $Q = \Omega \times (0, T)$. Define the bilinear forms

$$(w, \varphi)_Q = \int_0^T \left(w(t), \varphi(t)\right)_\Omega dt,$$

$$(w, \varphi)_{0,Q} = \int_0^T \left(w(t), \varphi(t)\right)_{0,\Omega} dt,$$

$$\mathcal{L}_Q(w, \varphi) = \int_0^T \mathcal{L}_\Omega\left(w(t), \varphi(t)\right) dt.$$

Introduce D_t, which is the derivative with respect to t and the integration operators

$$(I_t w)(t) = \int_0^t w(\theta)\, d\theta, \qquad (I_t^* w)(t) = \int_t^T w(\theta)\, d\theta.$$

Introduce the space $W_1^1(B)$ for $B = H^1$, H^{-1}, $H^{<-1>}$ with the norm

$$\|w\|_{W_1^1(B)} = \|w\|_{C(B)} + \|D_t w\|_{L_1(B)};$$

for $B = H,^{-1}$ by definition $(D_t w, \eta)_{0,Q} = -(w, D_t\eta)_{0,Q}$ for any $\eta \in W_1^1(H^1)$ such that $\eta|_{t=0,T} = 0$ (for $B = H^{<-1>}$ the duality relation $(\cdot,\cdot)_{0,Q}$ should be replaced by $(\cdot,\cdot)_Q$).

1.2. INITIAL-BOUNDARY VALUE PROBLEM

The initial-boundary value problem for second-order multidimensional hyperbolic equation basic in this paper is

$$\rho D_t^2 u + Lu = \rho f \text{ in } Q \tag{1.3}$$

$$u|_\Gamma = 0, \quad u|_{t=0} = u^{(0)}, \quad D_t u|_{t=0} = u^{(1)}, \tag{1.4}$$

where $\Gamma = \partial\Omega \times (0,T)$. We shall denote its solution by $P^{-1}\mathbf{d} \equiv P^{-1}(u^{(0)}, u^{(1)}, f)$.

By the generalized solution to problem (1.3) and (1.4) from the energy class (see reference 26, Chapter IV, Section 4), we call a function $u \in C(H^1)$, which has $D_t u \in C(H^0)$ and satisfies the integral identity

$$\mathcal{B}(u, \eta) \equiv -(D_t u, D_t\eta)_{0,Q} + \mathcal{L}_Q(u, \eta) = (u^{(1)}, \eta_0)_{0,\Omega} + (f, \eta)_{0,Q} \tag{1.5}$$

for any $\eta \in L_1(H^1)$ such that $D_t\eta \in L_1(H^0)$, $\eta|_{t=T} = 0$, and also satisfies the initial condition, $u|_{t=0} = u^{(0)}$. Here and in subsequent text, $\eta_0 = \eta|_{t=0}$.

Stipulate that in subsequent inequalities the constants $c, c_1, c_2, \ldots$ are nonnegative and not dependent either on T or on coefficients of Equation (1.3). The constants $c_0 > 0$ are absolute (i.e., they do not depend on anything).

Proposition 1.1. The estimate

$$\|D_t u\|_{C(H^0)} + \|u\|_{C(H^1)} \leq c_0 \Big(\|u^{(0)}\|_{H^1} + \|u^{(1)}\|_{H^0}$$

$$+ \|f\|_{L_1(H^0)} \Big). \tag{1.6}$$

is valid, that is, if $u^{(0)} \in H^0$, $u^{(1)} \in H^{-1}$, and $f \in L_1(H^0)$, then there exists a unique generalized solution to problem (1.3) and (1.4) from the energy class, and estimate (1.6) holds true. Estimate (1.6) remains valid if one substitutes $\|f\|_{W_1^1(H^{-1})}$ for $\|f\|_{L_1(H^0)}$.

Proof. There exist several methods to prove the given proposition (see reference 26, Chapter IV). For the assumptions made previously, the Fourier method is fairly convenient. With this method the solution is presented as a series,

$$u(t) = \sum_{k \geq 1} \tilde{u}_k(t)\, \mu_k, \qquad (1.7)$$

with the coefficients

$$\tilde{u}_k(t) = \tilde{u}_k^{(0)} \cos \sqrt{\lambda_A^{(k)}}\, t + \frac{\tilde{u}_k^{(1)}}{\sqrt{\lambda_A^{(k)}}} \sin \sqrt{\lambda_A^{(k)}} t$$

$$+ \frac{1}{\sqrt{\lambda_A^{(k)}}} \int_0^t \tilde{f}_k(\theta) \sin \sqrt{\lambda_A^{(k)}}\, (t - \theta)\, d\theta \quad (1.8)$$

where $\tilde{u}_k^{(0)}$, $\tilde{u}_k^{(1)}$, and $\tilde{f}_k(\theta)$ are the Fourier coefficients of the functions $u^{(0)}$, $u^{(1)}$, and $f(\theta)$, respectively. The Fourier method justification is presented in detail in reference 26, Chapter IV, Section 7. Here we note that the convergence of series (1.7) in the energy class of solutions and the validity of estimate (1.6) follow from the chain of inequalities ($0 \leq t \leq T$)

$$\|u(t)\|_{H^1}^2 + \|D_t u(t)\|_{H^0}^2 = \sum_{k \geq 1} \lambda_A^{(k)} \tilde{u}_k^2(t) + (D_t \tilde{u}_k)^2(t)$$

$$\leq c_0 \sum_{k \geq 1} \lambda_A^{(k)} \left(\tilde{u}_k^{(0)}\right)^2 + \left(\tilde{u}_k^{(1)}\right)^2 + \left(\int_0^t |\tilde{f}_k(\theta)|\, d\theta\right)^2$$

$$\leq c_0 \left(\|u^{(0)}\|_{H^1}^2 + \|u^{(1)}\|_{H^0}^2 + \|f\|_{L_1(H^0)}^2\right).$$

Here $\|f\|_{L_1(H^0)}$ appears (instead of $\|f\|_{L_2(Q)}$ as in reference 26) because by virtue of Minkowski's inverse inequality [37]

$$\sum_{k \geq 1} \left(\int_0^t |\tilde{f}_k(\theta)|\, d\theta\right)^2 \leq \left(\int_0^t \left(\sum_{k \geq 1} \tilde{f}_k^2(\theta)\right)^{1/2} d\theta\right)^2$$

$$= \|f\|_{L_1((0,t);H^0)}^2. \qquad (1.9)$$

To ensure the possibility of the norm substitution in estimate (1.6) which is indicated in the hypothesis, it is necessary to transform the last summand of the right-hand part of Equation (1.8) to the form

$$\frac{1}{\lambda_A^{(k)}} \left\{ \int_0^t (D_t \tilde{f}_k)(\theta) \left(1 - \cos \sqrt{\lambda_A^{(k)}} \, (t - \theta) \right) d\theta \right.$$

$$\left. + \tilde{f}_k(0) \left(1 - \cos \sqrt{\lambda_A^{(k)}} \, t \right) \right\}.$$

It is useful to supplement the proved proposition. Stipulate writing $g = D_t^{-1} f$ and $f = D_t^{(0)} g$ if (generalized) functions f and g are connected by the equality $(f, \eta)_{0,Q} = -(g, D_t \eta)_{0,Q}$ for the same η as in Equation (1.5) (sometimes we need additional η smoothness from the context). For $f \in L_1(H^0)$, we have $g = I_t f$ and $f = D_t g$, $g|_{t=0} = 0$.

Proposition 1.1'. For $u^{(0)} = u^{(1)} = 0$ the estimates

$$\|D_t u\|_{L_\infty(H^0)} + \|u\|_{C(H^1)} \leq c_0 \|D_t^{-1} f\|_{V(H^0)} \qquad (1.10)$$

$$\|D_t u\|_{C(H^0)} + \|u\|_{C(H^1)} \leq c_0 \|f\|_{V(H^{-1})} \qquad (1.11)$$

are valid.

Proof. To derive estimate (1.10), let $\tilde{g}_k(\theta)$ be the Fourier coefficients of the function $g = D_t^{-1} f$. Equation (1.8) remains valid if the integral within it is understood in the Stieltjes sense (i.e., $\tilde{f}_k(\theta) d\theta$ is replaced by $d\tilde{g}_k(\theta)$). Note that the functions $\tilde{u}_k$ are continuous on $[0, T]$.

Without loss of generality, let $g(t)$ be continuous from the right for $0 \leq t < T$ and continuous from the left for $t = T$. To prove estimate (1.10) it is enough to verify the inequality

$$\sum_{k \geq 1} (\text{var}_{[0,T]} \, \tilde{g}_k)^2 \leq (\text{var}_{[0,T]} \, g)^2. \qquad (1.12)$$

This inequality can be deduced from inequality (1.9) by limit transition. To do that, assume that $g(t) = g(T)$ for $t > T$ and construct the averages $g^{(\delta)}(t) = \delta^{-1} \int_0^\delta g(t + \theta) d\theta$, $\delta > 0$. It is clear that $\tilde{g}_k^{(\delta)}(t)$, the Fourier coefficients of $g^{(\delta)}(t)$, are counterparting averages for $\tilde{g}_k(t)$. By applying inequality (1.9) to $D_t g^{(\delta)}$, we obtain[42]

$$\sum_{1 \leq k \leq K} \left(\text{var}_{[0,T]} \, \tilde{g}_k^{(\delta)} \right)^2 \leq \left(\text{var}_{[0,T]} \, g^{(\delta)} \right)^2$$

$$\leq (\text{var}_{[0,T]} \, g)^2, \qquad K \geq 1.$$

By virtue of $\tilde{g}_k^{(\delta)}(t) \to \tilde{g}_k(t)$ for all $0 \le t \le T$ and

$$\mathrm{var}_{[0,T]}\,\tilde{g}_k^{(\delta)} \le \mathrm{var}_{[0,T]}\,\tilde{g}_k$$

we have

$$\mathrm{var}_{[0,T]}\,\tilde{g}_k^{(\delta)} \to \mathrm{var}_{[0,T]}\,\tilde{g}_k$$

as $\delta \to 0$, so inequality (1.12) is valid.

Estimate (1.11) can be proved in a similar way. The property $D_t u \in C(H^0)$ in it follows from the continuity of $D_t \tilde{u}_k$ on $[0,T]$ (which in the case of estimate (1.10), generally speaking, does not hold true).

In estimate (1.10) the previous given definition of the generalized solution is somewhat extended; here we assume that $D_t u \in L_\infty(H^0)$.

To analyze problem (1.3) and (1.4) in assumptions broader than in Proposition 1.1 (in particular, when the solution is discontinuous), we should considerably extend the notion of the solution. For the weak generalized solution to problem (1.3) and (1.4), let us call a function $u \in C(H^0)$ which has $I_t u \in C(H^1)$ and satisfies the integral identity

$$-(u, D_t\eta)_{0,Q} + \mathcal{L}_Q(I_t u, \eta) = \left(u^{(0)}, \eta_0\right)_{0,\Omega} + \left(u^{(1)}, (I_t^*\eta)_0\right)_{0,\Omega}$$
$$+ (f, I_t^*\eta)_{0,Q} \qquad (1.13)$$

for the same functions η as in identity (1.5).

Proposition 1.2. With $f_1 + D_t^{(0)} f_2 = f$,

$$\|u\|_{C(H^0)} + \|I_t u\|_{C(H^1)} \le c_0 \left(\left\|u^{(0)}\right\|_{H^0} + \left\|u^{(1)}\right\|_{H^{-1}}\right.$$
$$\left. + \|f_1\|_{L_1(H^{-1})} + \|f_2\|_{L_1(H^0)}\right) \qquad (1.14)$$

is a valid estimate.

Proof. Identity (1.3) means that $I_t u = P^{-1}(0, u^{(0)}, u^{(1)} + I_t f_1 + f_2)$. Hence, Proposition 1.2 follows from Proposition 1.1. It is useful to note that here series (1.7) converges in $C(H^0)$; the summand in Equation (1.8) corresponding to f_2 should be written in the form

$$\int_0^t \tilde{f}_{2,k}(\theta) \cos\sqrt{\lambda_A^{(k)}}\,(t-\theta)\,d\theta.$$

Remark 1.1. As follows from Equation (1.11), in estimate (1.14) the norm $\|f_1\|_{L_1(H^{-1})}$ can be weakened to $\|D_t^{-1} f_1\|_{V(H^{-1})}$.

To deduce error estimates in terms of data, it is necessary to know the smoothness of the generalized solution to problem (1.3) and (1.4) due to the data smoothness.

Introduce the space $F^{k,l}$ ($k \geq 0, l \geq 0$ are integers) with the norm

$$\|f\|_{F^{k,l}} = \|D_t^l f\|_{L_1(H^k)} + \sum_{0 \leq m < l} \|D_t^m f|_{t=0}\|_{H^{k+l-m-1}}.$$

Here it is assumed that $D_t^m f \in L_1(H^k)$ for $0 \leq m \leq l$. Elements of $F^{k,l}$ have the dominating mixed smoothness (of order k in x and of order l in t). Set $T_1 = \max\{T, 1\}$.

Proposition 1.3. For $k \geq 0$, $l \geq 0$, the estimate

$$\|D_t^{l-1} u\|_{C(H^{k+2})} + \|D_t^l u\|_{C(H^{k+1})} + \|D_t^{l+1} u\|_{C(H^k)}$$

$$+ T_1^{-1} \|D_t^{l+2} u\|_{L_1(H^{k-1})} \leq c \Big(\|u^{(0)}\|_{H^{k+l+1}}$$

$$+ \|u^{(1)}\|_{H^{k+l}} + \|f\|_{F^{k,l}} \Big). \tag{1.15}$$

is valid.

Remark 1.2. The estimate

$$\|D_t^{l+2} u\|_{L_r(H^{k-1})} \leq c_0 T^{1/r} \Big(\|u^{(0)}\|_{H^{k+l+1}} + \|u^{(1)}\|_{H^{k+l}}$$

$$+ \|f\|_{F^{k,l}} \Big) + \|D_t^l f\|_{L_r(H^{k-1})}$$

is also valid for $1 \leq r \leq \infty$.

Proof. Let $u^{(0)} \in H^{k+l+1}$, $u^{(1)} \in H^{k+l}$, $f \in F^{k,l}$. Introduce functions $u^{(m)}$ with the help of the recurrent equation

$$u^{(m)} = -Au^{(m-2)} + D_t^{m-2} f|_{t=0}, \qquad m \geq 2. \tag{1.16}$$

Then $$D_t^l u = P^{-1}(u^{(l)}, u^{(l+1)}, D_t^l f),$$

$$A^{k/2} D_t^l u = P^{-1} \Big(A^{k/2} u^{(l)}, A^{k/2} u^{(l+1)}, A^{k/2} D_t^l f \Big)$$

are valid equalities and can be successively justified by the Fourier method. Thus, by virtue of Proposition 1.1 the estimate

$$\|D_t^{l+1} u\|_{C(H^k)} + \|D_t^l u\|_{C(H^{k+1})} \leq c_0 \Big(\|u^{(l)}\|_{H^{k+1}}$$

$$+ \|u^{(l+1)}\|_{H^k} + \|D_t^l f\|_{L_1(H^k)} \Big)$$

is valid. Hence, with the help of Equation (1.16) we derive the estimate

$$\left\|D_t^{l+1}u\right\|_{C(H^k)} + \left\|D_t^l u\right\|_{C(H^{k+1})} \leq c_0\Big(\left\|u^{(0)}\right\|_{H^{k+l+1}}$$

$$+ \left\|u^{(1)}\right\|_{H^{k+l}} + \left\|f\right\|_{F^{k,l}}\Big). \qquad (1.17)$$

To estimate the remaining summand in the left-hand part of estimate (1.15) utilize the equations, which can also be justified by the Fourier method,

$$A^{k/2}D_t^{l+1}u + A^{k/2+1}D_t^{l-1}u = A^{k/2}D_t^{l-1}f, \qquad l \geq 1$$

$$(1.18)$$

$$A^{(k-1)/2}D_t^{l+2}u + A^{(k+1)/2}D_t^l u = A^{k-1/2}D_t^l f. \qquad (1.19)$$

They yield the estimates

$$\left\|D_t^{l-1}u\right\|_{C(H^{k+2})} \leq \left\|D_t^{l+1}u\right\|_{C(H^k)} + \left\|D_t^l f\right\|_{L_1(H^k)}$$

$$+ \left\|D_t^{l-1}f\big|_{t=0}\right\|_{H^k}, \qquad l \geq 1 \qquad (1.20)$$

$$\left\|D_t^{l+2}u\right\|_{L_1(H^{k-1})} \leq T\left\|D_t^l u\right\|_{C(H^{k+1})} + \left\|D_t^l f\right\|_{L_1(H^{k-1})}.$$

For $l = 0$ an equation similar to Equation (1.18) but with the additional summand $A^{k/2}u^{(1)}$ in its right-hand part is valid. Therefore, for $l = 0$ estimate (1.20) is also valid, but with $D_t^{l-1}f\big|_{t=0}$ replaced by $u^{(1)}$. Proposition 1.3 is proved.

Remark 1.2 follows from Equation (1.19) and estimate (1.15). Estimate (1.15) is valid, with

$$c = c_0 \max\left\{\left(\min_k \lambda_A^{(k)}\right)^{-1/2}, 1\right\};$$

if the last summand in its left-hand part is omitted, we can take $c = c_0$.

Proposition 1.3 and Remark 1.2 can be extended to the case of arbitrary integers k and l. In particular, Proposition 1.1 can be supplemented with the estimate (the case where $k = 1$, $l = -1$, and $u^{(0)} = 0$, $u^{(1)} = 0$)

$$\left\|D_t u\right\|_{L_\infty(H^0)} + \left\|u\right\|_{C(H^1)} \leq c_0\Big(\left\|D_t^{-1}f\right\|_{L_1(H^1)} + \left\|D_t^{-1}f\right\|_{L_\infty(H^0)}\Big).$$

$$(1.21)$$

This estimate remains valid if in both its parts we substitute $L_\infty(H^0)$ for $C(H^0)$.

2. THREE-LEVEL FINITE-ELEMENT METHOD WITH WEIGHT

2.1. ADDITIONAL NOTATION

Let $S_1 = S_{1,h}$ be a finite-dimensional subspace in H^1 (h is a parameter). Let s_1 and s_0 be the projectors onto S_1 in H^1 and H^0, respectively, that is,

$$\mathcal{L}_\Omega(w - s_1 w, \varphi) = 0, \quad (w - s_0 w, \varphi)_{0,\Omega} = 0 \quad \forall \varphi \in S_1. \tag{2.1}$$

The inequality

$$\max\{\|s_k w\|_{H^k}, \|w - s_k w\|_{H^k}\} \leq \|w\|_{H^k} \quad k = 0, 1. \tag{2.2}$$

is trivial.

Let us fix some basis in S_1 and let φ_h be the coefficient vector of the expansion of a function $\varphi \in S_1$ with respect to this basis. Let H_h be the space of vectors φ_h with the standard Euclidean norm $\|\cdot\|_h = (\cdot, \cdot)_h^{1/2}$, which is equal to the square root of the vector components' second power sum. Introduce the self-adjoint positive definite operators, B_h and L_h, acting in H_h such that

$$(B_h \varphi_h, \psi_h)_h = (\varphi, \psi)_{0,\Omega}, \quad (L_h \varphi_h, \psi_h)_h = \mathcal{L}_\Omega(\varphi, \psi) \quad \forall \varphi, \psi \in S_1.$$

For $w \in H^{-1}$ denote by w^h the vector from H_h such that $(w^h, \varphi_h)_h = (w, \varphi)_{0,\Omega}$ for all $\varphi \in S_h$.

Introduce on $[0, T]$ the mesh $\bar{\omega}^\tau$, with the nodes $0 = t_0 < t_1 < \ldots < t_M = T$ and the steps $\tau_m = t_m - t_{m-1}$. Let $\hat{S}_\tau$ be the space of functions which are continuous on $[0, T]$ and linear on each segment $[t_{m-1}, t_m]$, $1 \leq m \leq M$. The "hill" functions, $e_m \in \hat{S}_\tau$ ($0 \leq m \leq M$) such that $e_m(t_l) = 1$ for $l = m$ or $e_m(t_l) = 0$ for $l \neq m$, form a basis in S_h. For a function w continuous on $[0, T]$, introduce the function $s_t w \in \hat{S}_\tau$ with the property $s_t w(t_m) = w_m \equiv w(t_m)$ for $0 \leq m \leq M$. Define the averages $w_m^\tau = (w, e_m)_{(0,T)}/(1, e_m)_{(0,T)}$, where $(w, \eta)_{(0,T)} = \int_0^T w\eta \, dt$, and define $\tau_{\max} = \max_{1 \leq m \leq M} \tau_m$.

Let $\bar{S}_\tau$ be the space of functions constant on each half-interval $(t_{m-1}, t_m]$, with $1 \leq m \leq M$. For $w \in L_1(0, T)$ introduce the function $[w] \in \bar{S}_\tau$ such that $[w](t_m)$ is the mean value of w on $(t_{m-1}, t_m)(1 \leq m \leq M)$. Let $\tau \in \bar{S}_\tau$, $\tau(t_m) = \tau_m$ ($1 \leq m \leq M$). The identities

$$(D_t s_t w, D_t \eta)_{0,T} = (D_t w, D_t \eta)_{0,T} \quad \forall w \in W_1^1(0, T), \ \eta \in \hat{S}_\tau \tag{2.3}$$

$$([w], z)_{(0,T)} = (w, [z])_{(0,T)} = ([w], [z])_{(0,T)} \quad \forall w, z \in L_1(0, T) \tag{2.4}$$

are valid. We have

$$\|s_t w\|_{C(B)} = \|w\|_{C_\tau(B)} \equiv \max_{0 \le m \le M} \|w_m\|_B. \tag{2.5}$$

Introduce the mesh operators

$$\check{w}_m = w_{m-1}, \quad \hat{w}_m = w_{m+1}, \quad I_\tau^m w = \sum_{1 \le l \le m} w_l \tau_l \,;$$

$$\bar{\partial}_t w = \frac{w - \check{w}}{\tau}, \quad \partial_t w = \frac{\hat{w} - w}{\hat{\tau}}, \quad \hat{\partial}_t w = \frac{\hat{w} - w}{\bar{\tau}} \,;$$

$$\overset{\circ}{\partial}_t w = \frac{\hat{w} - \check{w}}{2\bar{\tau}}, \quad w^{(1/2)} = \frac{\check{w} + w}{2} \,,$$

where $\bar{\tau} = (\tau + \hat{\tau})/2$ and set $I_\tau^0 w = 0$, $\bar{\partial}_t w_0 = 0$, $w_0^{(1/2)} = w_0$.

In Section 2 and Sections 3–7 the mesh $\bar{\omega}^\tau$ is uniform (i.e., $\tau_m = T/M$ for all m) if the opposite is not indicated.

2.2.   THREE-LEVEL FINITE-ELEMENT METHOD

As an approximate solution to the initial-boundary value problem (1.3) and (1.4), we consider a function $v \in S = S_{1,h} \otimes \hat{S}_\tau$ satisfying the integral identity

$$\mathcal{B}_\sigma(v, \eta) \equiv (D_t v, D_t \eta)_{0,Q} - \left(\sigma - \frac{1}{6}\right) \mathcal{L}_Q(\tau^2 D_t v, D_t \eta) + \mathcal{L}_Q(v, \eta)$$

$$= (u^{(1)}, \eta_0)_{0,\Omega} + (f, \eta)_{0,Q} \qquad \forall \eta \in S, \ \eta|_{t=T} = 0 \tag{2.6}$$

and the initial condition $v|_{t=0} = v^{(0)}$ or $v|_{t=0} = s_1 u^{(0)}$. By definition, the function $v^{(0)} \in S_1$ satisfies the identity

$$\left(v^{(0)}, \varphi\right)_{0,\Omega} + \sigma^{(0)} \tau_1^2 \mathcal{L}_\Omega\left(v^{(0)}, \varphi\right) = \left(u^{(0)}, \varphi\right)_{0,\Omega} \qquad \forall \varphi \in S_1. \tag{2.7}$$

Here $\sigma, \sigma^{(0)}$ are parameters, $\sigma^{(0)} \ge \sigma - 1/4$ (this condition is not difficult to extend). These identities can be used for a nonuniform mesh, $\bar{\omega}^\tau$.

Let us fix $\varepsilon_0 \in (0, 1]$ and assume that $\sigma \ge 1/4 - (1 - \varepsilon_0^2)/(\tau^2 \alpha_h^2)$, where α_h is the least constant in the inequality $\|\varphi\|_{H^1} \le \alpha_h \|\varphi\|_{H^0}$ for all $\varphi \in S_1$. Then for $\varphi \in S_1$ we have

$$\varepsilon_0^2 \|\varphi\|_{H^0}^2 \le \|\varphi\|_{H_\tau^0}^2 \equiv \|\varphi\|_{H^0}^2 + \left(\sigma - \frac{1}{4}\right) \tau^2 \|\varphi\|_{H^1}^2. \tag{2.8}$$

Choose $\eta(x,t) = \varphi(x)e_m(t)$ with $\varphi \in S_1$ in identity (2.6). Simple calculations yield the identities

$$(\partial_t \bar{\partial}_t v_m, \varphi)_{0,\Omega} + \mathcal{L}_\Omega(\sigma\tau^2 \partial_t \bar{\partial}_t v_m + v_m, \varphi) = (f_m^\tau, \varphi)_{0,\Omega},$$

$$1 \leq m \leq M-1 \qquad (2.9)$$

$$(\partial_t v_0, \varphi)_{0,\Omega} + \mathcal{L}_\Omega\left(\sigma\tau^2 \partial_t v_0 + \frac{\tau}{2} v_0, \varphi\right) = \left(u^{(1)}, \varphi\right)_{0,\Omega}$$

$$+ \frac{\tau}{2} (f_0^\tau, \varphi)_{0,\Omega}. \qquad (2.10)$$

The union of identities (2.9) and (2.10) with any $\varphi \in S_1$ is equivalent to identity (2.6). It is easy to write identities (2.9), (2.10), and (2.7), respectively, in operator form, that is,

$$(B_h + \sigma\tau^2 L_h)\partial_t \bar{\partial}_t v_{h,m} + L_h v_{h,m} = f_m^{h,\tau}, \qquad 1 \leq m \leq M-1 \quad (2.11)$$

$$(B_h + \sigma\tau^2 L_h)\partial_t v_{h,0} + \frac{\tau}{2} L_h v_{h,0} = u^{(1),h} + \frac{\tau}{2} f_0^{h,\tau} \qquad (2.12)$$

$$\left(B_h + \sigma^{(0)}\tau^2 L_h\right) v_h^{(0)} = u^{(0),h}. \qquad (2.13)$$

It is clear that the finite-element method under consideration is a three-level one.

If $(f, \eta)_{0,Q} = -(f_2, D_t\eta)_{0,Q}$ for all $\eta \in S$, $\eta|_{t=T} = 0$, then

$$f_m^{h,\tau} = \partial_t[f_2^h]_m, \qquad 1 \leq m \leq M-1$$

$$f_0^{h,\tau} = \frac{2}{\tau} \partial_t[f_2^h]_1. \qquad (2.14)$$

In accordance with Equations (2.11) and (2.12), to find $v_{h,m+1}$ for $0 \leq m \leq M-1$ we need to solve an equation with operator $B_h + \sigma\tau^2 L_h$. This operator is positive definite (and consequently nondegenerate) for $\sigma > -1/(\tau^2 \alpha_h^2)$; thus, under the given condition Equations (2.11) and (2.12) are uniquely solvable. To unify calculations, the convenient choice of $\sigma^{(0)}$ is $\sigma^{(0)} = \sigma$ (compare Equation [2.13] with Equations [2.11] and [2.12]).

In the case of a nonuniform mesh, $\bar{\omega}^\tau$, it is not difficult to verify that Equation (2.11) should be replaced by

$$B_h \hat{\partial}_t \bar{\partial}_t v_{h,m} + \sigma L_h \hat{\partial}_t (\tau^2 \bar{\partial}_t v_h)_m + L_h v_{h,m} = f_m^{h,\tau}, \qquad 1 \leq m \leq M-1,$$

$$(2.11')$$

and in Equations (2.12) and (2.13) we should take $\tau = \tau_1$.

When defining an approximate solution one often starts from identity (2.9) (see references 5, 18, and others). The advantages of using identity (2.6) are, in particular, that the problems of v_1 defining and $\bar{\omega}^\tau$ nonuniformity are removed, and the derivation of a priori error estimates is considerably simplified (see Theorem 3.1 proof).

Introduce the level energetic norm such that

$$\|v\|_{H_E}^2 = \|\bar{\partial}_t v\|_{H_\tau^0}^2 + \|v^{(1/2)}\|_{H^1}^2.$$

For the function v establish the counterpart of Propositions 1.1 and 1.2.

Theorem 2.1. Let the function $v \in S$ satisfy identity (2.6) and the function $v_0 \in S_1$ be arbitrary.

(1) Let $(f, \eta)_{0,Q} = (f_1, \eta)_{0,Q} - (f_2, D_t\eta)_{0,Q}$ for all $\eta \in S$, $\eta|_{t=T} = 0$. Then the estimate

$$\max\left\{\|v\|_{C_\tau(H_\tau^0)}, \|I_t v\|_{C_\tau(H^1)}\right\} \le \|v_0\|_{H_\tau^0} + 2\|u^{(1)}\|_{H^{-1}}$$
$$+ 4\|f_1\|_{L_1(H^{-1})} + 2\varepsilon_0^{-1}\|f_2\|_{L_1(H^0)} \quad (2.15)$$

is valid.

(2) The estimate

$$\|v\|_{C_\tau(H_E)} \le \|v_0\|_{H^1} + \varepsilon_0^{-1}\|u^{(1)}\|_{H^0}$$
$$+ 2\varepsilon_0^{-1}\|f\|_{L_1(H^0)} \quad (2.16)$$

is also valid. The indicated estimate remains valid while $\|f\|_{L_1(H^0)}$ is replaced by $2\varepsilon_0\|f\|_{W_1^1(H^{-1})}$.

Remark 2.1. The inequality $\|v^{(0)}\|_{H_\tau^0} \le \varepsilon_0^{-1}\|u^{(0)}\|_{H^0}$ holds true (for $\sigma^{(0)} \ge 0$ the multiplier ε_0^{-1} can be omitted); it follows from identity (2.7), with $\varphi = v^{(0)}$, and from inequality (2.8). Also recall that $\|s_1 u^{(0)}\|_{H^1} \le \|u^{(0)}\|_{H^1}$.

Remark 2.2. In estimate (2.15) the norm of f_1 can be weakened to $\|D_t^{-1} f_1\|_{V(H^{-1})}$. In estimate (2.16) the norm of f can be weakened to $\|D_t^{-1} f\|_{V(H^0)}$ or replaced by $2\varepsilon_0\|f\|_{V(H^{-1})}$.

Proof.

(1) In identity (2.6) choose the function η in the form $\eta = s_t I_t^* \xi$, with $\xi \in S$. The equalities

$$D_t\eta = -[\xi],$$
$$(v, \eta)_{(0,T)} = (D_t I_t v, \eta)_{(0,T)} = (I_t v, [\xi])_{(0,T)}$$

are valid. Also, using identity (2.4) and the equality $[I_t v] = [s_t I_t v] - (\tau^2/12) D_t v,^\dagger$ change identity (2.6) to the form

$$\left(D_t v, [\xi]\right)_{0,Q} + \left(\sigma - \frac{1}{4}\right) \mathcal{L}_Q(\tau^2 D_t v, [\xi])$$

$$+ \mathcal{L}_Q([s_t I_t v], [\xi]) = \left(u^{(1)}, (I_t^* \xi)_0\right)_{0,\Omega}$$

$$+ (f_1, s_t I_t^* \xi)_{0,Q} + (f_2, [\xi])_{0,Q}.$$

Compare this with identity (1.13).

Let $1 \leq l \leq M$. Choose $\xi_m = (-1)^{m-l} \xi_l$ for $l < m \leq M$. Then $[\xi](\cdot, t) = (s_t I_t^* \xi)(\cdot, t) = 0$ for $t_l < t \leq T$, and thus in the last identity it is possible to replace Q by $Q^l = \Omega \times (0, t_l)$. Again applying identity (2.4) and the equality $(I_t^* \xi)_m = (I_t \xi)_l - (I_t \xi)_m,\ 0 \leq m \leq l$, we obtain

$$\left(D_t v, \xi\right)_{0,Q^l} + \left(\sigma - \frac{1}{4}\right) \mathcal{L}_{Q^l}(\tau^2 D_t v, \xi) + \mathcal{L}_{Q^l}(s_t I_t v, [\xi])$$

$$= \left(u^{(1)}, (I_t \xi)_l\right)_{0,\Omega} + (f_1, (I_t \xi)_l - s_t I_t \xi)_{0,Q^l}$$

$$+ \left(f_2, [\xi]\right)_{0,Q^l}. \tag{2.17}$$

Finally, choose $\xi_m = v_m$ for $0 \leq m \leq l$ and, taking into account equality (2.5), derive the inequality

$$\frac{1}{2} \left(\|v_l\|_{H_\tau^0}^2 + \|(I_t v)_l\|_{H^1}^2 \right) \leq \frac{1}{2} \|v_0\|_{H_\tau^0}^2$$

$$+ \left(\|u^{(1)}\|_{H^{-1}} + 2 \|f_1\|_{L_1(H^{-1})} \right) \|I_t v\|_{C_\tau(H^1)}$$

$$+ \|f_2\|_{L_1(H^0)} \|v\|_{C_\tau(H^0)}. \tag{2.18}$$

Using arbitrariness in the choice of l and the linear character of the dependence of v on $u^{(1)}, f_1, f_2$, we complete the proof of estimate (2.15).

(2) Choose $\varphi = (\partial_t v)_m^{(1/2)}$ in identity (2.9). Since

$$(\partial_t \bar{\partial}_t v)\varphi = \frac{1}{2} \partial_t \{(\bar{\partial}_t v)^2\},$$

$$v\varphi = \frac{1}{2} \partial_t \left\{ (v^{(1/2)})^2 - \frac{\tau^2}{4} \left(\bar{\partial}_t v\right)^2 \right\}$$

---

†This equality follows from the well-known trapezoidal quadrature rule, with a remainder term expressed by the second derivative in an intermediate point.

the energy conservation law

$$\frac{1}{2}\,\partial_t\big(\|v\|_{H_E}^2\big) = \Big(f^\tau, (\partial_t v)^{(1/2)}\Big)_{0,\Omega} \qquad (2.19)$$

is valid. It is not difficult to verify that

$$\big(\|v\|_{H_E}^2\big)_1 = (\partial_t v_0, \partial_t v_0)_{0,\Omega}$$
$$+ \mathcal{L}_\Omega\Big(\sigma\tau^2 \partial_t v_0 + \frac{\tau}{2}\,v_0, \partial_t v_0\Big) + \mathcal{L}_\Omega\Big(v_0, v_1^{(1/2)}\Big).$$

Therefore, taking into account identity (2.10), it follows from law (2.19) that

$$\big(\|v\|_{H_E}^2\big)_l = \mathcal{L}_\Omega(v_0, v_1^{(1/2)}) + (u^{(1)}, \partial_t v_0)_{0,\Omega}$$
$$+ \frac{\tau}{2}\,(f_0^\tau, \partial_t v_0)_{0,\Omega} + 2I_\tau^{l-1}\big(f^\tau, (\partial_t v)^{(1/2)}\big)_{0,\Omega},$$
$$1 \le l \le M. \quad (2.20)$$

Remembering the definition of f^τ, we have

$$\big(\|v\|_{H_E}^2\big)_l \le \|v_0\|_{H^1}\|v_1^{(1/2)}\|_{H^1}$$
$$+ \big(\|u^{(1)}\|_{H^0} + 2\|f\|_{L_1(H^0)}\big)\|\bar{\partial}_t v\|_{C_\tau(H^0)}.$$

Hence, similar to the first item of the proof, inequality (2.16) is valid. Summarizing by parts, for $v_0 = 0$ we obtain the chain of relations

$$\frac{\tau}{2}\,(f_0^\tau, \partial_t v_0)_{0,\Omega} + 2I_\tau^{l-1}(f^\tau, (\partial_t v)^{(1/2)})_{0,\Omega}$$

$$= -2I_\tau^{l-1}(\bar{\partial}_t f^\tau, v^{(1/2)})_{0,\Omega} + 2(f_{l-1}^\tau, v_l^{(1/2)})_{0,\Omega}$$

$$- (f_0^\tau, v_1^{(1/2)})_{0,\Omega} \le \big(2I_\tau^{l-1}\,\|\bar{\partial}_t f^\tau\|_{H^{-1}}$$

$$+ 2\,\|f_{l-1}^\tau\|_{H^{-1}} + \|f_0^\tau\|_{H^{-1}}\big)\,\|v^{(1/2)}\|_{C_\tau(H^1)}$$

$$\le \big(4\|D_t f\|_{L_1(H^{-1})} + 3\|f\|_{C(H^{-1})}\big)\,\|v^{(1/2)}\|_{C_\tau(H^1)}.$$

The possibility indicated in Theorem 2.1, item (2), of replacement of the norm of f follows from this chain.

Let us justify Remark 2.2. For example, recall estimate (2.15). As is evident from the proof of inequality (2.18), it is possible to replace

$$\|f_1\|_{L_1(H^{-1})}$$

by $\qquad \|(f_1)^\tau\|_{L_{1,\tau}(H^{-1})} \equiv \frac{\tau}{2}\|(f_1)_0^\tau\|_{H^{-1}} + I_\tau^{M-1}\|(f_1)^\tau\|_{H^{-1}}$

in inequality (2.18). For $f_1 = D_t^{-1}g$, taking into account equalities (2.14), we have $\|(f_1)^\tau\|_{L_{1,\tau}(H^{-1})} \leq \|g\|_{V(H^{-1})}$.

Remark 2.3. The case of $\sigma = 1/4$ stands out, because for this case estimate (2.15) is valid for any nonuniform mesh, $\bar{\omega}^\tau$.[†] As a consequence, the same is valid (looking ahead) for the corresponding a priori error estimate (3.4) and the estimates in terms of data (see Theorem 4.1 and Theorem 4.3, item (1)[‡]).

In identity (2.6) the summand with the multiplier $\sigma - 1/6$ breaks the formal projection principle of the approximate solution construction (compare with identity [1.5]). In accordance with Theorem 2.1, the appropriate choice of σ enables us to achieve absolute stability (i.e., for any τ, h), whereas for $\sigma = 1/6$ the stability is only conditional.

The choice of $\sigma = 1/4$ is recommended by some,[5,18,22,39] but it has a drawback: for $\sigma = 1/4$ the stability in the norm $\|\cdot\|_{C_\tau(H^1)}$ does not follow from estimate (2.16) (also see Theorems 4.2, 5.2, and 5.3, and Section 8, in subsequent text).

Lemma 2.1. Let $\varepsilon_0 \in (0,1]$. The inequality

$$\frac{1}{2}\varepsilon_0^2(\|v_{m-1}\|_{H^1}^2 + \|v_m\|_{H^1}^2) \leq (\|v\|_{H_E}^2)_m \qquad \forall v_{m-1}, v_m \in S_1 \tag{2.21}$$

is fulfilled if and only if

$$\sigma \geq \frac{1+\varepsilon_0^2}{4} - \frac{1}{\tau^2\alpha_h^2}. \tag{2.22}$$

Corollary 2.1. Under condition (2.22) we have

$$(\varepsilon_0/\sqrt{2})\|v\|_{C_\tau(H^1)} \leq \|v\|_{C_\tau(H_E)}.$$

[†]The proof remains valid due to disappearance of the second summand in the left-hand part of Equation (2.17); also, $\|v\|_{C_\tau(H_\tau^0)} = \|v\|_{C_\tau(H^0)}$ for $\sigma = 1/4$.

[‡]In these theorems τ^2 should be replaced by τ_{max}^2.

Proof. Let us perform the following transformations:

$$\|v\|_{H_E}^2 - \frac{1}{2}\,\varepsilon_0^2\left(\|\breve{v}\|_{H^1}^2 + \|v\|_{H^1}^2\right) = \tau^{-2}\|v\|_{H^0}^2$$

$$+ \left(\sigma - \frac{1}{2}\,\varepsilon_0^2\right)\|v\|_{H^1}^2 - 2\left\{\tau^{-2}(v,\breve{v})_{0,\Omega} + \left(\sigma - \frac{1}{2}\right)\mathcal{L}_\Omega(v,\breve{v})\right\}$$

$$+ \tau^{-2}\|\breve{v}\|_{H^0}^2 + \left(\sigma - \frac{1}{2}\,\varepsilon_0^2\right)\|\breve{v}\|_{H^1}^2$$

$$= \left(\begin{bmatrix} C_{1,h} & C_{2,h} \\ C_{2,h} & C_{1,h} \end{bmatrix}\begin{bmatrix} v_h \\ -\breve{v}_h \end{bmatrix}, \begin{bmatrix} v_h \\ -\breve{v}_h \end{bmatrix}\right)_h = (C_h w_h, w_h)_h$$

where $C_{1,h} = \tau^{-2}B_h + (\sigma - \frac{1}{2}\,\varepsilon_0^2)L_h$, $C_{2,h} = \tau^{-2}B_h + (\sigma - \frac{1}{2})L_h$. Thus, inequality (2.21) is equivalent to the inequality $C_h \geq 0$, that is, to a nonnegative definiteness of C_h in $H_h \times H_h$.[†]

Let us consider the eigenvalue problem, $L_h\varphi_h = \lambda B_h\varphi_h$ in H_h. Denote by $\lambda_k > 0$ the eigenvalues of this problem; emphasize that $\alpha_h = \max_k \lambda_k^{1/2}$. The $(B_h\cdot,\cdot)_h$ orthogonal eigenvectors of the considered problem form the basis in H_h. Expanding v_h and $\breve{v}_h$ with respect to this basis, we can reduce the condition $C_h \geq 0$ to the condition of a nonnegative definiteness of 2×2 matrices, that is,

$$\begin{bmatrix} c_{1,k} & c_{2,k} \\ c_{2,k} & c_{1,k} \end{bmatrix} \geq 0 \quad \text{for all } k,$$

where $c_{1,k} = \tau^{-2} + (\sigma - 1/2\,\varepsilon_0^2)\lambda_k$, $c_{2,k} = \tau^{-2} + (\sigma - 1/2)\lambda_k$, or, what is the same, to the condition $|c_{1,k}| \leq c_{2,k}$ for all k. For $\varepsilon_0 \in (0,1]$, the latter condition (and therefore initial inequality [2.21]) is equivalent to condition (2.22).

Now we establish the counterpart of estimate (1.21) for the function v. It is utilized not in the main theorems but for discussing the broadest error estimates in the energetic norm.

Theorem 2.2. Let function $v \in S$ satisfy the identity $\mathcal{B}_\sigma(v,\eta) = -(f_2, D_t\eta)_{0,Q}$ for all $\eta \in S$, $\eta|_{t=T} = 0$ while $v_0 = 0$.

If condition (2.22) is fulfilled, then the estimate

$$\max\left\{(\varepsilon_0^2/\sqrt{2})\|\bar{\partial}_t v\|_{C_\tau(H^0)},\ \varepsilon_0\|v\|_{C_\tau(H^1)}\right\}$$

$$\leq 2\varepsilon_0^{-1}\|L_h^{1/2}B_h^{-1}f_2^h\|_{L_1(H_h)} + (\varepsilon_0/\sqrt{2})\|f_2\|_{L_\infty(H^0)} \quad (2.23)$$

is valid.

[†]Compare this with reference 35.

Proof. Let us introduce the auxiliary function $z_m = I_\tau^m v$, $0 \le m \le M$. Since according to the hypothesis $u^{(1)} = 0$ and $v_0 = 0$, with the help of equalities (2.14) Equation (2.12) takes the form $(B_h + \sigma\tau^2 L_h)\bar{\partial}_t v_{h,1} = [f_2^h]_1$. Applying to Equation (2.11) the operator $I_\tau(\cdot)$, and using the obtained equation for $\bar{\partial}_t v_{h,1}$ and equalities (2.14), derive the equality

$$(B_h + \sigma\tau^2 L_h)\bar{\partial}_t v_{h,m} + L_h \check{z}_{h,m} = [f_2^h]_m, \qquad 1 \le m \le M. \tag{2.24}$$

We have $\bar{\partial}_t z = v$, so equality (2.24), the condition $v_0 = 0$, and the definition of the operator I_τ yield the following problem for the function z,

$$(B_h + \sigma\tau^2 L_h)\partial_t \bar{\partial}_t z_h + L_h z_h = \widehat{[f_2^h]} \tag{2.25}$$

$$(B_h + \sigma\tau^2 L_h)\partial_t z_{h,0} = \tau[f_2^h]_1, \qquad z_{h,0} = 0. \tag{2.26}$$

Applying the operator $L_h B_h^{-1}$ to Equation (2.25) and multiplying the result scalarly by $\mathring{\partial}_t z_h$, we obtain

$$\left(L_h \partial_t \bar{\partial}_t z_h, \mathring{\partial}_t z_h\right)_h + \sigma\tau^2 \left(B_h^{-1} L_h \partial_t \bar{\partial}_t z_h, L_h \mathring{\partial}_t z_h\right)_h$$
$$+ \left(B_h^{-1} L_h z_h, L_h \mathring{\partial}_t z_h\right)_h = \left(L_h B_h^{-1}[\hat{f}_2], \mathring{\partial}_t z_h\right)_h.$$

Hence, similarly to (2.19) the following equality holds true

$$\frac{1}{2}\partial_t(\|z\|_{H_E^2}^2) = \left(L_h B_h^{-1}[\hat{f}_2], \mathring{\partial}_t z_h\right)_h \tag{2.27}$$

where

$$\|z\|_{H_E^2}^2 \equiv \left\|L_h^{1/2}\bar{\partial}_t z_h\right\|_h^2 + \left(\sigma - \frac{1}{4}\right)\tau^2 \left\|B_h^{-1/2} L_h \bar{\partial}_t z_h\right\|_h^2$$
$$+ \left\|B_h^{-1/2} L_h z_h^{(1/2)}\right\|_h^2.$$

Since $z_1^{(1/2)} = (\tau/2)\bar{\partial}_t z_1$, we have

$$(\|z\|_{H_E^2})_1^2 = \left(L_h B_h^{-1}(B_h + \sigma\tau^2 L_h)\bar{\partial}_t z_{h,1}, \bar{\partial}_t z_{h,1}\right)_h. \tag{2.28}$$

Applying operator I_τ to equality (2.27) and using relations (2.28) and (2.26), we obtain (for $1 \le l \le M$)

$$(\|z\|_{H_E^2})_l = \tau\left(L_h B_h^{-1}[f_2^h]_1, \bar{\partial}_t z_{h,1}\right)_h + 2I_\tau^{l-1}\left(L_h B_h^{-1}[\hat{f}_2^h], \mathring{\partial}_t z_h\right)_h.$$

Then the estimate

$$\|z\|^2_{H^2_E} \leq 2\Big(I_\tau\big\|L_h^{1/2}B_h^{-1}[f_2^h]\big\|_h\Big)\big\|L_h^{1/2}\bar\partial_t z_h\big\|_{C_\tau(H_h)}$$

is valid, and because $\big\|B_h^{-1/2}L_h^{1/2}\big\| = \big\|L_h^{1/2}B_h^{-1/2}\big\| \leq \alpha_h$, the estimate

$$\varepsilon_0\|v\|_{C_\tau(H^1)} = \varepsilon_0\|\bar\partial_t z\|_{C_\tau(H^1)} \leq \|z\|_{C_\tau(H^2_E)}$$

$$\leq 2\varepsilon_0^{-1}\big\|L_h^{1/2}B_h^{-1}f_2^h\big\|_{L_1(H_h)} \tag{2.29}$$

is also valid.

It remains only to estimate $\bar\partial_t v$. Under the condition $\varepsilon_0 \in (0,1]$ we can verify that the inequality

$$\frac{1}{2}\,\varepsilon_0^2\Big(\big\|B_h^{-1/2}L_h z_{h,m-1}\big\|^2_h + \big\|B_h^{-1/2}L_h z_{h,m}\big\|^2_h\Big) \leq \Big(\|z\|^2_{H^2_E}\Big)_m$$

$$\forall z_{m-1}, z_m \in S_1 \tag{2.30}$$

is equivalent to the operator inequality $L_h B_h^{-1}C_h \geq 0$ in the same way as in Lemma 2.1; therefore it is fulfilled if and only if condition (2.22) holds true.

First, turning to equality (2.24) and then inequalities (2.30) and (2.8), we derive

$$\varepsilon_0\left(\big\|B_h^{1/2}\bar\partial_t v_h\big\|^2_h + \sigma\tau^2\big\|L_h^{1/2}\bar\partial_t v_h\big\|^2_h\right)^{1/2}$$

$$\leq \left\|B_h^{-1/2}\big(-L_h\check z_h + [f_2^h]\big)\right\|_h \leq \frac{\sqrt{2}}{\varepsilon_0}\|z\|_{H^2_E} + \left\|B_h^{-1/2}[f_2^h]\right\|_h.$$

This inequality and Equation (2.29) yields Equation (2.23).

3. SECOND-ORDER ACCURACY A PRIORI ERROR ESTIMATES

Let us stipulate that the rate of the best approximation by elements of $S_{1,h}$ in the H^1 norm is characterized by the inequality

$$\|w - s_1 w\|_{H^1} \leq \beta_h \|w\|_{H^2} \qquad \forall w \in H^2. \tag{3.1}$$

By virtue of the well-known Nitsche-(Aubin-Oganesjan-Rukhovets) technique, the estimate (with $\lambda = 1,2$)

$$\|w - s_1 w\|_{H^0} \leq \beta_h \|w - s_1 w\|_{H^1} \leq \beta_h^\lambda \|w\|_{H^\lambda} \tag{3.2}$$

is valid. Remember that if we introduce the auxiliary function Ψ such that $A\Psi = w - s_1 w$, then by virtue of the properties of s_1

$$\|w - s_1 w\|_{H^0}^2 = (w - s_1 w, A\Psi)_{0,\Omega} = \mathcal{L}_\Omega(w - s_1 w, \Psi)$$

$$= \mathcal{L}_\Omega(w - s_1 w, \Psi - s_1 \Psi) \le \|w - s_1 w\|_{H^1}\|\Psi - s_1 \Psi\|_{H^1}$$

$$\le \beta_h \|w - s_1 w\|_{H^1}\|A\Psi\|_{H^0}.$$

Hence, estimate (3.2) is valid.

Note that the indicated technique of deriving the L_2-norm error estimate from the energetic norm error estimate is also effective in the parabolic case.[31,43] But in the situation under consideration it seems to be useless, and in subsequent text corresponding error estimates are given independently from one another.

In the finite-element method, instead of Equation (3.1) the inequality

$$\|w - s_1 w\|_{H^{<1>}} \le c h_{\max} \|w\|_{H^{<2>}}$$

$$\forall w \in H^{<2>} \equiv W_2^2(\Omega) \cap \overset{\circ}{W}_2^1(\Omega) \quad (3.3)$$

is typical where $h_{\max}$ is the maximal diameter of finite elements for discretization of Ω. Of course, other situations are possible (see specifically Section 5). From elliptic equation theory it is well known[27] that under the proper smoothness conditions on a_{ij} and $\partial\Omega$ and for $\|a\|_{L_{p_n^1}(\Omega)} \le N$ with $p_n^1 = \max\{2, n/2\}$ the problem $Lw = g$ with $g \in L_2(\Omega)$ is solvable in $H^{<2>}$. Then $\|w\|_{H^{<2>}} \le c\|g\|_\Omega \le c_1\|w\|_{H^2}$, and inequality (3.3) yields Equation (3.1) with $\beta_h = c_2 h_{\max}$.

To begin, let us derive the a priori error estimates (i.e., those in terms of an appropriate smoothness of the solution) of the order $O(\beta_h^2 + \tau^2)$. Do that on the basis of stability Theorem 2.1. It is assumed that $|\sigma|\tau^2 \le N(\beta_h^2 + \tau^2)$; of course, this condition is broader than $|\sigma| \le N$.

Theorem 3.1. For the approximate solution defined by identity (2.6) a priori error estimates

$$\|u - v\|_{C_\tau(H^0)} + \|s_1 u - v\|_{C_\tau(H_\tau^0)} + \|s_1 I_t u - s_t I_t v\|_{C(H^1)}$$

$$\le c\Big\{\|s_1 u^{(0)} - v_0\|_{H_\tau^0} + (\beta_h^2 + \tau^2)\Big(\|u^{(0)}\|_{H^2}$$

$$+ \|u^{(1)}\|_{H^1} + \|D_t^2 u\|_{L_1(H^1)} + \|D_t Au\|_{L_1(H^0)}\Big)\Big\} \quad (3.4)$$

$$\|\bar{\partial}_t(u - v)\|_{C_\tau(H^0)} + \|s_1 u - v\|_{C_\tau(H_E)} \le c\Big\{\|s_1 u^{(0)} - v_0\|_{H^1}$$

$$+ (\beta_h^2 + \tau^2)\Big(\|u^{(1)}\|_{H^2} + \|D_t^2 Au\|_{L_1(H^0)}\Big)\Big\} \quad (3.5)$$

are valid. The condition $u \in L_1(H^2)$ is assumed in the given estimates.

Proof. Let $\eta \in S$, $\eta|_{t=T} = 0$. By virtue of properties (2.1) and (2.3) of operators s_1, s_t we have

$$\mathcal{B}_\sigma(s_1 s_t u, \eta) = - (D_t s_1 u,\, D_t \eta)_{0,Q}$$
$$- \left(\sigma - \frac{1}{6}\right)\mathcal{L}_Q(\tau^2 D_t u,\, D_t \eta) + \mathcal{L}_Q(s_t u, \eta).$$

The identity $\mathcal{B}_\sigma(v, \eta) = \mathcal{B}(u, \eta)$ follows from identities (1.5) and (2.6). Therefore, the function $r = s_1 s_t u - v \in S$ satisfies the identity

$$\mathcal{B}_\sigma(r, \eta) = \mathcal{B}_\sigma(s_1 s_t u, \eta) - \mathcal{B}(u, \eta) = \left(D_t u - s_1 D_t u\right.$$
$$\left. - \left(\sigma - \frac{1}{6}\right)\tau^2 D_t A u,\, D_t \eta\right)_{0,Q} - \mathcal{L}_Q(u - s_t u, \eta).$$

First derive two estimates for the function r. Write the equality $\mathcal{L}_Q(u - s_t u, \eta) = (A(u - s_t u), \eta)_{0,Q}$. According to Theorem 2.1, item (1), and with the help of the equality $\|Aw\|_{H^{-1}} = \|w\|_{H^1}$, we obtain

$$\max\left\{\|r\|_{C_\tau(H^0_\tau)},\, \|I_t r\|_{C_\tau(H^1)}\right\} \le \|r_0\|_{H^0_\tau} + 4\|u - s_t u\|_{L_1(H^1)}$$
$$+ 2\varepsilon_0^{-1}\left(\|D_t u - s_1 D_t u\|_{L_1(H^0)} + \left|\sigma - \frac{1}{6}\right|\tau^2\|D_t A u\|_{L_1(H^0)}\right).$$

To derive another estimate for r, transform the identity for r

$$\mathcal{B}_\sigma(r, \eta) = -(u^{(1)} - s_1 u^{(1)} - \left(\sigma - \frac{1}{6}\right)\tau^2 A u^{(1)},\, \eta_0)_{0,\Omega}$$
$$-(D_t^2 u - s_1 D_t^2 u - \left(\sigma - \frac{1}{6}\right)\tau^2 D_t^2 A u + A(u - s_t u), \eta)_{0,\Omega}.$$

Now according to Theorem 2.1, item (2), we obtain

$$\|r\|_{C_\tau(H_E)} \le \|r_0\|_{H^1}$$
$$+ \varepsilon_0^{-1}\left(\|u^{(1)} - s_1 u^{(1)}\|_{H^0} + \left|\sigma - \frac{1}{6}\right|\tau^2\|A u^{(1)}\|_{H^0}\right)$$
$$+ 2\varepsilon_0^{-1}\left(\|D_t^2 u - s_1 D_t^2 u\|_{L_1(H^0)}\right.$$
$$\left. + \left|\sigma - \frac{1}{6}\right|\tau^2\|D_t^2 A u\|_{L_1(H^0)} + \|A u - s_t A u\|_{L_1(H^0)}\right).$$

To pass from the estimated norms of r to the norms of the functions in the left-hand parts of estimates (3.4) and (3.5), it is enough to utilize the inequalities[†]

$$\|\bar{\partial}_t^k w\|_{C_\tau(H^0)} \leq \|(D_t^k w)_0\|_{H^0} + \|D_t^{k+1} w\|_{L_1(H^0)}, \qquad k = 0, 1$$

$$\|s_1 w\|_{H^1} \leq \|w\|_{H^1}$$

and to notice that

$$\|u - s_1 u\|_{C_\tau(H^0)} + \|s_1 I_t u - s_t I_t s_1 s_t u\|_{C(H^1)}$$

$$\leq \|u^{(0)} - s_1 u^{(0)}\|_{H^0} + \|D_t u - s_1 D_t u\|_{L_1(H^0)}$$

$$+ \|I_t u - s_t I_t u\|_{C(H^1)} + \|u - s_t u\|_{L_1(H^1)}.$$

To wind up derivation of estimates (3.4) and (3.5) it remains to make use of the well-known inequality $\|w - s_t w\|_{L_q(H^k)} \leq \tau^2 \|D_t^2 w\|_{L_q(H^k)}$ $(1 \leq q \leq \infty)$ and inequality (3.2).

Error estimates (3.4) and (3.5) do not include all the "derivatives" $A^{k/2} D_t^l u$ $(k \geq 0, l \geq 0)$ of the order $k + l = 3$ (in the case of estimate [3.4]) or $k + l = 4$ (in the case of estimate [3.5]), but only some of them (with $k \leq 2, l \leq 2$). This fact, on the one hand, is significant in Section 4 when the estimates in the function classes of the right-hand parts are derived. On the other hand, it shows "superconvergence" of $I_t v$ to $s_1 I_t u$ in estimate (3.4) and $\bar{\partial}_t v$ to $\bar{\partial}_t u$ as well as v to $s_1 u$ in estimate (3.5).

Let us study the properties of the function $v^{(0)}$.

Lemma 3.1.

(1) If $|\sigma^{(0)}|\tau^2 \leq N(\beta_h^2 + \tau^2)$, then the estimate

$$\|s_1 u^{(0)} - v^{(0)}\|_{H_\tau^0} \leq c(\beta_h^2 + \tau^2)\|u^{(0)}\|_{H^2} \qquad (3.6)$$

is valid;

(2) if one of the conditions

$$0 \leq \sigma^{(0)} \leq N, \qquad |\sigma^{(0)}|\tau\alpha_h \leq N \qquad (3.7)$$

is fulfilled, then the estimate

$$\|s_1 u^{(0)} - v^{(0)}\|_{H_\tau^0} \leq c(\beta_h + \tau)\|u^{(0)}\|_{H^1} \qquad (3.8)$$

is also valid;

[†]For $k = 0$ one could replace $C_\tau(H^0)$ by $C(H^0)$.

(3) if $\sigma^{(0)} \geq 0$, $\beta_h \alpha_h \leq N$, then the estimate

$$\|v^{(0)}\|_{H^1} \leq 2\|s_0 u^{(0)}\|_{H^1} \leq 2(1+N)\|u^{(0)}\|_{H^1}. \qquad (3.9)$$

is valid. If in addition $\sigma^{(0)} \tau^2 \leq N(\beta_h^2 + \tau^2)$, then the estimate

$$\|s_1 u^{(0)} - v^{(0)}\|_{H^1} \leq c(\beta_h + \tau)\|u^{(0)}\|_{H^2} \qquad (3.10)$$

is valid.

Proof. Let us denote the bilinear form in the left-hand part of identity (2.7) by $\mathcal{J}(v^{(0)}, \varphi)$. Let $r^{(0)} = s_1 u^{(0)} - v^{(0)}$ and $\varphi \in S_1$.

(1) Since $\mathcal{J}(s_1 u^{(0)}, \varphi) = (s_1 u^{(0)}, \varphi)_{0,\Omega} + \sigma^{(0)} \tau^2 \mathcal{L}_\Omega(u^{(0)}, \varphi)$, we have

$$\mathcal{J}(r^{(0)}, \varphi) = -(u^{(0)} - s_1 u^{(0)}, \varphi)_{0,\Omega}$$
$$+ \sigma^{(0)} \tau^2 \mathcal{L}_\Omega(u^{(0)}, \varphi). \qquad (3.11)$$

Making use of Remark 2.1 (for $r^{(0)}$ as $v^{(0)}$) we obtain

$$\|r^{(0)}\|_{H_\tau^0} \leq \varepsilon_0^{-1} \left(\|u^{(0)} - s_1 u^{(0)}\|_{H^0} \right.$$
$$\left. + |\sigma^{(0)}| \tau^2 \|A u^{(0)}\|_{H^0} \right). \qquad (3.12)$$

This inequality and estimate (3.2) yield estimate (3.6).

(2) Inequality (3.11) also yields an estimate similar to estimate (3.12), but with the replacement of $\|A u^{(0)}\|_{H^0}$ by $\alpha_h \|u^{(0)}\|_{H^1}$. Hence, under the condition $|\sigma^{(0)}| \tau \alpha_h \leq N$ estimate (3.8) is valid.

In the case of $0 \leq \sigma^{(0)} \leq N$ identity (3.11) with $\varphi = r^{(0)}$ yields

$$\|r^{(0)}\|_{H_\tau^0} \leq (\mathcal{J}(r,r))^{1/2}$$
$$\leq \left(\|u^{(0)} - s_1 u^{(0)}\|_{H^0}^2 + \sigma^{(0)} \tau^2 \|u^{(0)}\|_{H^1}^2 \right)^{1/2}.$$

Thus, in this case estimate (3.8) is also valid.

(3) If $\sigma^{(0)} > 0$, then according to identity (2.7) we have

$$\mathcal{J}(v^{(0)} - s_0 u^{(0)}, \varphi) = -\sigma^{(0)} \tau^2 \mathcal{L}_\Omega(s_0 u^{(0)}, \varphi)$$
$$\leq \sigma^{(0)} \tau^2 \|s_0 u^{(0)}\|_{H^1} \|\varphi\|_{H^1}.$$

Hence, for $\varphi = v^{(0)} - s_0 u^{(0)}$ we obtain $\|v^{(0)} - s_0 u^{(0)}\|_{H^1} \leq \|s_0 u^{(0)}\|_{H^1}$. Thus, $\|v^{(0)}\|_{H^1} \leq 2\|s_0 u^{(0)}\|_{H^1}$ for $\sigma^{(0)} \geq 0$.[†]

Right-hand inequality (3.9) is actually proved in Lemma 2 in reference 43; according to the operators s_0 and s_1 properties (see inequality [2.2]) we can write

$$\|s_0 u^{(0)}\|_{H^1} \leq \|s_1 u^{(0)}\|_{H^1} + \|s_0 u^{(0)} - s_1 u^{(0)}\|_{H^1} \leq \|u^{(0)}\|_{H^1}$$
$$+ \alpha_h \|u^{(0)} - s_1 u^{(0)}\|_{H^0} \leq (1 + \beta_h \alpha_h)\|u^{(0)}\|_{H^1}.$$

Using identity (3.11), for $\sigma^{(0)} \geq 0$ we have

$$\|r^{(0)}\|_{H^1} \leq \min\left\{\alpha_h, \frac{1}{\sqrt{\sigma^{(0)}\tau}}\right\} (\mathcal{J}(r^{(0)}, r^{(0)}))^{1/2}$$
$$\leq (\beta_h^2 \alpha_h + \sqrt{\sigma^{(0)}}\tau)\|u^{(0)}\|_{H^2}.$$

(Compare this inequality with Equation [3.12].) Taking into account other assumptions in item (3), we derive estimate (3.10).

Corollary 3.1. If $\beta_h \alpha_h \leq N$, then

$$\left\|L_h^{1/2} B_h^{-1} f_2^h\right\|_{L_1(H_h)} \leq \|s_0 f_2\|_{L_1(H^1)} \leq (1 + N)\|f_2\|_{L_1(H^1)}.$$

This is of interest in connection with Theorem 2.2.

4. FRACTIONAL-ORDER ERROR ESTIMATES IN NONSMOOTH DATA CLASSES

4.1. FUNCTION SPACES

Introduce some function spaces in addition to those considered in Section 1. Let B_0, B_1 be Banach compatible spaces (see reference 1, Section 2.3), for example, let B_1 be imbedded into B_0. Denote by $(B_0, B_1)_{\theta,q} = \bar{B}_{\theta,q}$ the Banach spaces constructed with the help of the $K_{\theta,q}$ real interpolation method ($0 < \theta < 1$, $1 \leq q < \infty$ and $0 \leq \theta \leq 1$, $q = \infty$ (see Section 3.1 in reference 11).

Interpolation inequality (the interpolation theorem for linear operators) is valid, that is,

$$\|R\|_{(B_0,B_1)_{\theta,q} \to (C_0,C_1)_{\theta,q}} \leq \|R\|_{B_0 \to C_0}^{1-\theta} \|R\|_{B_1 \to C_1}^{\theta} \qquad (4.1)$$

[†]In the case of $\sigma^{(0)} = 0$, the equality $v^{(0)} = s_0 u^{(0)}$ is taken into account.

where $\|R\|_{B\to C}$ is the norm of a linear operator $R : B \to C$. In this section $C_0 = C_1$, $q = \infty$; in this case $\|\cdot\|_{(C_0,C_1)_{\theta,\infty}} = \|\cdot\|_{C_0}$.

Introduce the spaces $H^{(\kappa)}(\kappa \geq -1)$ such that: (1) $H^{(\kappa)} = H^\kappa$ for the integer κ; (2) $H^{(\kappa)} = (H^{[\kappa]}, H^{[\kappa]+1})_{\kappa-[\kappa],\infty}$ for the noninteger κ. Here $[\kappa]$ is the integral part of the number κ.

Let B be a Banach space. Denote by $WH_1^\lambda(B)(0 < \lambda \leq 1)$ the space consisting of functions $w \in WH_1^0(B) \equiv L_1(B)$ with the finite norm

$$\|w\|_{WH_1^\lambda(B)} = T^{-\lambda}\|w\|_{L_1(B)} + \sup_{0<\gamma<T} \gamma^{-\lambda}\|w(t+\gamma) - w(t)\|_{L_1((0,T-\gamma);B)}.$$

The space $WH_1^1(B)$ can be identified with $V(B)$ in some sense.[42]

Introduce the spaces $F^{(\kappa,\lambda)}(\kappa \geq -1, \lambda \geq -1)$ with the norm

$$\|f\|_{F^{(\kappa,\lambda)}} = \|D_t^{[\lambda]}f\|_{WH_1^{\lambda-[\lambda]}(H^{(\kappa)})} + \sum_{0\leq m<[\lambda]} \|D_t^m f|_{t=0}\|_{H^{(\kappa+\lambda-m-1)}}.$$

Here, for $\lambda \geq 1$ it is assumed that $D_t^m f \in L_1(H^{(\kappa)})$ for $0 \leq m < [\lambda]$, and for $-1 \leq \lambda < 0$ it is assumed that f is a linear continuous functional on the space of functions η with the norm $\|\eta\|_{C(H^{-\kappa_0})} + \|D_t\eta\|_{L_\infty(H^{-\kappa_0})}$ and such that $\eta|_{t=T} = 0$ (with $\kappa_0 = \min\{[\kappa],0\}$). If $\kappa \geq 0, \lambda \geq 0$ are integers, then the spaces $\bar{F}^{(\kappa,\lambda)}$ and $F^{\kappa,\lambda}$ (see Section 1) are identical.

Introduce the spaces $\bar{F}^{(\kappa,l)}$ ($\kappa \geq -1$, $l \geq 0, l$ is an integer) with the norm

$$\|f\|_{\bar{F}^{(\kappa,\lambda)}} = \|D_t^{l-1}f\|_{WH_1^1(H^{(\kappa)})} + \sum_{0\leq m<l-1} \|D_t^m f|_{t=0}\|_{H^{(\kappa+l-m-1)}}.$$

Here we should make the same assumptions on f as previously for $\lambda = l - 1$. Note that $\|f\|_{\bar{F}^{(\kappa,l)}} \leq 2\|f\|_{F^{(\kappa,l)}}$, and the space $\bar{F}^{(\kappa,l)}$ is significantly broader than $F^{(\kappa,l)}$. The inequality

$$\|f\|_{F^{(\kappa,\lambda_0)}} \leq cT_1^{\lambda_1-\lambda_0}\|f\|_{F^{(\kappa,\lambda_1)}} \tag{4.2}$$

is valid.

Also introduce the spaces $H^{<\kappa>} = (H^{<k>}, H^{<k+1>})_{\kappa-\lambda,\infty}$ for noninteger $\kappa \in (-1,2)$ (here $k = [\kappa]$) and the spaces $F^{<\kappa,\lambda>}(-1 \leq \kappa \leq 0, \lambda = 0,1)$ and $\bar{F}^{<\kappa,l>}(-1 \leq \kappa \leq 0, l = 0,1)$ resulting from the spaces $F^{(\kappa,\lambda)}$ and $\bar{F}^{<\kappa,l>}$, respectively, when in their definitions $H^{(\kappa)}$ is replaced by $H^{<\kappa>}$ (and in the definition of D_t^{-1}, $(\cdot,\cdot)_{0,Q}$ is replaced by $(\cdot,\cdot)_Q$) (see Section 1).

The norms in $H^{(\kappa)}$ and $H^{<\kappa>}$ are equivalent: (1) for $0 \leq \kappa \leq 1$; (2) for $1 < \kappa \leq 2$, provided that $\|w\|_{H^{<2>}} \leq c\|w\|_{H^2}$ for all $w \in H^2$ (see Section 3) and $\|D_i a_{ij}\|_{L_{p_n}(\Omega)} + \|a\|_{L_{\max\{2,n/2\}}(\Omega)} \leq N$ for $1 \leq j \leq n$. The space $H^{(\kappa)}$ for noninteger $\kappa > 0$ is closely connected with the Nikol'skiĭ space $H_2^\kappa(\Omega)$.[30,38] As an example, under appropriate requirements on $\partial\Omega$ the first space coincides with the second one for $0 < \kappa < 1/2$ or with its subspace $\{w \in H_2^\kappa(\Omega) | w|_{\partial\Omega} = 0\}$ for $1/2 < \kappa < 2$, $\kappa \neq 1, 3/2$. The space $H^{(1/2)}$ includes discontinuous piecewise differentiable functions (under rather weak requirements on $n - 1$-dimensional surfaces of discontinuity).[36,45] The similar property is valid for $F^{(1/2,0)}$, $F^{(0,1/2)}$ (but not for $(H^0, H^1)_{1/2,q}$ with $q < \infty$, which explains the choice of $q = \infty$ made previously).

Let $\delta(\cdot)$ be the Dirac delta function. Take note that $\|w(x)\delta(t - t^{(0)})\|_{\bar{F}(\kappa,0)} \leq 2\|w\|_{H^{(\kappa)}}$ for $t^{(0)} \in (0, T)$ $(\kappa \geq -1)$. If $n = 1$, then $\|\delta(x - x^{(0)})\|_{H^{<-1/2>}} \leq c_1$, $\|\delta(x - x^{(0)}, t - t^{(0)})\|_{\bar{F}^{<-1/2,0>}} \leq 2c_1$. Also, $\|z(t)\delta(x - \mu(t))\|_{F^{(-1/2,0)}} \leq c_1\|z\|_{L_1(0,T)}$ for measurable $\mu : (0, T) \to \bar{\Omega}$ (the constant c_1 in all the inequalities is one and the same). For $n > 1$ the last estimates are transferred on the δ functions concentrated in x on some $n - 1$-dimensional sets in Ω.

Proposition 4.1. Let $0 < \theta < 1$, the numbers k_0, k_1, l_0, l_1 be integers, and $-1 \leq k_0 < k_1$, $-1 \leq l_0 < l_1$. The inequalities (imbeddings)

$$\|w\|_{(H^{k_0}, H^{k_1})_{\theta,\infty}} \leq c\|w\|_{H^{((1-\theta)k_0+\theta k_1)}} \tag{4.3}$$

$$\|f\|_{(F^{(k_0,l_1)}, F^{(k_1,l_1)})_{\theta,\infty}} \leq c\|f\|_{F^{((1-\theta)k_0+\theta k_1,l_1)}} \tag{4.4}$$

$$\|f\|_{(F^{<-1,l>}, F^{<0,l>})_{\theta,\infty}} \leq c\|f\|_{F^{<\theta-1,l>}}, \qquad l = 0, 1 \tag{4.5}$$

$$\|f\|_{(F^{(\kappa,l_0)}, F^{(\kappa,l_1)})_{\theta,\infty}} \leq c\|f\|_{F^{(\kappa,l_\theta)}}, \qquad \kappa \geq 0 \tag{4.6}$$

are valid, with $l_\theta = (1 - \theta)l_0 + \theta l_1$. The constant c in inequalities (4.3)–(4.6) does not depend on θ.

Remark 4.1. For the integer l_θ and $0 < \theta \leq 1$ in Equation (4.6) it is possible to strengthen $F^{(\kappa,l_\theta)}$ to $\bar{F}^{(\kappa,l_\theta)}$.

The proof of Proposition 4.1 is beyond the scope of this paper.

4.2. FRACTIONAL-ORDER ERROR ESTIMATES

Let us formulate and prove fractional-order error estimates.

Theorem 4.1. Let $v_0 = v^{(0)}$ and $|\sigma^{(0)}|\tau^2 \leq N(\beta_h^2 + \tau^2)$ (see Equation [2.7]). The error estimate,

$$\|u - v\|_{C_\tau(H^0)} + \|s_1 I_t u - s_t I_t v\|_{C(H^1)} \leq c\big(T_1(\beta_h^2 + \tau^2)\big)^{\alpha/3}\|\mathbf{d}\|_\alpha$$

$$\equiv c\big(T_1(\beta_h^2 + \tau^2)\big)^{\alpha/3}\Big(\big\|u^{(0)}\big\|_{H^{(\alpha)}}$$

$$+ \big\|u^{(1)}\big\|_{H^{(\alpha-1)}} + \|f\|_{F^{(\alpha_1,\alpha_2)}}\Big), \tag{4.7}$$

is valid ($0 \leq \alpha \leq 3$). The numbers α_1, α_2 are such that $\alpha_1 + \alpha_2 = \alpha - 1$ and any one of the following conditions is fulfilled: (a) $\alpha_1 = 0$, $-1 \leq \alpha_2 \leq 2$; (b) $0 < \alpha_1 \leq 1$, $0 < \alpha_2 \leq 1$; (c) $-1 \leq \alpha_1 \leq 2$, $\alpha_2 = 0$ (see Figure 1).

Theorem 4.2. Let $v_0 = s_1 u^{(0)}$. The error estimate

$$\left\| \bar{\partial}_t (u - v) \right\|_{C_\tau(H^0)} + \left\| (s_1 u - v)^{(1/2)} \right\|_{C_\tau(H^1)}$$

$$\leq c \left(T_1 (\beta_h^2 + \tau^2) \right)^{(\alpha-1)/3} \times$$

$$\left(\left\| u^{(0)} \right\|_{H^{(\alpha)}} + \left\| u^{(1)} \right\|_{H^{(\alpha-1)}} + \left\| f \right\|_{F^{(\alpha_1, \alpha_2)}} \right) \quad (4.8)$$

is valid ($1 \leq \alpha \leq 4$). The numbers α_1, α_2 are such that $\alpha_1 + \alpha_2 = \alpha - 1$, and any one of the following conditions is fulfilled: (a) $-1 \leq \alpha_1 < 0$, $\alpha_2 = 1$; (b) $\alpha_1 = 0$, $0 \leq \alpha_2 \leq 3$; (c) $0 < \alpha_1 \leq 1$, $0 < \alpha_2 \leq 2$; (d) $1 < \alpha_1 \leq 2$, $0 < \alpha_2 \leq 1$; (e) $0 < \alpha_1 \leq 3$, $\alpha_2 = 0$ (see Figure 2). Under condition (2.22) we can omit the operator $(\cdot)^{(1/2)}$ in estimate (4.8).

Proof. To prove Theorem 4.1, use Proposition 1.2 and Theorem 2.1, item (1). Taking into account properties of the operators s_1 and s_t, we have

$$\| u - v \|_{C(H^0)} + \| s_1 I_t u - s_t I_t v \|_{C(H^1)} \leq \| u \|_{C(H^0)}$$

$$+ \| I_t u \|_{C(H^1)} + \varepsilon_0^{-1} \| v \|_{C_\tau(H^0_\tau)} + \| I_t v \|_{C_\tau(H^1)}$$

$$\leq c \left(\left\| u^{(0)} \right\|_{H^0} + \left\| u^{(1)} \right\|_{H^{-1}} + \| f \|_{F^{(k,-1-k)}} \right),$$

$$k = -1, 0. \quad (4.9)$$

A priori error estimate (3.4) (and Lemma 3.1, item (1)) and Proposition 1.3 yield the estimate

$$\| u - v \|_{C_\tau(H^0)} + \| s_1 I_t u - s_t I_t v \|_{C(H^1)} \leq c T_1 (\beta_h^2 + \tau^2) \times$$

$$\left(\left\| u^{(0)} \right\|_{H^3} + \left\| u^{(1)} \right\|_{H^2} + \| f \|_{F^{k,2-k}} \right), \quad k = 0, 1, 2. \quad (4.10)$$

For $\alpha = 0$, error estimate (4.7) immediately follows from estimate (4.9), and for $\alpha = 3$ it follows from estimate (4.10). For $0 < \alpha < 3$ we will derive estimate (4.7) from the same one for $\alpha = 0, 3$, with the help of the interpolation theory of linear operators.

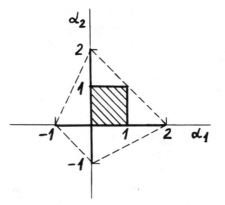

FIGURE 1. The set of indices α_1, α_2 from Theorem 4.1.

In the case of $u^{(1)} = 0$, $f = 0$, it is enough to consider the operators

$$R_1 u^{(0)} = u - v : \ H^\alpha \to C_\tau(H^0),$$

$$R_2 u^{(0)} = s_1 I_t u - s_t I_t v : \ H^\alpha \to C(H^1),$$

with $\alpha = 0, 3$ and to utilize interpolation inequality (4.1) and inequality (4.3). The case $u^{(0)} = 0$, $f = 0$ is considered in the same way.

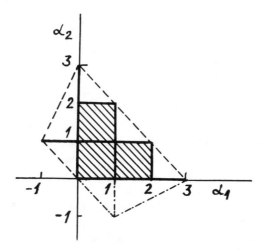

FIGURE 2. The set of indices α_1, α_2 from Theorem 4.2.

Now let $u^{(0)} = 0$, $u^{(1)} = 0$. We will derive estimate (4.7) gradually again with the help of interpolation inequality (4.1). If $\alpha_1 = 0$, $-1 < \alpha_2 < 2$, then we need to use estimate (4.9) for $k = 0$, estimate (4.10) for $k = 0$, and inequality (4.6) for $\kappa = 0$, $l_0 = -1$, $l_1 = 2$. If $-1 < \alpha_1 < 2$, $\alpha_2 = 0$, we can use estimate (4.9) for $k = -1$, estimate (4.10) for $k = 2$, and inequality (4.4) for $k_0 = -1$, $k_1 = 2$, $l_1 = 0$.

For $0 < \alpha_1 < 1$, $\alpha_2 = 1$, estimate (4.7) follows from the same one already proved for $\alpha_1 = 0$, $\alpha_2 = 1$, estimate (4.10) for $k = 1$, and inequality (4.4) for $k_0 = 0$, $k_1 = 1$, $l_1 = 1$.

Finally, for $0 < \alpha_1 \leq 1$, $0 < \alpha_2 < 1$, estimate (4.7) follows from the same one for $\alpha_2 = 0, 1$ and inequality (4.6) for $\kappa = \alpha_1$, $l_0 = 0$, $l_1 = 1$. Theorem 4.1 is proved.

Proof. To prove Theorem 4.2, using Proposition 1.1 and Theorem 2.1, item (2) we have

$$\|\bar{\partial}_t(u - v)\|_{C_\tau(H^0)} + \|(s_1 u - v)^{(1/2)}\|_{C_\tau(H^1)} \leq \|D_t u\|_{L_\infty(H^0)}$$

$$+ \|u\|_{C(H^1)} + 2\varepsilon_0^{-1}\|v\|_{C_\tau(H_E)} \leq c\Big(\|u^{(0)}\|_{H^1}$$

$$+ \|u^{(1)}\|_{H^0} + \|f\|_{F^{(k,-k)}}\Big), \qquad k = -1, 0. \qquad (4.11)$$

A priori error estimate (3.5) and Proposition 1.3 yield the estimate

$$\|\bar{\partial}_t(u - v)\|_{C_\tau(H^0)} + \|(s_1 u - v)^{(1/2)}\|_{C_\tau(H^1)} \leq cT_1\big(\beta_h^2 + \tau^2\big) \times$$

$$\Big(\|u^{(0)}\|_{H^4} + \|u^{(1)}\|_{H^3} + \|f\|_{F^{(k,3-k)}}\Big), \qquad k = 0, 1, 2, 3.$$

$$(4.12)$$

For $\alpha = 1$, error estimate (4.8) results from estimate (4.11), and for $\alpha = 4$ it results from estimate (4.12).

Estimate (4.8) for $1 < \alpha < 4$ is derived from the indicated estimates with the help of interpolation theory similar to the method used in Theorem 4.1. The case $u^{(1)} = 0$, $f = 0$ (or $u^{(0)} = 0$, $f = 0$) is considered almost in the same way as in Theorem 4.2. In the case of $u^{(0)} = 0$, $u^{(1)} = 0$, estimate (4.8) can be derived with the help of inequalities (4.11) and (4.12), for example, in the following order: (1) for $\alpha_1 = 0$, $0 < \alpha_2 < 3$; (2) for $\alpha_2 = 0$, $0 < \alpha_1 < 3$; for $\alpha_2 = 1$, $-1 < \alpha_1 < 2$, and for $\alpha_2 = 2$, $0 < \alpha_1 < 1$; (3) for $0 < \alpha_2 < 1$, $0 < \alpha_1 \leq 2$; and (4) for $1 < \alpha_2 < 2$, $0 < \alpha_1 \leq 1$. Under condition (2.22) we can omit the operator $(\cdot)^{(1/2)}$ in estimates (4.11) and (4.12), and therefore in estimate (4.8). Theorem 4.2 is proved.

In Figures 1 and 2 we see sets of the smoothness indices α_1 and α_2 from Theorems 4.1 and 4.2, respectively. These sets, J_k, are the thick segments and the shaded squares ($k = 0$ for Theorem 4.1, $k = 1$ for Theorem 4.2).

Theorems 4.1 and 4.2 can be extended to the case of interpolation spaces (seemingly of the $F^{(\alpha_1,\alpha_2)}$ type) with the indices α_1, α_2 from broader sets, namely from the convex hulls of J_k. In order to do that it is necessary to again apply function space interpolation theory.

The complement of the set J_k up to its convex hull consists of several right triangles (see Figures 1 and 2), with their legs belonging to J_k. Therefore, to reduce the case of indices from every such triangle to the already considered cases of indices from the set J_k, it is convenient to perform the interpolation with respect to α_1, α_2 in the direction of the triangle hypotenuse from one of its legs to the other.

As to the triangles with hypotenuses lying on straight lines $\alpha_1 + \alpha_2 =$ const , here formally we may not apply interpolation theory. This case reduces to justification of decomposition of the elements f belonging to the spaces with the indices α_1, α_2 from those triangles into the sum of the form

$$f = f_1 + f_2, \qquad f_1 \in F^{(\bar\alpha_1, \alpha_1 - \bar\alpha_1 + \alpha_2)}, \quad f_2 \in F^{(\alpha_1 + \alpha_2 - \bar\alpha_2, \bar\alpha_2)}$$

where $\bar\alpha_i = [\alpha_i](i = 1, 2)$ for $\alpha_1 - [\alpha_1] + \alpha_2 - [\alpha_2] \leq 1$ or $\bar\alpha_i = [\alpha_i] + 1 (i = 1, 2)$ for $\alpha_1 - [\alpha_1] + \alpha_2 - [\alpha_2] > 1$. This approach was realized for proving the results of the paper[41]. Therein, for the one-dimensional case ($n = 1$) the sets of indices $\alpha_1, \alpha_2 \in [0, 2 + k)$ such that $\alpha_1 + \alpha_2 < 2 + k$ were covered. In connection with these questions see reference 4 and Chapters 5 and 6 in reference 30; note that in the present paper some moments complicate the situation in comparison with reference 4.

Due to estimate (1.21) and Theorem 2.2 (and Corollary 3.1) Theorem 4.2 can be extended also to the case of indices α_1, α_2 from the triangle with the vertices $(0,0)$, $(0,3)$, $(1,-1)$ (see Figure 2).

Below we need the inequalities

$$\|w\|_{\bar{B}_{\theta,\infty}} \leq \|w\|_{(\bar{B}_{\theta_0,\infty}, \bar{B}_{\theta_1,\infty})_{\eta,\infty}} \tag{4.13}$$

$$\|w\|_{(X_0, X_1)_{\eta,\infty}} \leq c_0 \kappa \{(\theta - \theta_0)(\theta_1 - \theta)\}^{-1} \|w\|_{\bar{B}_{\theta,\infty}} \tag{4.14}$$

where $0 \leq \theta_0 < \theta < \theta_1 \leq 1$, $\eta = (\theta - \theta_0)/(\theta_1 - \theta_0)$, and X_k is an intermediate Banach space with respect to the pair (B_0, B_1) such that $\|w\|_{X_k} \leq \kappa \|w\|_{B_0}^{1-\theta_k} \|w\|_{B_1}^{\theta_k}$, ($k = 0, 1$). It is known that the space $X_k = \bar{B}_{\theta_k, q}$ satisfies the last inequality, with $\kappa = \kappa(\theta_k, q)$ for all $1 \leq q \leq \infty$.

Besides that, the inequality

$$\|w\|_{(B_0,\bar{B}_{\theta_1},q)_{\theta/\theta_1},\infty} \le c_0(\min\{1-\theta,\ \theta_1-\theta\})^{-1/q}\|w\|_{\bar{B}_\theta,\infty} \qquad (4.15)$$

is valid, where $0 < \theta < \theta_1 < 1$ for $1 \le q < \infty$ or $0 < \theta < \theta_1 \le 1$ for $q = \infty$ (in the latter case the multiplier depending on q should be omitted). All the enumerated inequalities follow from the proof of the reiteration theorem (see Section 3.5 in reference 11); to derive Equation (4.15) we also need Holmstedt's formula (see reference 11, Section 3.6).

Starting from Theorems 4.1 and 4.2 we can obtain additional results.

Theorem 4.3.

(1) Theorem 4.1 remains valid for one or several of these operations:

    (a) for $0 \le \alpha \le 3/2$ add the summand $\|I_t u - s_1 I_t v\|_{C(H^1)}$ into the left-hand part of error estimate (4.7);

    (b) weaken the norm $\|f\|_{F^{(\alpha_1,\alpha_2)}}$ to $\|f\|_{\bar{F}}^{(\alpha_1,\alpha_2)}$ for integer $\alpha_2 \ge 0$;

    (c) for $0 \le \alpha \le 1$, $\alpha_2 = 0$ replace

$$\|u^{(1)}\|_{H^{(\alpha-1)}} + \|f\|_{F^{(\alpha_1,\alpha_2)}}$$

    by $\|u_\rho^{(1)}\|_{H^{<\alpha-1>}} + \|f_\rho\|_{\bar{F}^{<\alpha_1,\alpha_2>}}.$

    (for the definition of the operator $(\cdot)_\rho$ see Section 1.1);

    (d) for $\alpha_1 \le 0$ or $\alpha_2 \ge 1$, strengthen $\|u - v\|_{C_\tau(H^0)}$ to $\|u - v\|_{C(H^0)}$;

    (e) for $1 \le \alpha \le 3$, first choose $v_0 = s_1 u^{(0)}$; second, for $v_0 = v^{(0)}$ and under one of conditions (3.7), or for $v_0 = s_1 u^{(0)}$, add the summand $\|s_1 u - v\|_{C_\tau(H^0)}$ into the left-hand part of error estimate (4.7).

(2) Theorem 4.2 remains valid for one or several of the following operations:

    (a) for $1 \le \alpha \le 5/2$, add the summand

$$\|(u - v)^{(1/2)}\|_{C_\tau(H^1)}$$

    into the left-hand part of error estimate (4.8);

    (b) weaken the norm $\|f\|_{F^{(\alpha_1,\alpha_2)}}$ to $\|f\|_{\bar{F}^{(\alpha_1,\alpha_2)}}$ for integer $\alpha_2 \ge 0$;

    (c) for $1 \le \alpha \le 2$, $\alpha_2 = 1$, replace $\|f\|_{F^{(\alpha_1,\alpha_2)}}$ by

$$\|f_\rho\|_{\bar{F}^{<\alpha_1,\alpha_2>}};$$

(d) for $\alpha_1 \leq 0$ or $\alpha_2 \geq 1$ under condition (2.22), strengthen $\|(\cdot)^{(1/2)}\|_{C_\tau(H^1)}$ to $\|\cdot\|_{C(H^1)}$;

(e) for $1 \leq \alpha \leq 5/2$ and under all conditions of Lemma 3.1, item (3), choose $v_0 = v^{(0)}$ (the condition $\sigma^{(0)} \geq \sigma - 1/4$ is not required).

Proof.

(1a) The estimate $\|I_t u - s_1 I_t u\|_{C(H^1)} \leq \beta_h^\alpha \|I_t u\|_{C(H^{1+\alpha})}$ ($\alpha = 0, 1$) follows from the properties of s_1. Majorizing the norms of $I_t u$ by data norms according to Propositions 1.2 and 1.3 (with $k = l = 0$), we obtain the estimate

$$\|I_t u - s_1 I_t u\|_{C(H^1)}$$
$$\leq c\beta_h^\alpha \left(\|u^{(0)}\|_{H^{(\alpha)}} + \|u^{(1)}\|_{H^{(\alpha-1)}} + \|f\|_{F^{(\alpha_1,\alpha_2)}} \right) \quad (4.16)$$

for $\alpha = 0, 1$ and for $(\alpha_1, \alpha_2) = (-1, 0), (0, -1), (0, 0)$. Applying interpolation inequality (4.1) and Proposition 4.1, we set Equation (4.16) also for $0 < \alpha < 1$ and for the same (α_1, α_2) as in Theorem 4.1.

The right-hand part of Equation (4.16) for $0 \leq \alpha \leq 1$ can be majorized by the right-hand part of Equation (4.7). In addition, the right-hand part of Equation (4.16) for $\alpha = 1$ can be majorized (with the help of inequality [4.2]) by the right-hand part of estimate (4.7) for any $1 < \alpha \leq 1/5$; by that, item (1a) is proved.

(b) For $\alpha_1 \geq 0$, $\alpha_1 \geq 1$, it is possible to extend $F^{(\alpha_1,\alpha_2)}$ to $\bar{F}^{(\alpha_1,\alpha_2)}$ due to Remark 4.1. In the general case we can approximate the function f by its averages with respect to t and perform limit transition relying on Remarks 1.1 and 2.2.

(c) Let $\alpha_2 = 0$. The possibility of replacing $\|u^{(1)}\|_{H^{(\alpha-1)}} + \|f\|_{F^{(\alpha_1,\alpha_2)}}$ by $\|u_\rho^{(1)}\|_{F^{<\alpha-1>}} + \|f_\rho\|_{F^{<\alpha_1,\alpha_2>}}$ is obvious for $\alpha_1 = -1, 0$ (see Section 1), and then for $-1 < \alpha_1 < 0$ it follows from interpolation theory. Finally, it is possible to extend $F^{<\alpha_1,\alpha_2>}$ to $\bar{F}^{<\alpha_1,\alpha_2>}$ as in item (b).

(d) A priori error estimate (3.4) remains valid if in its left-hand part we replace $C_\tau(H^0)$ by $C(H^0)$, whereas in the right-hand part we add the summand, $\|D_t^2 u\|_{L_\infty(H^0)}$, to the norms of u. But then according to the proof of Theorem 4.1, for $\alpha_1 \leq 0$ or $\alpha_2 \geq 1$ we can also replace $C_\tau(H^0)$ by $C(H^0)$ in estimate (4.7).

This replacement is also possible for the rest of α_1, α_2, but at the expense of the norm of f strengthening (see

Remark 1.2). For example, for $\alpha_1 > 0, \alpha_2 = 0$, it is enough to add the summand $\|f\|_{L_\infty(H^0)}$ to the norm of f in estimate (4.7).

(e) According to the proof of Theorem 4.1 it is enough to consider the cases $\alpha = 1, 3$; moreover, for $\alpha = 3$ it is enough to use error estimate (3.4) to a greater extent. For $\alpha = 1$ it is necessary to notice, first, that when $v_0 = v^{(0)}$ is replaced by $v_0 = s_1 u^{(0)}$, the corresponding change of v in the norm $\| \cdot \|_{C_\tau(H^0_\rho)} + \|I_t \cdot \|_{C_\tau(H^1)}$ does not exceed the quantity $c(\beta_h + \tau)\|u^{(0)}\|_{H^1}$ (which follows from Theorem 2.1, item (1), and Lemma 3.1, item (2)). Second, the estimate

$$\|u - s_1 u\|_{C(H^0)} \le \beta_h \|u\|_{C(H^1)}$$

$$\le c\beta_h \left(\left\|u^{(0)}\right\|_{H^1} + \left\|u^{(1)}\right\|_{H^0} + \|f\|_{L_1(H^0)} \right)$$

is valid (see estimate [3.2] and Proposition 1.1).

(2a) In a manner similar to estimate (4.16), we can derive the estimate

$$\|u - s_1 u\|_{C(H^1)} \le c\beta_h^{\alpha-1} \times$$

$$\left(\left\|u^{(0)}\right\|_{H^{(\alpha)}} + \left\|u^{(1)}\right\|_{H^{(\alpha-1)}} + \|f\|_{F^{(\alpha_1,\alpha_2)}} \right) \quad (4.17)$$

$(1 \le \alpha \le 2)$ where $\alpha_1 + \alpha_2 = \alpha - 1$ and any one of these conditions is fulfilled: (i) $-1 \le \alpha_1 \le 0, \alpha_2 = 0$; (ii) $\alpha_1 = 0, 0 \le \alpha_2 \le 1$; (iii) $0 \le \alpha_1 \le 1, \alpha_2 = 0$. From the indicated estimate we obtain item (2a) (which is like item (1a)), excluding the case $\alpha_1, \alpha_2 \in (0, 1)$.

Consider this last case. Let $u^{(0)} = u^{(1)} = 0$, and t is any number in $[0, T]$. From Propositions 1.2 and 1.3, and with the help of interpolation inequality (4.1) and Proposition 4.1, we derive the estimate

$$\|u(t)\|_{(H^1,H^3)_{\theta,\infty}} \le c\|f\|_{F^{(2\theta-l,l)}} \quad \text{for } 0 < \theta < 1, l = 0, 1.$$

Hence, with the help of Equation (4.13) we have

$$\|u(t)\|_{(H^1,H^3)_{(\alpha-1)/2,\infty}} \le c\|f\|_{F^{(\alpha_1,\alpha_2)}} \quad (4.18)$$

for $1 < \alpha < 3$ and $\alpha_1, \alpha_2 \in (0, 1)$.

The space $(H^1, H^3)_{\theta,\infty}$ is imbedded in H^2 for $1/2 < \theta < 1$; thus, according to Equation (4.14) we obtain

$$\|w\|_{H^{(\gamma)}} \le c_\theta \|w\|_{(H^1,H^3)_{\theta(\gamma-1),\infty}}, \quad 1 \le \gamma \le 2. \quad (4.19)$$

Using the estimate $\|u(t) - s_1 u(t)\|_{H^1} \leq \beta_h^{\gamma-1} \|u(t)\|_{H(\gamma)}$ for $1 \leq \gamma \leq 2$, choosing $\theta(\gamma - 1) = (\alpha - 1)/2$, $\theta = 3/4$, and successively applying inequalities (4.19) and (4.18), we derive the estimate

$$\|u - s_1 u\|_{C(H^1)} \leq c\beta_h^{2(\alpha-1)/3} \|f\|_{F^{(\alpha_1, \alpha_2)}}$$

for $1 < \alpha \leq 5/2$ and $\alpha_1, \alpha_2 \in (0, 1)$. Item 2(a) is proved. Items 2(b)–(e) can be proved similarly to items 1(b)–(e).

Let us discuss items (1a) and (2a) in the proved Theorem. It is naturally impossible to generalize item (1a) for $\alpha > 1/5$ or item (2a) for $\alpha > 5/2$ only under condition (3.1). But even in this situation we can obtain additional results for some concrete spaces S_1. In order do that, in the error estimates we should pass over from the H^1 norm to its mesh counterpart, and use corresponding results on superconvergence for the elliptic case[31,25,45,46] and others. For $n = 1$ this will be done in Section 5. The case of $n = 2, 3$ we will consider at the end of Section 7.

5. THE $W_{2,h}^1$ AND C_h ERROR ESTIMATES FOR THE ONE-DIMENSIONAL CASE

5.1. THE $W_{2,h}^1$ ERROR ESTIMATES

Let us consider the one-dimensional case ($n = 1$) when $\Omega = (0, X)$ and $Lw = -D(a_{11}Dw) + aw$. Introduce the nonuniform mesh $\bar{\omega}^h$ formed by the nodes $0 = x^{(0)} < x^{(1)} < \ldots < x^{(n_1)} = X$, with steps $h_k = x^{(k)} - x^{(k-1)}$. Let $h = (h_1, \ldots, h_{n_1})$ and $h_{\max} = \max_{1 \leq k \leq n_1} h_k$, $h_{\min} = \min_{1 \leq k \leq n_1} h_k$.

Introduce the space $S_{1,h,\tilde{L}}$ which consists of the functions φ continuous on $[0, X]$, equal zero for $x = 0$ and X, and satisfying the equation

$$\tilde{L}\varphi(x) = 0, \qquad \tilde{L}\varphi \equiv -D(\tilde{a}_{11}D\varphi) + \tilde{a}\varphi \tag{5.1}$$

on each interval $\Omega_k = (x^{(k-1)}, x^k)$ $(1 \leq k \leq n)$. Consider simultaneously the following three variants: (A) $\tilde{L} = L$; (B) $\tilde{L}(\cdot) = -D(a_{11}D(\cdot))$; (C) $\tilde{L} = -D \cdot D$. In variant (C) we assume that $\|Da_{11}\|_{L_\infty(\Omega)} \leq N$ (more generally, we assume that $\|Da_{11}\|_{L_\infty(\bar{x}^{(k-1)}, \bar{x}^{(k)})} \leq N$ for $1 \leq k \leq \bar{n}_1$, where $0 = \bar{x}^{(0)} < \bar{x}^{(1)} < \ldots < \bar{x}^{(n_1)} = X$ and $\bar{x}^{(k)} \in \bar{\omega}^h$; thereby a_{11} may have points of discontinuity but they should belong to the mesh).

Variant (C) is the most standard when $S_{1,h,\tilde{L}} = \hat{S}_{1,h}$ is the space of functions continuous on $[0, X]$, equal zero for $x = 0, X$, and linear on each interval $(x^{(k-1)}, x^k)$ $(1 \leq k \leq n_1)$. For the choice of $S_{1,h} = S_{1,h,\tilde{L}}$

the approximation of each summand in the left-hand part of Equation (1.3), as well as the approximation of its right-hand part, depends on the coefficients of the operator, $\tilde{L}$. In variant (A) these approximations are very cumbersome, but in variant (B) they are already visible. Let us emphasize that in variants (A) and (B) the coefficients ρ, a_{11}, a satisfy only rather weak conditions indicated at the beginning of Section 1.1, that is, in essence it is enough for them to be measurable and bounded. Variant (B) is studied in detail for stationary problems in reference 1 and is considered for parabolic problems in references 40 and 43.

For a function φ defined on $\bar{\omega}^h$ and equal to zero for $x = 0$ and X, first let us denote by $\tilde{s}_1 \varphi \in S_{1,h,\tilde{L}}$ the function with the property $(\tilde{s}_1 \varphi)(x) = \varphi(x)$ for $x \in \bar{\omega}^h$; for brevity let $s = \tilde{s}_1$ in variant (C). Second, let us introduce the norms

$$\|\varphi\|_{H_h^{<1>}} = \left(\sum_{1 \le k \le n_1} (\delta\varphi)^2_{k-1/2} h_k \right)^{1/2}, \qquad \|\varphi\|_{C_h} = \max_{x \in \bar{\omega}_h} |\varphi(x)|$$

where $\delta\varphi_{k-1/2} = (\varphi_k - \varphi_{k-1})/h_k$.

Let the space $\tilde{H}^1$ and the bilinear form $\tilde{\mathcal{L}}_\Omega(\cdot, \cdot)$ result from H^1 and $\mathcal{L}_\Omega(\cdot, \cdot)$ when the coefficients of the operator L are replaced by the corresponding coefficients of $\tilde{L}$.

Lemma 5.1. In variants (A), (B), and (C), operator $\tilde{s}_1$ is a projector in $\tilde{H}^1$ onto $S_{1,h,\tilde{L}}$, and the estimates

$$\|w - s_1 w\|_{H^1} \le \|w - \tilde{s}_1 w\|_{H^1} \le c h_{\max}^{\lambda-1} \|w\|_{H^\lambda} \qquad (5.2)$$

$$\|w - s_1 w\|_{H^0} + \|w - \tilde{s}_1 w\|_{H^0} \le c h_{\max}^\lambda \|w\|_{H^\lambda} \qquad (5.3)$$

$$\|s_1 w - \tilde{s}_1 w\|_{H^1} \le c h_{\max}^\lambda \|w\|_{H^\lambda} \qquad (5.4)$$

$$c_1 \|\varphi\|_{H_h^{<1>}} \le \|\varphi\|_{H^1} \le c_2 \|\varphi\|_{H_h^{<1>}} \qquad \forall \varphi \in S_{1,h,\tilde{L}} \qquad (5.5)$$

are valid for $\lambda = 1, 2$.

Corollary 5.1. In variants (A), (B), and (C) we can set $\beta_h = c_1 h_{\max}$, and with such β_h, Theorems 4.1–4.3 are valid; besides that, in Theorem 4.3, item (1e), operator s_1 can be replaced by $\tilde{s}_1$.

Proof. Set $r = w - \tilde{s}_1 w$ for $w \in H^1$. Let $\varphi \in S_{1,h,\tilde{L}}$. Utilizing integration by parts, the property $r(x) = 0$ for $x \in \bar{\omega}^h$, and property (5.1), we have

$$\tilde{\mathcal{L}}_\Omega(r, \varphi) = \sum_{1 \le k \le n_1} \tilde{\mathcal{L}}_{\Omega_k}(r, \varphi) = \sum_{1 \le k \le n_1} (r, \tilde{L}\varphi)_{\Omega_k} = 0. \qquad (5.6)$$

Thus $\tilde{s}_1$ is a projector in $\tilde{H}^1$ onto $S_{1,h,\tilde{L}}$.

The property $r(x) = 0$ for $x \in \bar{\omega}^h$ implies the inequality $|r(x)| \le \|Dr\|_{L_1(\Omega_k)}$ for $x \in \Omega_k$ ($1 \le k \le n_1$). Hence, by virtue of Hölder's inequality and the equivalence of norms in $H^{<1>}$ and $\tilde{H}^1$ we obtain

$$\|r\|_\Omega \le ch_{\max}\|r\|_{H^1}. \tag{5.7}$$

Since $\tilde{s}_1$ is a projector, first, $\|r\|_{\tilde{H}^1} \le c\|w\|_{H^1}$. Second, using property (5.6) and the inequality $\|w\|_{H^1} \le c\|w\|_{H^2}$ and also the condition on Da_{11} in variant (C), we have

$$\|r\|_{\tilde{H}^1}^2 = \tilde{\mathcal{L}}_\Omega(r,r) = \tilde{\mathcal{L}}_\Omega(r,w) = \sum_{1\le k\le n_1}(r,\tilde{L}w)_{\Omega_k} \le c\|r\|_\Omega\|w\|_{H^2}. \tag{5.8}$$

Substituting inequality (5.8) into estimate (5.7), we obtain the estimate $\|r\|_\Omega \le ch_{\max}^2\|w\|_{H^2}$. Substituting this estimate into inequality (5.8), we obtain the estimate $\|r\|_{\tilde{H}^1} \le ch_{\max}\|w\|_{H^2}$. Utilizing inequality (3.2), we complete the proof of estimate (5.2) and (5.3).

To establish estimate (5.4) write the identity

$$\mathcal{L}_\Omega(s_1w - \tilde{s}_1w, \varphi) = \mathcal{L}_\Omega(r,\varphi). \tag{5.9}$$

In variant (B), according to identity (5.6) we have $\mathcal{L}_\Omega(r,\varphi) = (ar,\varphi)_\Omega$. In variant (C) notice that due to $D\varphi(x) = \delta\varphi_{k-1/2}$ for $x \in \Omega_k$ ($1 \le k \le n_1$) it follows that

$$(a_{11}Dr, D\varphi)_\Omega = \sum_{1\le k\le n_1}(a_{11}, Dr)_{\Omega_k}\delta\varphi_{k-1/2} = -\sum_{1\le k\le n_1}(Da_{11}, rD\varphi)_{\Omega_k}$$

(see references 45 and 46). Therefore, in all variants for $\varphi = s_1w - \tilde{s}_1w$, identity (5.9) yields the estimate $\|s_1w - \tilde{s}_1w\|_{H^1} \le c\|r\|_\Omega$ and, by virtue of estimate (5.3), also estimate (5.4).

Finally, the operators s and $\tilde{s}_1$ are projectors, the norms in $H^{<1>}$ and $\tilde{H}^1$ are equivalent, and $\varphi = \tilde{s}_1s\varphi$, so we have

$$\|s\varphi\|_{H^{<1>}} \le \|\varphi\|_{H^{<1>}} \le c\|\varphi\|_{\tilde{H}^1} \le c\|s\varphi\|_{\tilde{H}^1} \le c_1\|s\varphi\|_{H^{<1>}}.$$

Hence, by virtue of the equality $\|s\varphi\|_{H^{<1>}} = \|\varphi\|_{H_h^{<1>}}$, inequality (5.5) holds true.

Remark 5.1. In reference 41, for the general case the first summand in estimate (4) should be corrected by the corresponding

summand from estimate (4.8), and the modification of estimate
(4) indicated in reference 41, Remark 2, should be corrected by
the same one as given in Theorem 4.3, item (2a).

Theorem 5.1. Let any one of variants (A), (B), or (C) be fulfilled.

(1) If the summand $\|I_t(u - v)\|_{C_\tau(H_h^{<1>})}$ is added to the left-
hand part of estimate (4.7), and simultaneously for $\alpha_2 > 0$
the summand $ch_{max}^2\|f\|_{F^{(0,0)}}$ is added to its right-hand part,
Theorem 4.1 remains valid;

(2) If the summand $\|(u - v)^{(1/2)}\|_{C_\tau(H_h^{<1>})}$ is added to the left-
hand part of estimate (4.8), and simultaneously for $\alpha_2 > 1$
the summand $ch_{max}^2\|f\|_{F^{(0,1)}}$ is added to its right-hand part,
Theorem 4.2 remains valid. Also, Theorem 4.2 in its initial
form as well as in the indicated one is true for $v_0 = \tilde{s}_1 u^{(0)}$.

Proof.

(1) The estimate

$$\|s_1 I_t u - \tilde{s}_1 I_t u\|_{C(H^1)} \le ch_{max}^{1+\alpha} \times$$
$$\left(\left\|u^{(0)}\right\|_{H^{(\alpha)}} + \left\|u^{(1)}\right\|_{H^{(\alpha-1)}} + \|f\|_{F^{(\alpha_1,\alpha_2)}}\right) \quad (5.10)$$

is valid ($0 \le \alpha \le 1$) where α_1, α_2 satisfy any one of the
conditions: (a) $\alpha_1 = \alpha - 1$, $\alpha_2 = 0$; (b) $\alpha_1 = 0$, $\alpha_2 = \alpha - 1$.
For $\alpha = 0, 1$ this estimate follows from estimate (5.4)
and Propositions 1.2 and 1.3 ($k = l = 0$). Then for
$0 < \alpha < 1$, estimate (5.10) is easily obtained with the
help of interpolation theory for linear operators.

Since according to estimate (5.5)

$$\|I_t(u - v)\|_{C_\tau(H_h^{<1>})} \le c\left(\|s_1 I_t u - I_t v\|_{C_\tau(H^1)}\right.$$
$$\left. + \|s_1 I_t u - \tilde{s}_1 I_t u\|_{C_\tau(H^1)}\right), \quad (5.11)$$

item (1) follows from Theorem 4.1 and estimate (5.10).

(2) Similar to estimate (5.10), the estimate

$$\|s_1 u - \tilde{s}_1 u\|_{C(H^1)} \le ch_{max}^\alpha\left(\left\|u^{(0)}\right\|_{H^{(\alpha)}} + \left\|u^{(1)}\right\|_{H^{(\alpha-1)}}\right.$$
$$\left. + \|f\|_{F^{(\alpha_1,\alpha_2)}}\right) \quad (5.12)$$

is valid ($1 \le \alpha \le 2$) where α_1, α_2 satisfy any one of the
conditions: (a) $\alpha_1 = \alpha - 2$, $\alpha_2 = 1$; (b) $\alpha_1 = 0$, $\alpha_2 = \alpha - 1$;
(c) $\alpha_1 = \alpha - 1$, $\alpha_2 = 0$.

Estimate (5.11) remains valid if we omit all operators I_t. Thus, item (2) follows from Theorem 4.2 and estimate (5.12) (for $0 \leq \alpha_2 \leq 1$, $\alpha_1 \geq 0$ it is necessary to take into account that for integer α_1, α_2 the right-hand part of estimate (5.12) is majorized by the right-hand part of estimate (4.8); then apply interpolation theory).

The substitution of $v_0 = \tilde{s}_1 u^{(0)}$ for $v_0 = s_1 u^{(0)}$ is possible. The corresponding change of the function, v, in the norm $\| \cdot \|_{C_\tau(H_E)}$ would not exceed the quantity $\|s_1 u^{(0)} - \tilde{s}_1 u^{(0)}\|_{H^1} \leq c h_{max}^\alpha \|u^{(0)}\|_{H^{(\alpha)}}$ $(1 \leq \alpha \leq 2)$, according to Theorem 2.1, item (2), and estimate (5.4); for $1 < \alpha < 2$, interpolation theory is used.

5.2.   THE C_h ERROR ESTIMATES

Let us introduce the interpolation spaces $H^{(k+0.5);1} = (H^k, H^{k+1})_{0.5;1}$, $F^{(k+0.5,l);1} = (F^{(k,l)}, F^{(k+1,l)})_{0.5;1}$, $F^{(k,l+0.5);1} = (F^{(k,l)}, F^{(k,l+1)})_{0.5;1}$ for integers $k \geq -1$, $l \geq -1$.

Theorem 5.2.   In variants (A), (B), and (C) for $v_0 = \tilde{s}_1 u^{(0)}$ the error estimate

$$\left\|(u-v)^{(1/2)}\right\|_{C_\tau(C_h)} \leq c T_1 \left(h_{max}^2 + \tau^2\right) \times$$

$$\left(\left\|u^{(0)}\right\|_{H^{(7/2);1}} + \left\|u^{(1)}\right\|_{H^{(5/2);1}} + T_1^\gamma \|f\|_{F^{(\alpha_1,\alpha_2);1}}\right) \qquad (5.13)$$

is valid. The numbers α_1, α_2 are such that $\alpha_1 + \alpha_2 = 5/2$, $\alpha_1 \geq 0$, $\alpha_2 \geq 0$, and one of α_1, α_2 is an integer (the corresponding points are marked with light circles on Figure 3); $\gamma = 1/2$ for $\alpha_2 = 5/2$, otherwise $\gamma = 0$. Under condition (2.22) we can omit operator $(\cdot)^{(1/2)}$ in Equation (5.13).

Proof.   For $\Omega = (0, X)$ the inequality

$$\|w\|_{C(\bar{\Omega})} \leq c\|w\|_{L_2(\Omega)}^{1/2} \|w\|_{W_2^1(\Omega)}^{1/2}. \qquad (5.14)$$

is well known. This yields (see reference 11, Section 3.5) the imbedding,

$$\|w\|_{C(\bar{\Omega})} \leq c_1 \|w\|_{H^{(1/2);1}}. \qquad (5.15)$$

Note that $\|(u - v)^{(1/2)}\|_{C_h} \leq \|(\tilde{s}_1 u - v)^{(1/2)}\|_{C(\bar{\Omega})}$. By virtue of interpolation inequality (4.1), with $(C_0, C_1)_{\theta,q} = H^{(1/2);1}$,

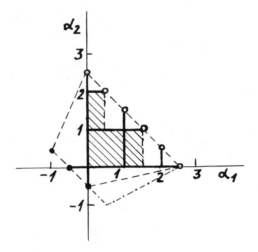

FIGURE 3. The set of indices α_1, α_2 from Theorems 5.2 and 5.3.

estimate (5.13) follows from, first, Theorems 4.1 and 4.3, item (1e), for $\alpha = 3$ (see Corollary 5.1) and second, Theorem 5.1, item (2) for $\alpha = 4$ and from inequality (4.2).

Theorem 5.3. Consider variants (A), (B), and (C). Let $v_0 = v^{(0)}$ (and then $\sigma^{(0)} \geq 0$, $h_{\max} \leq Nh_{\min}$) for $1/2 < \alpha \leq 1$ or $v_0 = \tilde{s}_1 u^{(0)}$ for $1 \leq \alpha < 7/2$. The error estimate

$$\left\| (u - v)^{(1/2)} \right\|_{C_\tau(C_h)} \leq c_{\alpha_1, \alpha_2} \left(T_1 \left(h_{\max}^2 + \tau^2 \right) \right)^{(\alpha - 1/2)/3}$$

$$\left(\left\| u^{(0)} \right\|_{H^{(\alpha)}} + \left\| u^{(1)} \right\|_{H^{(\alpha-1)}} + T_1^\gamma \| f \|_{F^{(\alpha_1, \alpha_2)}} \right) \qquad (5.16)$$

is valid ($1/2 < \alpha < 7/2$). The numbers α_1, α_2 are such that $\alpha_1 + \alpha_2 = \alpha - 1$ and any one of the following conditions is fulfilled: (a) $-1/2 < \alpha_1 < 5/2$, $\alpha_2 = 0$; (b) $\alpha_1 = 0$, $-1/2 < \alpha_2 < 5/2$; (c) $0 < \alpha_1 < 1/2$, $0 < \alpha_2 \leq 2$; (d) $1/2 \leq \alpha_1 < 3/2$, $0 < \alpha_2 \leq 1$; (e) $\alpha_1 = 1$, $1 < \alpha_2 < 3/2$; (f) $\alpha_1 = 2$, $0 < \alpha_2 < 1/2$. In addition, $\gamma = 0$ in the cases (a) and (d)–(f), $\gamma = \max\{(\alpha_2 - 3/2)/2, 0\}$ in the (b) case, $\gamma = \max\{(1/2 - \alpha_1)(\alpha_2 - 1)/2, 0\}$ in the (c) case. Under condition (2.22) we can omit operator $(\cdot)^{(1/2)}$ in estimate (5.16).

Proof. In a manner similar to imbedding (5.15), the imbedding

$$\|w\|_{C(\bar{Q})} \leq c_1 \|w\|_{(C(H^0), C(H^1))_{1/2, 1}}$$

is valid. By virtue of the interpolation inequality (4.1) from

Propositions 1.1, 1.2, and also from Theorem 2.1, items (1) and (2), it follows that

$$\left\|(u-v)^{(1/2)}\right\|_{C_\tau(C(\bar\Omega))} \leq \|u\|_{C(\bar Q)} + \left\|v^{(1/2)}\right\|_{C_\tau(C(\bar\Omega))}$$

$$\leq c\left(\left\|u^{(0)}\right\|_{H^{(1/2);1}} + \left\|u^{(1)}\right\|_{H^{(-1/2);1}} + \|f\|_{F^{(\alpha_1,\alpha_2);1}}\right) \quad (5.17)$$

where $(\alpha_1,\alpha_2) = (-1,1/2), (-1/2,0), (0,-1/2)$ (marked with dark circles on Figure 3). In Equation (5.17) it is assumed that $v_0 = v^{(0)}$ and $\sigma^{(0)} \geq 0$, $h_{max} \leq N h_{min}$ (see item (3) of Lemma 3.1).

By virtue of multiplicative inequality (5.14) from Theorem 4.1 and Theorems 4.2 and 4.3, item (2a) (see also Corollary 5.1) we obtain the estimate (with $1 \leq \alpha \leq 5/2$)

$$\left\|(u-v)^{(1/2)}\right\|_{C_\tau(C(\bar\Omega))} \leq c(T_1(h^2_{max}+\tau^2))^{(\alpha-1/2)/3}\|\mathbf{d}\|_\alpha \quad (5.18)$$

where the numbers α_1, α_2 are the same as in Theorem 4.1. The indicated estimate is valid for $v_0 = \tilde{s}_1 u^{(0)}$ (in this connection, see Corollary 5.1 and Theorem 5.1, item (2)); if $\alpha = 1$, it is valid also for $v_0 = v^{(0)}$.

Error estimate (5.16) follows from estimate (5.17), Theorem 5.2, and estimate (5.18) (in which we can limit ourselves to the cases $\alpha = 1$, $u^{(1)} = 0$, $f = 0$, and $(\alpha_1,\alpha_2) = (0,3/2)$, $u^{(0)} = 0$, $u^{(1)} = 0$) with the help of interpolation theory. Namely, we need to use inequalities (4.1), (4.14), and (4.15), and Proposition 4.1. It is convenient to implement the interpolation with respect to f in the following order. First, we consider the case of α_1, α_2 satisfying conditions (a) and (b), then the conditions (c)–(f) such that one of the numbers is integer, and, finally, satisfying the conditions (c)–(f) and noninteger.

The set of indices α_1, α_2 from Theorem 5.3 is shown on Figure 3 (thick segments and dashed area). As in Section 4, this set can be extended to its convex hull.

Remark 5.2. From the proof of Theorem 5.3 it follows that in estimate (5.16) we have $c_{\alpha_1,\alpha_2} = O\left((7/2-\alpha)^{-1}\right)$, if $\alpha_1+\alpha_2 \to 5/2$ and one of the numbers α_1, α_2 is integer. Consequently, under the hypothesis of Theorem 5.2 the error estimate

$$\left\|(u-v)^{(1/2)}\right\|_{C_\tau(C_h)} \leq cT_1^{1+\gamma}(h^2_{max}+\tau^2)|\ln\{T_1^{1+\gamma}(h^2_{max}+\tau^2)\}|\,\|\mathbf{d}\|_\alpha$$

is valid for $T_1^{1+\gamma}(h^2_{max}+\tau^2) \leq 1/2$; this supplements Theorems 5.2 and 5.3. Under condition (2.22) we can omit operator $(\cdot)^{(1/2)}$ in this estimate.

Error estimates in the uniform norm are obtained in references 7, 15, 17, 20, and 39.

6. SECOND-ORDER HYPERBOLIC EQUATIONS OF GENERAL FORM

In this section we consider the initial-boundary value problem for the second-order hyperbolic equation of general form,

$$\rho D_t^2 u + Lu + L_0 D_t u + L_1 u = \rho f \tag{6.1}$$

where $L_0 D_t u = D_t(a_i D_t u) + a_0 D_t u$, $L_1 u = b_i D_i u + b_0 u$. As done previously, let us confine ourselves to the case that coefficients are independent of t. Introduce the vector functions $\mathbf{a} = (a_1, \ldots, a_n)$, $\mathbf{b} = (b_1, \ldots, b_n)$, and assume that

$$\|\mathbf{a}\|_{L_\infty(\Omega)} + \|a_0\|_{L_\infty(\Omega)} + \|\mathbf{b}\|_{L_\infty(\Omega)} + \|b_0\|_{L_{p_n}(\Omega)}$$

$$+ \|\operatorname{div}\mathbf{a}\|_{L_\infty(\Omega)} + \|\operatorname{div}\mathbf{b}\|_{L_{p_n}(\Omega)} \leq N$$

where, for example, $\operatorname{div}\mathbf{a} = D_i a_i$.

The generalized solution to problem (6.1) and (1.4) from the energy class is defined as in Section 1.2, with this modification of identity (1.5),

$$\bar{\mathcal{B}}(u, \eta) \equiv \mathcal{B}(u, \eta) + \mathcal{L}_{0,Q}(D_t u, \eta) + (L_1 u, \eta)_Q = (u^{(1)}, \eta_0)_{0,\Omega} + (f, \eta)_{0,Q}. \tag{6.2}$$

Here and in subsequent text, $\mathcal{L}_{0,G}(w, \eta) = -(a_i w_i, D_i \eta)_G + (a_0 w, \eta)_G$, $G = \Omega, Q$.

Propositions 1.1 and 1.2 can be extended to the case under consideration with the help of methods described in reference 26; but the constant, c_0, in estimates (1.6) and (1.14) should be replaced by $c_1 \exp(c_2 T)$.

The approximate solution, $v \in S$, to problem (6.1) and (1.4) we define as in Section 2, item (2.2), but just with the replacement of integral identity (2.6) by (σ_b is a parameter, $|\sigma_b| \leq N$),

$$\bar{\mathcal{B}}_\sigma(v, \eta) \equiv \mathcal{B}_\sigma(v, \eta) + \mathcal{L}_{0,Q}(D_t v, \eta) + \mathcal{L}_{1,\sigma_b,Q}(v, \eta) = (u^{(1)}, \eta_0)_{0,\Omega}$$

$$+ (f, \eta)_{0,Q} \qquad \forall \eta \in S, \ \eta|_{t=T} = 0 \tag{6.3}$$

where $\mathcal{L}_{1,\sigma_b,Q}(v, \eta) = -(\sigma_b - 1/6)(\tau^2 L_1 D_t v, D_t \eta)_Q + (L_1 v, \eta)_Q$.

Let us introduce the operators $L_{0,h}$ and $L_{1,h}$, acting in H_h, with the help of the identities

$$(L_{0,h} w_h, \varphi_h)_h = \mathcal{L}_{0,\Omega}(w, \varphi), \quad (L_{1,h} w_h, \varphi_h)_h = (L_1 w, \varphi)_\Omega$$

$$\forall w, \varphi \in S_1.$$

Then identity (6.3) can be transformed to operator form (compare with Equations [2.11] and [2.12]),

$$(B_h + \sigma\tau^2 L_h + \sigma_b\tau^2 L_{1,h})\partial_t\bar{\partial}_t v_{h,m} + L_{0,h}\overset{\circ}{\partial}_t v_{h,m} + (L_h + L_{1,h})v_{h,m}$$
$$= f_m^{h,\tau}, \qquad 1 \le m \le M - 1 \qquad (6.4)$$

$$(B_h + \sigma\tau^2 L_h + \sigma_b\tau^2 L_{1,h})\partial_t v_{h,0} + \frac{\tau}{2} L_{0,h}\partial_t v_{h,0}$$
$$+ \frac{\tau}{2}(L_h + L_{1,h})v_{h,0} = u^{(1),h} + \frac{\tau}{2} f_0^{h,\tau}. \qquad (6.5)$$

We shall use the following version of the finite difference counterpart of Gronwall's lemma (a similar statement was proved in reference 2).

Lemma 6.1. Let functions d_0, d_1, d_2 be defined and nonnegative on $\omega^\tau = \bar{\omega}^\tau\backslash\{t_0\}$, and let $0 < \varepsilon_1 < 1$. If the function $w \ge 0$ satisfies the inequality

$$w^2 \le \bar{w}_0^2 + I_\tau\left(d_0 w^2 + d_1 w\check{w} + d_2\check{w}^2\right) \qquad (6.6)$$

on $\bar{\omega}^\tau$ and $d_0\tau \le 1 - \varepsilon_1$, then the estimate

$$w \le |\bar{w}_0|\exp\{(2\varepsilon_1)^{-1}I_\tau(d_0 + 2d_1 + d_2)\}. \qquad (6.7)$$

holds true.

Proof. Denote by z^2 the right-hand side of inequality (6.6); we assume that $z \ge 0$. Then $w \le z$ and

$$\bar{\partial}_t(z^2) = d_0 w^2 + d_1 w\check{w} + d_2\check{w}^2 \le d_0 z^2 + d_1 z\check{z} + d_2\check{z}^2.$$

Hence, under the condition $d_0\tau \le 1 - \varepsilon_1$ we have

$$z^2 \le \frac{\tau d_1}{1 - \tau d_0} z\check{z} + \frac{1 + \tau d_2}{1 - \tau d_0}\check{z}^2 \le \varepsilon_1^{-1}\tau d_1 z\check{z} + \left(1 + \varepsilon_1^{-1}(d_0 + d_2)\right)\check{z}^2.$$

Solving this quadratic inequality for z, we derive

$$z \le \left\{\varepsilon_1^{-1}\tau d_1 + \left(1 + \varepsilon_1^{-1}\tau(d_0 + d_2)\right)^{1/2}\right\}\check{z}$$
$$\le \left(1 + (2\varepsilon_1)^{-1}\tau(d_0 + 2d_1 + d_2)\right)\check{z}.$$

The obtained recurrent inequality yields the estimate

$$z \le z_0\exp\{(2\varepsilon_1)^{-1}I_\tau(d_0 + 2d_1 + d_2)\},$$

and because $w \le z$ and $z_0 = |\bar{w}_0|$, yields also estimate (6.7).

Note that in the constraint on τ only the coefficient, d_0 (but not d_1 or d_2), is present; subsequently this will be essential. Let us establish the stability of the approximate solution.

Theorem 6.1. Let $\tau \leq \tau^0$, where $\tau^0 = \tau^0(\varepsilon_0, N) > 0$ is small enough. Theorem 2.1 is valid also for the function v satisfying identity (6.3) provided that the right-hand parts of estimates (2.15) and (2.16) are multiplied by $c_1 \exp(c_2 T)$.

For $1/2 \mathrm{div}\, \mathbf{a} + a_0 \geq 0$, $\sigma_b = 0$, the indicated result is valid for any τ (i.e., $\tau^0 = T$). If, moreover, $\mathbf{b} = 0$, $b_0 = 0$, then it is possible to take $c_2 = 0$.

Remark 6.1. Let $\mathbf{a} = 0$, and let $\tilde{a}_0 \equiv a_0/\rho = \mathrm{const}$, or $\|D\tilde{a}_0\|_{L_{p_n}(\Omega)} \leq N$, $\beta_h \alpha_h \leq N$. As a result,

$$\|L_h^{1/2} B_h^{-1} L_{0,h} \varphi_h\|_h \leq c\|\varphi\|_{H^1}$$

for all $\varphi \in S_1$. Then the previous assertion refers not only to Theorem 2.1, but also to Theorem 2.2. To ensure the property $\tau^0 = T$, the condition $\tilde{a}_0 = \mathrm{const}$ is additionally required.

Proof.

(1) Let us turn to the proof of Theorem 2.1, item (1). Choose the functions η, ξ as indicated there. Transform the summands added to $\mathcal{B}_\sigma(v, \eta)$ in identity (6.3) as

$$\mathcal{L}_{0,Q}(D_t v, \eta) = -(a_i v, D_i[\xi])_Q + (a_0 v, [\xi])_Q$$
$$- \mathcal{L}_{0,\Omega}(v_0, \eta_0)$$
$$= \left(\frac{\mathrm{div}\, \mathbf{a}}{2} + a_0, [v]^2\right)_{Q^l} - \mathcal{L}_{0,\Omega}(v_0, (I_t v)_l)$$

where the equality $-(a_i[v], D_i v)_{Q^l} = \left(\mathrm{div}\, \mathbf{a}/2, [v]^2\right)_{Q^l}$ is taken into account;

$$\mathcal{L}_{1,\sigma_b,Q}(v, \eta) = \left(\sigma - \frac{1}{4}\right)(\tau^2 L_1 D_t v, [v])_{Q^l}$$
$$+ \left(L_1[s_t I_t v], [v]\right)_{Q^l}$$
$$= I_\tau^l\left\{\left(\sigma - \frac{1}{4}\right)(\tau^2 L_1 \bar{\partial}_t v, v^{(1/2)})_\Omega\right.$$
$$\left. + \left(L_1\left(\check{I}_t v + \frac{\tau}{2} v^{(1/2)}\right), v^{(1/2)}\right)_\Omega\right\}$$

$$= I_\tau^l \left\{ \left(L_1 v_{\sigma_b}, \tau v^{(1/2)} \right)_\Omega + \left(L_1 \check{I}_t v, v^{(1/2)} \right)_\Omega \right\}$$

$$= I_\tau^l \Big\{ \left((b_0 - \operatorname{div} \mathbf{b}) v_{\sigma_b}, \tau v^{(1/2)} \right)_\Omega$$

$$- \left(b_i v_{\sigma_b}, D_i \left(\tau v^{(1/2)} \right) \right)_\Omega + \left(L_1 \check{I}_t v, v^{(1/2)} \right)_\Omega \Big\}$$

where $v_{\sigma_b} = \sigma_b v - (\sigma_b - 1/2)\check{v}$, $\check{I}_t v = \dot{I}_t v$.

Now we can write the estimate (see inequalities [1.1])

$$-\mathcal{L}_{0,Q}(D_t v, \eta) \le d_a I_\tau^l \left(\|v\|_\Omega^2 + \|\check{v}\|_\Omega^2 \right) + \frac{\varepsilon}{4} \|D(I_t v)_l\|_\Omega^2$$

$$+ \varepsilon^{-1} \left(\|\mathbf{a}\|_{L_\infty(\Omega)} + c^{(1)} \|a_0\|_{L_\infty(\Omega)} \right)^2 \|v_0\|_\Omega^2$$

$$d_a \equiv \frac{1}{2} \left\| \min\left\{ \frac{\operatorname{div} \mathbf{a}}{2} + a_0, 0 \right\} \right\|_{L_\infty(\Omega)}$$

where the inequality $\alpha\beta \le (\varepsilon/4)\alpha^2 + \varepsilon^{-1}\beta^2$ for any $\varepsilon > 0$ is used, and also the estimate

$$- \mathcal{L}_{1,\sigma_b,Q}(v, \eta) \le d_b I_\tau^l \left\{ \left(|\sigma_b| \, \|v\|_\Omega + \left| \sigma - \frac{1}{2} \right| \|\check{v}\|_\Omega \right) \times \right.$$

$$\left. \left(\|DI_t v\|_\Omega + \|D\check{I}_t v\|_\Omega \right) + \left(\|v\|_\Omega + \|\check{v}\|_\Omega \right) \|D\check{I}_t v\|_\Omega \right\}$$

$$d_b \equiv c^{(0)} \left(\|b_0\|_{L_{p_n}(\Omega)} + \|\operatorname{div} \mathbf{b}\|_{L_{p_n}(\Omega)} \right) + \|\mathbf{b}\|_{L_\infty(\Omega)}$$

where the equality $\tau v^{(1/2)} = I_t v - \check{I}_t v$ is used.

Taking into account that the norms $\| \cdot \|_{H^0}$ and $\| \cdot \|_\Omega$ as well as $\| \cdot \|_{H^1}$ and $\|D \cdot \|_\Omega$ are equivalent and choosing ε small enough, with the help of Lemma 6.1 we derive the required version of estimate (2.15). As can be seen from the proof, in the special cases marked in the Theorem formulation the conditions on τ do not arise or we have $c_2 = 0$.

(2) Let us turn to the proof of Theorem 2.1, item (2). Here it is necessary to add the summands $\Phi^{(0)}(v) + I_\tau^{l-1}\Phi(v)$ to the right-hand part of equality (2.19), where

$$\Phi^{(0)}(v) = -\frac{\tau}{2} \mathcal{L}_{0,\Omega}(\partial_t v_0, \partial_t v_0)$$

$$- \left(\sigma_b \tau^2 L_1 \partial_t v_0 + \frac{\tau}{2} L_1 v_0, \partial_t v_0 \right)_\Omega$$

$$\Phi(v) = -\mathcal{L}_{0,\Omega}\big(\mathring{\partial}_t v, \mathring{\partial}_t v\big) - \big(\sigma_b \tau^2 L_1 \partial_t \bar{\partial}_t v + L_1 v, \mathring{\partial}_t v\big)_\Omega.$$

Let us perform such transformations,

$$\Phi^{(0)}(v) = -\frac{\tau}{2}\left(\frac{\operatorname{div}\mathbf{a}}{2} + a_0, (\partial_t v_0)^2\right)_\Omega$$

$$-2\sigma_b\tau\left(L_1 v_1^{(1/2)}, \partial_t v_0\right)_\Omega$$

$$+4\left(\sigma_b - oneforth\right)\left(L_1 v_0, v_1^{(1/2)} - v_0\right)_\Omega$$

$$\Phi(v) = -\left(\frac{\operatorname{div}\mathbf{a}}{2} + a_0, \left(\overset{\circ}{\partial}_t v\right)^2\right)_\Omega$$

$$-\left(L_1\partial_t v_{\sigma_b}, \tau\partial_t\left(v^{(1/2)}\right)\right)_\Omega - \left(L_1 v^{(1/2)}, (\partial_t v)^{(1/2)}\right)_\Omega.$$

Introduce the level norm such that $\|v\|^2_{\bar{H}^1} = \|\bar{\partial}_t v\|^2_\Omega + \|Dv^{(1/2)}\|^2_\Omega$. Taking into account the inequality $\tau|\partial_t w| \le |\hat{w}| + |w|$ we obtain as in item (1) the estimates

$$\Phi^{(0)}(v) \le c_0\bar{d}_0\tau\left(\|v\|^2_{\bar{H}^1}\right)_1 + c_0\left|\sigma_b - 1/4\right|\left|\sigma_b - \frac{1}{4}\right|d_b c^{(1)} \times$$

$$\left\{\varepsilon\left(\|v\|^2_{\bar{H}^1}\right)_1 + (1+\varepsilon^{-1})\|Dv_0\|^2_\Omega\right\}, \qquad \varepsilon > 0,$$

$$I_\tau^{l-1}\Phi(v) \le c_0 I_\tau^l\left\{\bar{d}_0\|v\|^2_{\bar{H}^1} + \bar{d}_1\|\check{v}\|_{\bar{H}^1}\left(\|v\|_{\bar{H}^1} + \|\check{v}\|_{\bar{H}^1}\right)\right\}$$

where $\bar{d}_0 = d_0 + |\sigma_b|d_b$, $\bar{d}_1 = d_a + (|\sigma_b - 1/2| + 1/2)d_b$.

Utilize the inequality $\|v\|^2_{\bar{H}^1} \le c_3\|v\|^2_{H_E}$. Under the condition $c_0 c_3 \bar{d}_0\tau \le 1/2$, for $l = 1$ we initially obtain the estimate

$$\left(\|v\|^2_{H_E}\right)_1 \le c\left(\|v_0\|_{H^1} + \|u^{(1)}\|_{H^0} + \|f\|_{L_1(H^0)}\right).$$

Then, using Lemma 6.1 we obtain the required version of estimate (2.16). We shall not dwell on the justification of Remark 2.1.

Let us establish a priori error estimates of second-order accuracy.

Theorem 6.2. Let $\tau \le \tau^0$. Theorem 3.1 is valid also for the function v satisfying identity (6.3) with the following changes: (a) the right-hand parts of estimates (3.4) and (3.5) should be multiplied by $\exp\{c_2 T\}$; (b) in estimate (3.4) it is necessary to add $\|u\|_{L_1(H^2)}$ to the sum of the norms of u (for $\mathbf{b} = 0$, $b_0 = 0$ it is not necessary); (c) in estimate (3.5) it is necessary to add to the sum of the norms of u $\|D_t u\|_{L_1(H^2)} + \|D_t^2 u\|_{L_\infty(H^{k_a})} + \|D_t^3 u\|_{L_1(H^{k_a})}$, where $k_a = 1$ in the general case, or $k_a = 0$ for $\mathbf{a} = 0$. (For $\mathbf{a} = 0$, $a_0 = 0$, only $\|D_t u\|_{L_1(H^2)}$ can be added). Here, the quantities τ^0, c_2 are the same as in Theorem 6.1.

Remark 6.2. If $\mathbf{a} = 0$, $\|D\tilde{a}_0\|_{L_{p_n}(\Omega)} \leq N$, $\beta_h \alpha_h \leq N$ and condition (2.22) is fulfilled, then in item (c) it is enough to add $\|D_t u\|_{L_1(H^2)} + \|D_t^2 u\|_{L_\infty(H^0)}$.

Proof.

(1) The function $r = s_1 s_t u - v$ satisfies the identity

$$\bar{\mathcal{B}}_\sigma(r, \eta) = \left(\mathcal{B}_\sigma(s_1 s_t u, \eta) - \mathcal{B}(u, \eta)\right) - \mathcal{I}^{(1)}(u, \eta)$$

$$\forall \eta \in S, \ \eta|_{t=T} = 0$$

$$\mathcal{I}^{(1)}(u, \eta) \equiv \mathcal{L}_{0,Q}\left(D_t(u - s_1 s_t u), \eta\right) + \left(\sigma_b - \frac{1}{6}\right) \times$$

$$(\tau^2 L_1 D_t s_1 s_t u, D_t \eta)_Q + \left(L_1(u - s_1 s_t u), \eta\right)_Q.$$

The first summand on the right-hand part of the indicated identity for r was considered in the proof of Theorem 3.1. Transform $\mathcal{I}^{(1)}(u, \eta)$, which is the sum of additional summands. Using the formula $D_t(u - s_1 s_t u) = [D_t(u - s_1 s_t u)] + D_t(u - s_t u)$, identity (2.3), and integrating by parts, we obtain

$$\mathcal{I}^{(1)}(u, \eta) = -\left(a_i[Dt(u - s_1 u)], D_i \eta\right)_Q$$

$$+ \left(a_0[D_t(u - s_1 u)], \eta\right)_Q - \left((a_0 + \operatorname{div} \mathbf{a})(u - s_t u)\right.$$

$$+ a_i D_i(u - s_t u), D_t \eta\right)_Q + \left(\sigma_b - \frac{1}{6}\right)(\tau^2 L_1 D_t s_1 u, D_t \eta)_Q$$

$$+ \left((b_0 - \operatorname{div} \mathbf{b})(u - s_1 s_t u), \eta\right)_Q - \left(b_i(u - s_1 s_t u), D_i \eta\right)_Q.$$

Combining similar terms, we obtain

$$\mathcal{I}^{(1)}(u, \eta) = \left(\bar{f}^{(1)}, \eta\right)_Q + \left(\bar{f}_i, D_i \eta\right)_Q + \left(\bar{f}^{(2)}, D_t \eta\right)_Q.$$

In addition, according to the conditions on the coefficients of Equation (6.1) and inequalities (1.1), we have for $\bar{\mathbf{f}} = (\bar{f}_1, \ldots, \bar{f}_n)$

$$\|\bar{f}^{(1)}\|_{L_1(H^{<-1>})} + \|\operatorname{div}\bar{\mathbf{f}}\|_{L_1(H^{<-1>})} + \|\rho^{-1}\bar{f}^{(2)}\|_{L_1(H^0)}$$

$$\leq c\Big(\|\bar{f}^{(1)}\|_{L_1\left(L_{2p_n/p_n+2}(\Omega)\right)} + \|\bar{\mathbf{f}}\|_{L_1(H^0)} + \|\bar{f}^{(2)}\|_{L_1(H^0)}\Big)$$

$$\leq c_1\Big(\big\|[D_t(u - s_1 u)]\big\|_{L_1(H^0)} + \|u - s_1 s_t u\|_{L_1(H^0)}$$

$$+ \|u - s_t u\|_{L_1(H^1)} + \Big|\sigma_b - \frac{1}{6}\Big|\tau^2 \|D_t s_1 u\|_{L_1(H^1)}\Big). \quad (6.8)$$

From the triangle inequality for norms and the imbedding from H^1 into H^0, it follows that

$$\|u - s_1 s_t u\|_{L_1(H^0)} \leq \|u - s_1 u\|_{L_1(H^0)}$$

$$+ c\|s_1(u - s_t u)\|_{L_1(H^1)}.$$

Since the operators $[\cdot], s_1$ are projectors in $L_1(H^0), H^1$, respectively, and H^2 is imbedded into H^1, estimate (6.8) can be continued as follows,

$$\|\bar{f}^{(1)}\|_{L_1(H^{<-1>})} + \|\operatorname{div}\bar{\mathbf{f}}\|_{L_1(H^{<-1>})} + \|\rho^{-1}\bar{f}^{(2)}\|_{L_1(H^0)}$$

$$\leq c\big(\|D_t u - s_1 D_t u\|_{L_1(H^0)} + \|u - s_1 u\|_{L_1(H^0)}$$

$$+ \|u - s_t u\|_{L_1(H^1)} + \tau^2\|A D_t u\|_{L_1(H^0)}\big).$$

For $\mathbf{b} = 0$, $b_0 = 0$ in estimate (6.8), the term $\|u - s_1 s_t u\|_{L_1(H^0)}$ disappears; therefore, in the last estimate the term $\|u - s_1 u\|_{L_1(H^0)}$ can be omitted. As for the rest, the derivation of the required version of estimate (3.4) remains unchanged.

(2) To derive the required version of estimate (3.5), let us transform $\mathcal{I}^{(1)}(u, \eta)$ in a somewhat different way than previously:

$$\mathcal{I}^{(1)}(u, \eta) = -\big(a_i[D_t(u - s_1 u)], D_i\eta\big)_Q$$

$$+ \big(a_0[D_t(u - s_1 u)], \eta\big)_Q + \big(L_0 D_t(u - s_t u), \eta\big)_Q$$

$$- \Big(\sigma_b - \frac{1}{6}\Big)\big(\tau^2 L_1 D_t^2 s_1 u, \eta\big)_Q$$

$$- \Big(\sigma_b - \frac{1}{6}\Big)\big(\tau^2 L_1 s_1 u^{(1)}, \eta_0\big)_\Omega$$

$$+ \big((b_0 - \operatorname{div}\mathbf{b})(u - s_1 u), \eta\big)_Q - \big(b_i(u - s_1 u), D_i\eta\big)_Q$$

$$+ \big(L_1 s_1(u - s_t u), \eta\big)_Q.$$

Combining similar terms, we obtain

$$\mathcal{I}^{(1)}(u, \eta) = (\bar{u}^{(1)}, \eta_0)_{0,\Omega}$$
$$+ \left(\bar{f}^{(0)} + \bar{f}^{(1)} + \bar{f}^{(2)}, \eta\right)_Q + \left(\bar{f}_i, D_i\eta\right)_Q$$

where, for example, $\bar{f}^{(1)} = L_0 D_t(u - s_t u)$, $\bar{f}^{(2)} = (b_0 - \text{div } \mathbf{b})(u - s_1 u)$. Furthermore, according to the conditions on the coefficients of Equation (6.1) and inequalities (6.1) (see also Remark 2.2 and its justification), we have

$$\Psi_1(u) \equiv \|\bar{u}^{(1)}\|_{H^0} + \|\rho^{-1}\bar{f}^{(0)}\|_{L_1(H^0)}$$

$$\leq c\left(\left|\sigma_b - \frac{1}{6}\right|\tau^2\|s_1 u^{(1)}\|_{H^1} + \|\,[D_t(u - s_1 u)]\,\|_{L_1(H^0)}\right.$$

$$+ \left|\sigma_b - \frac{1}{6}\right|\tau^2\|s_1 D_t^2 u\|_{L_1(H^1)} + \|s_1(u - s_t u)\|_{L_1(H^1)}\right)$$

$$\Psi_2(u) \equiv \left\|(\bar{f}^{(1)})^\tau\right\|_{L_{1,\tau}(H^0)} \leq c\left\|(D_t(u - s_t u))^\tau\right\|_{L_{1,\tau}(H^{k_a})}$$

$$\Psi_3(u) \equiv \left\|\bar{f}^{(2)}\right\|_{W_1^1(H^{<-1>})} + \|\text{div }\bar{\mathbf{f}}\|_{V(H^{<-1>})}$$

$$\leq c\left(\|D_t(u - s_1 u)\|_{L_1(H^0)} + \|u^{(0)} - s_1 u^{(0)}\|_{H^0}\right.$$

$$+ \left\|D_t^2(u - s_1 u)\right\|_{L_1(H^0)} + \left\|u^{(1)} - s_1 u^{(1)}\right\|_{H^0}\right).$$

As previously, the given estimates yield

$$\Psi_1(u) + \Psi_2(u) + \Psi_3(u) \leq c\left(\left\|u^{(0)} - s_1 u^{(0)}\right\|_{H^0}\right.$$

$$+ \tau^2\|u^{(1)}\|_{H^1} + \|u^{(1)} - s_1 u^{(1)}\|_{H^0}$$

$$+ \|D_t u - s_1 D_t u\|_{L_1(H^0)} + \tau^2\|D_t^2 u\|_{L_1(H^1)}$$

$$+ \|u - s_t u\|_{L_1(H^1)} + \tau^2\left(\|D_t^2 u\|_{L_\infty(H^{k_a})}\right.$$

$$+ \left.\|D_t^3 u\|_{L_1(H^{k_a})}\right) + \left.\|D_t^2 u - s_1 D_t^2 u\|_{L_1(H^0)}\right).$$

Otherwise, the derivation of the required version of estimate (3.5) remains unchanged (we also need the imbedding from H^2 into H^1).

To justify Remark 6.2 we should write the equality $(L_0 D_t(u - s_t u), \eta)_Q = -(L_0(u - s_t u), D_t \eta)_Q$ and use Theorem 2.2, Remark 6.1, and Corollary 3.1.

Using Theorems 6.1 and 6.2 (and generalizing Proposition 1.3), for finite element methods (6.4) and (6.5) it is possible to derive the extension of Theorems 4.1 and 4.2. (In the latter theorem, for $0 < \alpha_1 \leq 3$, $0 \leq \alpha_2 < 1$ we should consider that $\mathbf{a} = 0$, $a_0 = 0$ or that only $\mathbf{a} = 0$; but then the norm of f should be somewhat strengthened.) In the case where the coefficients of Equation (6.1) and $\partial\Omega$ have some smoothness, and also $\mathbf{a}|_{\partial\Omega} = 0$, $\mathbf{b}|_{\partial\Omega} = 0$, the essential difficulties do not appear. (If $u^{(0)} = 0$, $u^{(1)} = 0$, and $\alpha_1 \leq 0$, it is enough that $\mathbf{a}|_{\partial\Omega} = 0$). If the last condition on $\mathbf{a}, \mathbf{b}$ is not fulfilled, then the transition to other data spaces is required, and the situation becomes more complicated.

7. FINITE-ELEMENT METHOD WITH THE SPLITTING OPERATOR

In this section we consider the initial-boundary value problem (1.3) and (1.4) when ρ is a constant and Ω is a rectangular parallelepiped. Without loss of generality let us set $\rho(x) \equiv 1$, $\Omega = (0,1)^n$; then $H^0 = L_2(\Omega)$ and $L = A$.

Introduce the mesh in x_i with the step h_i, and choose $S_{1,h} = \hat{S}_{1,h_1} \otimes \ldots \otimes \hat{S}_{1,h_n}$ with $h = (h_1, \ldots, h_n)$. Denote by H_h the space of vectors $\varphi_{h,\mathbf{i}}$ with $\mathbf{i} = (i_1, \ldots, i_n)$, $1 \leq i_k \leq h_k^{-1} - 1$ for all $k = 1, \ldots, n$, having the scalar product $(\varphi_h, \eta_h)_h = \sum_{\mathbf{i}} \varphi_{h,\mathbf{i}} \eta_{h,\mathbf{i}} h_1 \ldots h_n$. The formula $\varphi(x_{\mathbf{i}}) = \varphi_{h,\mathbf{i}}$, where $x_{\mathbf{i}} = (i_1 h_1, \ldots, i_n h_n)$, establishes one-to-one correspondence between $S_{1,h}$ and H_h (compare this with Section 2.1). Set $B_k \varphi_h = \varphi_h - (h_k^2/6)\Lambda_k \varphi_h$, $(\Lambda_k \varphi_h)_{\mathbf{i}} = -(\varphi_{h,\mathbf{i}+\chi_k} - 2\varphi_{h,\mathbf{i}} + \varphi_{h,\mathbf{i}-\chi_k})/h_k^2$, where χ_k is the unit vector of kth direction in $\mathbf{R}^n$ and $\varphi_{h,\mathbf{i}}|_{i_k=0,h_k^{-1}} = 0$. Let B_h, L_h, and $(\cdot)^h$ correspond to the space H_h introduced here (in accordance with Section 2.1).

Let us construct the finite-element method with the splitting operator. In equations (2.11)–(2.13) choose $\sigma = \sigma^{(0)} = 0$ and replace the operator $B_h = B_1 \ldots B_n$ by $\hat{B}_h = (B_1 + \tau\kappa_1\Lambda_1) \ldots (B_n + \tau\kappa_n\Lambda_n)$. Then for the approximate solution $y \in S$ and the function $y^{(0)} \in S_1$ we obtain

$$\hat{B}_h \partial_t \bar{\partial}_t y_{h,m} + L_h y_{h,m} = f_m^{h,\tau}, \qquad 1 \leq m \leq M - 1 \qquad (7.1)$$

$$\hat{B}_h \partial_t y_{h,0} + \frac{\tau}{2} L_h y_{h,0} = u^{(1),h} + \frac{\tau}{2} f_0^{h,\tau} \qquad (7.2)$$

$$\hat{B}_h y_h^{(0)} = u^{(0),h}. \qquad (7.3)$$

Introduce the bilinear forms $\mathcal{R}_G(w, \varphi) = \tau^2 \kappa_i (D_i w, D_i \varphi)_G + \mathcal{R}_G^{(1)}(w, \varphi)$,

$$\mathcal{R}_G^{(1)}(w, \varphi) = \sum_{2 \leq k \leq n} \tau^{2k} \sum_{1 \leq i_1 < \ldots < i_k \leq n} \kappa_{i_1} \ldots \kappa_{i_k} (D_{i_1} \ldots D_{i_k} w, D_{i_1} \ldots D_{i_k} \varphi)_G.$$

The equivalent projection formulation of equations (7.1)–(7.3) is: the function $y \in S$ satisfies the identity

$$\hat{B}(y, \eta) \equiv B_0(y, \eta) - R_Q(D_t y, D_t \eta) = (u^{(1)}, \eta_0)_\Omega + (f, \eta)_Q$$
$$\forall \eta \in S, \ \eta|_{t=T} = 0 \quad (7.4)$$

and the function $y^{(0)} \in S_1$ satisfies the identity

$$(y^{(0)}, \varphi)_\Omega + R_\Omega(y^{(0)}, \varphi) = (u^{(0)}, \varphi)_\Omega \quad \forall \varphi \in S_1. \quad (7.5)$$

Here the bilinear form $B_0(\cdot, \cdot)$ is the special case of $B_\sigma(\cdot, \cdot)$ for $\sigma = 0$ (see identity [2.6]).

Let us subject the parameters κ_i to the condition $\kappa_i = \kappa_i^{(1)} + \kappa_0$, where

$$\frac{1}{4} a_{ij}(x)\xi_j \xi_i \le \kappa_i^{(1)} \xi_i^2 \quad \forall x \in \Omega, \ \forall \xi \in \mathbf{R}^n$$

$$\varepsilon_0^2 + (c^{(0)})^2 \|a\|_{L_{p_n/2}(\Omega)} \le \kappa_0.$$

Introduce the norms (the second one is defined for each level) such that

$$\|w\|_{\mathring{H}_\tau^0}^2 = \|w\|_\Omega^2 + \varepsilon_0^2 \tau^2 \|Dw\|_\Omega^2 + R_\Omega^{(1)}(w, w),$$

$$\|y\|_{\tilde{H}_E}^2 = \|\bar{\partial}_t y\|_{\mathring{H}_\tau^0}^2 + \|y^{(1/2)}\|_{H^1}^2,$$

and the self-adjoint operator, $\tilde{B}_h$, acting in H_h such that $(\tilde{B}_h w_h, w_h)_h = \|w\|_{\mathring{H}_\tau^0}^2$ for all $w \in S_{1,h}$. If we use inequality (1.2) and subsequently Remark 2.1, then we obtain the inequality

$$\min\{\sqrt{2}\varepsilon_0, \sqrt{\nu_0/2}\}\|y\|_{C_\tau(H^{<1>})} \le \|y\|_{C_\tau(\tilde{H}_E)}.$$

The following result on the absolute stability of method (7.1) and (7.2) holds true.

Theorem 7.1.

(1) If the function $y \in S$ satisfies the identity

$$\hat{B}(y, \eta) = (u^{(1)}, \eta_0)_\Omega + (f_1, \eta)_Q + (f_{2,h}, D_t \eta_h)_{L_2(H_h)} \quad (7.6)$$

for all $\eta \in S$, $\eta|_{t=T} = 0$, then the estimate

$$\max\{\|y\|_{C_\tau(\tilde{H}_\tau^0)}, \|I_t y\|_{C_\tau(H^1)}\} \le \|\hat{B}_h^{1/2} y_0\|_h + 2\|u^{(1)}\|_{H^{-1}}$$
$$+ 4\|f_1\|_{L_1(H^{-1})} + 2\|\tilde{B}_h^{-1/2} f_{2,h}\|_{L_1(H_h)} \quad (7.7)$$

is valid.

(2) If the function $y \in S$ satisfies the identity

$$\hat{B}(y, \eta) = (u_h^{(1)}, \eta_{h,0})_h + (f_h, \eta_h)_{L_2(H_h)} + (f_1, \eta)_Q \qquad (7.8)$$

for all $\eta \in S$, $\eta|_{t=T} = 0$, then the estimate

$$\|y\|_{C_\tau(\tilde{H}_E)} \leq \|y_0\|_{H^1} + \|\tilde{B}_h^{-1/2} u_h^{(1)}\|_h$$

$$+ 2\|\tilde{B}_h^{-1/2} f_h\|_{L_1(H_h)} + 4\|f_1\|_{W_1^1(H^{-1})} \qquad (7.9)$$

is valid.

Remark 7.1. The inequality $\|\hat{B}_h^{1/2} y_h^{(0)}\|_h \leq \|u^{(0)}\|_\Omega$ is valid; it immediately follows from identity (7.5) for $\varphi = y^{(0)}$. Furthermore, $\|\hat{B}_h^{-1/2} w^h\|_h \leq \|w\|_\Omega$.

Proof. Introduce the norm such that

$$\|w\|_{\hat{H}_\tau^0}^2 = \|w\|_\Omega^2 + \tau^2 \left(\kappa_i \|D_i w\|_\Omega^2 - \frac{1}{4} \|w\|_{H^1}^2\right) + \mathcal{R}_\Omega^{(1)}(w, w).$$

Then $\|\tilde{B}_h^{1/2} w_h\|_h = \|w\|_{\hat{H}_\tau^0} \leq \|w\|_{\hat{H}_\tau^0} \leq \|\hat{B}_h^{1/2} w_h\|_h$ for all $w \in S_1$. Turn to the proof of Theorem 2.1.

(1) Identity (7.6) yields the inequality (instead of inequality [2.18])

$$\frac{1}{2}\left(\|y_l\|_{\hat{H}_\tau^0}^2 + \|(I_t y)_l\|_{H^1}^2\right) \leq \frac{1}{2}\|y_0\|_{\hat{H}_\tau^0}^2$$

$$+ \left(\|u^{(1)}\|_{H^{-1}} + 2\|f_1\|_{L_1(H^{-1})}\right)\|I_t v\|_{C_\tau(H^1)}$$

$$+ \|\tilde{B}_h^{-1/2} f_{2,h}\|_{L_1(H_h)}\|y\|_{C_\tau(\tilde{H}_\tau^0)}.$$

Estimate (7.7) follows from this inequality.

(2) Identity (7.8) for $f_1 = 0$ yields the equality (instead of law [2.19])

$$\frac{1}{2} \partial_t\left(\|\bar{\partial}_t y\|_{\hat{H}_\tau^0}^2 + \|y^{(1/2)}\|_{H^1}^2\right) = (f_h^\tau, (\partial_t y_h)^{(1/2)})_h.$$

The equation

$$\|\bar{\partial}_t y_1\|_{\hat{H}_\tau^0}^2 + \|y_1^{(1/2)}\|_{H^1}^2 = (\hat{B}_h \partial_t y_{h,0}, \partial_t y_{h,0})_h$$

$$+ \frac{\tau}{2} \mathcal{L}_\Omega(y_0, \partial_t y_0) + \mathcal{L}_\Omega(y_0, y_1^{(1/2)})$$

is valid. As in the proof of Theorem 2.1, we obtain the inequality

$$\left\|\bar\partial_t y_l\right\|^2_{\mathring{H}^0_\tau} + \left\|y_l^{(1/2)}\right\|^2_{H^1} \le \|y_0\|_{H^1}\|y_1^{(1/2)}\|_{H^1}$$

$$+ \left(\|\tilde B_h^{-1/2}u_h^{(1)}\|_h + 2\|\tilde B_h^{-1/2}f_h\|_{L_1(H_h)}\right)\|\bar\partial_t y\|_{C_\tau(\tilde H^0_\tau)}.$$

Estimate (7.9) for $f_1 \not\equiv 0$ follows from this inequality. The case $f_1 \not\equiv 0$ is analyzed in just the same way as in the proof of Theorem 2.1.

To derive error estimates in this section we need additional assumptions. Let a_{ij} be continuous on $\bar\Omega$ for $1 \le i \le n$, $1 \le j \le n$ and $\|D_i a_{ij}\|_{L_\infty(\Omega)} + \|a\|_{L_{p_n}(\Omega)} \le N$ for $1 \le j \le n$. Then from the results of reference 44, §1, it follows that for $|h| \le h_0$ the estimate,

$$\|\Lambda_h(s_1 w)_h\|_h \le c\|w\|_{H^{<2>}} \le c_1\|w\|_{H^2} \qquad (7.10)$$

is valid, where the operator Λ_h is such that $(\Lambda_h w_h, \varphi_h)_h = (D_i w, D_i \varphi)_\Omega$ for all $w, \varphi \in S_1$ (thus $(-\Lambda_h)$ is the finite-element counterpart of the Laplace operator). Here and in subsequent text, the quantities c, c_k, h_0 depend on $n, N, c^{(0)}$ as well as on a majorant of continuity modules of a_{ij}. Moreover, the quantity $h_0 > 0$ is small enough. The condition $|h| \le h_0$ is assumed to be valid in subsequent text.

Let also $\kappa_i \le N$ $(1 \le i \le n)$, and for $n \ge 3$ the condition on the mesh steps be satisfied, that is, $\tau/h_i \le N_0 (1 \le i \le n)$, with the parameter $N_0 \ge 1$. It is convenient to set $N_0 = 1$ for $n < 3$.

Remark 7.2. The indicated condition on the mesh steps can be slightly weakened. For example, for $n = 3$ it is sufficient to require that $\tau/\max_{1 \le i \le 3} h_i \le N_0$.

Lemma 7.1. For any $w, \varphi \in S_1$ the estimate

$$\left((\hat B_h - B_h)w_h, \varphi_h\right)_h = \mathcal{R}_\Omega(w, \varphi)$$

$$\le cN_0^{n-2}\tau^2\,\|\Lambda_h w_h\|_h\,\|\tilde B_h^{1/2}\varphi_h\|_h \qquad (7.11)$$

holds true.

Proof. By definition of the bilinear form $\mathcal{R}_\Omega(w, \varphi)$ and the operators $\Lambda_h, \bar B_h$ we have

$$\mathcal{R}_\Omega(w, \varphi) \le \tau^2\|B_h^{-1/2}\Lambda_h w_h\|_h\,\|B_h^{1/2}\varphi_h\|_h$$

$$+ \mathcal{R}_\Omega^{(1)}(w, w)^{1/2}\,\mathcal{R}_\Omega^{(1)}(\varphi, \varphi)^{1/2}$$

$$\le (\tau^2\|B_h^{-1/2}\Lambda_h w_h\|_h\| + \mathcal{R}_\Omega^{(1)}(w, w)^{1/2})\|\tilde B_h^{1/2}\varphi_h\|_h.$$

To majorize $\mathcal{R}_\Omega^{(1)}(w, w)$ apply the inequalities

$$\|D_k z\|_{L_2(0,1)} \leq (c_0/h_k)\|z(x_k)\|_{L_2(0,1)} \qquad \forall z \in \hat{S}_{1,h_k}$$

$$\sum_{i \neq j} \|D_i D_j w\|_\Omega^2 \leq c_0 \|B_h^{-1/2} \Lambda_h w_h\|_h^2$$

and the condition $\tau/h_i \leq N_0$ for $n \geq 3$, therefore we derive Equation (7.11).

Corollary 7.1. The estimate

$$\|\tilde{B}_h^{-1/2}(\hat{B}_h - B_h)(s_1 w)_h\|_h \leq c N_0^{n-2} \tau^2 \|w\|_{H^2}.$$

is valid. Proof follows from estimate (7.11), also taking into account estimate (7.10).

Corollary 7.2. The estimate

$$\left\|\hat{B}_h^{1/2}(s_1 u^{(0)} - y^{(0)})_h\right\|_h \leq c\left(|h|^2 + N_0^{n-2}\tau^2\right)\|u^{(0)}\|_{H^2} \qquad (7.12)$$

is valid.

Proof. By definition of $y^{(0)}$ (see Equation [7.3]) we have

$$\left(\hat{B}_h\big(s_1 u^{(0)} - y^{(0)}\big)_h, \varphi_h\right)_h = \big(s_1 u^{(0)} - u^{(0)}, \varphi\big)_\Omega$$

$$+ \mathcal{R}_\Omega\big(s_1 u^{(0)}, \varphi\big) \leq \|u^{(0)} - s_1 u^{(0)}\|_\Omega \|\varphi\|_\Omega$$

$$+ \|\tilde{B}_h^{-1/2}(\hat{B}_h - B_h)(s_1 u^{(0)})_h\|_h \|\tilde{B}_h^{1/2}\varphi_h\|_h.$$

It remains to apply inequality (3.2) (here it is valid, with $\beta_h = c|h|$) and the previous Corollary as well as the inequality $\|\varphi\|_\Omega + \|\tilde{B}_h^{1/2}\varphi\|_h \leq 2\|\hat{B}_h^{1/2}\varphi_h\|_h$ and to set $\varphi = s_1 u^{(0)} - y^{(0)}$.

If $\tau/h_i \leq N_0$ $(1 \leq i \leq n)$ for $n \geq 2$, then the right-hand part of estimate (7.12) can be replaced by $c(|h| + N_0^{n-1}\tau)\|u^{(0)}\|_{H^1}$.

Let us establish a priori error estimates of the second order.

Theorem 7.2. For the solution to Equations (7.1) and (7.2), the a priori error estimates

$$\|u - y\|_{C_\tau(H^0)} + \|s_1 u - y\|_{C_\tau(\tilde{H}_\tau^0)} + \|s_1 I_t u - I_t y\|_{C(H^1)}$$

$$\leq c\left\{\left\|\hat{B}_h^{1/2}(s_1 u^{(0)} - y^{(0)})_h\right\|_h + \left(|h|^2 + N_0^{n-2}\tau^2\right)\left(\|u^{(0)}\|_{H^2}\right.\right.$$

$$\left.\left. + \|u^{(1)}\|_{H^1} + \|D_t^2 u\|_{L_1(H^1)} + \|D_t A u\|_{L_1(H^0)}\right)\right\} \qquad (7.13)$$

$$\left\|\bar{\partial}_t(u - y)\right\|_{C_\tau(H^0)} + \|s_1 u - y\|_{C_\tau(\tilde{H}_E)} \leq c\left\{\left\|s_1 u^{(0)} - y^{(0)}\right\|_{H^1}\right.$$

$$\left. + \left(|h|^2 + N_0^{n-2}\tau^2\right)\left(\|u^{(1)}\|_{H^2} + \|D_t^2 A u\|_{L_1(H^0)}\right)\right\} \qquad (7.14)$$

are valid. The condition $u \in L_1(H^2)$ is assumed in the given estimates.

Proof. Let $\eta \in S$, $\eta|_{t=T} = 0$. Identities (7.4) and (1.5) yield the identity for $s_1 s_t u - y$, that is,

$$\hat{B}_\tau(s_1 s_t u - y, \eta) = B_0(s_1 s_t u, \eta) - R_Q(D_t s_1 s_t u, D_t \eta) - B(u, \eta).$$

By virtue of identities (2.3) and (7.11) we can write

$$\hat{B}_\tau(s_1 s_t u - y, \eta) = B_0(s_1 s_t u, \eta) - B(u, \eta)$$
$$- \left((\hat{B}_h - B_h)(s_1 D_t u)_h, D_t \eta_h \right)_{L_2(H_h)}.$$

The error estimates derivation is based on Theorem 7.1 and is accomplished almost in the same way as in the proof of Theorem 3.1. The new term is only the last summand in the expression for $\hat{B}_\tau(s_1 s_t u - y, \eta)$. Its contribution into error estimate (7.13), according to Theorem 7.1, item (1), and Corollary 7.1, is represented by the quantity

$$\left\| \tilde{B}_h^{-1/2}(\hat{B}_h - B_h)(s_1 D_t u)_h \right\|_{L_1(H_h)} \leq c N_0^{n-2} \tau^2 \left\| D_t A u \right\|_{L_1(H^0)}.$$

Integration by parts yields

$$-\left((\hat{B}_h - B_h)(s_1 D_t u)_h, D_t \eta_h \right)_{L_2(H_h)}$$
$$= \left((\hat{B}_h - B_h)(s_1 u^{(1)})_h, \eta_h \right)_h$$
$$+ \left((\hat{B}_h - B_h)(s_1 D_t^2 u)_h, \eta_h \right)_{L_2(H_h)}.$$

Therefore, the contribution of the singled-out summand into error estimate (7.14) is majorized (similarly) by the quantity

$$c N_0^{n-2} \tau^2 \left(\|u^{(1)}\|_{H^2} + \|D_t^2 A u\|_{L_1(H^0)} \right).$$

Let us present the error estimates in the data classes. As it seen from the method of proof of Theorems 4.1–4.3, these estimates follow from Theorems 7.1, 7.2 (and Corollary 7.2).

Theorem 7.3. For the function y (instead of v) Theorems 4.1, 4.2, and 4.3 $\left(\text{except the choice of } y_0 = s_1 u^{(0)} \text{ in item (1e) and} \right.$ except item (2e)$\big)$ are valid. But it is necessary to replace $\beta_h^2 + \tau^2$ by $|h|^2 + N_0^{n-2} \tau^2$ and $v^{(0)}$ by $y^{(0)}$, and also to omit the operator,

$(\cdot)^{(1/2)}$. This latter action is necessary in Theorems 4.2 and 4.3, item (2a).

Remark 7.3. If the condition $\tau/h_i \le N_0$ $(1 \le i \le n)$ for $n \ge 2$ is fulfilled and $\beta_h^2 + \tau^2$ is replaced by $|h|^2 + N_0^{2(n-1)}\tau^2$, then the choice of $y_0 = s_1 u^{(0)}$ is possible in Theorem 4.3, item (1e).

Let us dwell on some supplements to Theorem 7.3 concerning the simplest choice of the initial values and the superconvergence of the gradient. For a function w continuous on $\bar{\Omega}$, $w|_{\partial\Omega} = 0$, introduce the function $sw \in S_1$ such that $sw(x_i) = w(x_i)$ for all i. Let $n = 2, 3$.

First, in the variant of Theorem 4.2 for y it is possible to replace $y_0 = s_1 u^{(0)}$ by $y_0 = su^{(0)}$ for $2 \le \alpha \le 4$. Second, in the variants of Theorems 4.1 and 4.2 for y it is possible to replace the operator s_1 by s for $1 \le \alpha \le 3$ and $2 \le \alpha \le 4$, respectively (for some values of α_1, α_2). In both cases, additional smoothness and summability conditions on coefficients a_{ij}, a, as well as slight modifications of the data norms, are required. Moreover, the condition $a_{ij} = 0$ for all $i \ne j$ is assumed in the second case. The derivation of these supplements is based on results of references 45 and 46. The details can be found in references 47 and 51.

Further results on methods with splitting operators for hyperbolic equations in the class of nonsmooth data are given in references 47–51. The literature on splitting methods is quite extensive (see reference 29); the finite-element splitting methods were introduced in reference 16. But error estimates of these methods for the problem under consideration are known largely for the case when the solution possesses sufficiently high smoothness. For $n = 2$ the case of moderate smoothness is discussed in references 22 and 23.

8.   TWO-LEVEL FINITE-ELEMENT METHOD

Initial-boundary value problems (1.3) and (1.4) can be rewritten as the system of first-order equations with respect to t,[28,6,8]

$$\rho D_t \dot{u} + Lu = \rho f, \quad D_t u - \dot{u} = 0$$

with the boundary and initial conditions

$$u|_\Gamma = 0, \quad \dot{u}|_{t=0} = \dot{u}^{(0)} \equiv u^{(1)}, \quad u|_{t=0} = u^{(0)}.$$

Let us consider as an approximate solution to the indicated system the vector function $\mathbf{y} = (\dot{y}, y) \in S \times S$, satisfying the integral identities,

$$\mathcal{B}^{(1)}(\mathbf{y}, \dot{\eta}) \equiv (D_t \dot{y}, \eta)_{0,Q} + \mathcal{L}_Q(y, \dot{\eta}) = (f, \dot{\eta})_{0,Q} \qquad \forall \dot{\eta} \in \bar{S} = S_1 \times \bar{S}_\tau,$$

$$(8.1)$$

$$(D_t y - \dot{y}, \eta)_{0,Q} = 0 \qquad \forall \eta \in \bar{S}, \tag{8.2}$$

and the initial conditions $\dot{y}|_{t=0} = s_0 \dot{u}^{(0)}$, $y|_{t=0} = s_k u^{(0)}$ ($k = 0$ or 1).

It is easy to transform identities (8.1) and (8.2) to the operator form

$$B_h \bar{\partial}_t \dot{y}_{h,m} + L_h y_{h,m}^{(1/2)} = [f^h]_m, \quad \bar{\partial}_t y_{h,m} - \dot{y}_{h,m}^{(1/2)} = 0, \qquad 1 \le m \le M. \tag{8.3}$$

The initial conditions can be written in the form

$$B_h \dot{y}_{h,0} = \dot{u}^{(0),h}; \quad B_h y_{h,0} = u^{(0),h}, \qquad k = 0;$$

$$L_h y_{h,0} = (A u^{(0)})^h, \qquad k = 1. \tag{8.4}$$

Evidently, the given projection (Bubnov–)Galerkin–Petrov method is a two-level one. In fact, it is the Crank–Nicholson–Galerkin method traditionally widely used in the parabolic case. Obviously, identity (8.2) can be replaced by the equivalent identity,

$$\mathcal{L}_Q(D_t y - \dot{y}, \eta) = 0 \quad \forall \eta \in \bar{S}. \tag{8.5}$$

This method is closely connected with the three-level method, with weight $\sigma = 1/4$ from Section 2. Let us exclude the function $\dot{y}$ from Equations (8.3). In order to do that, first apply the operators $\bar{\tau}^{-1}(\hat{\tau}(\cdot))^{(1/2)}$ and $\hat{\partial}_t$ to both Equations (8.3), respectively, and sum up the results. Second, for $m = 1$ apply the operators $(\tau_1/2)(\cdot)$ and B_h to the same equations, respectively, and again sum up the results. So we obtain equations for y,

$$B_h \hat{\partial}_t \bar{\partial}_t y_{h,m} + \frac{1}{4} L_h \hat{\partial}_t (\tau^2 \bar{\partial}_t y_h)_m + L_h y_{h,m} = \langle f^h \rangle_m^\tau,$$

$$1 \le m \le M - 1 \tag{8.6}$$

$$(B_h + \frac{1}{4} \tau_1^2 L_h) \partial_t y_{h,0} + \frac{\tau_1}{2} L_h y_{h,0} = B_h \dot{y}_{h,0} + \frac{\tau_1}{2} [f^h]_1 \tag{8.7}$$

where $\langle f^h \rangle_m^\tau = \bar{\tau}^{-1}(\hat{\tau}[\cdot])_m^{(1/2)}$ is the mean value on (t_{m-1}, t_m). They differ from Equations (2.11′) and (2.12) with $\sigma = 1/4$ only in right-hand parts. Moreover, since $B_h \dot{y}_{h,0} = u^{(1),h}$ (see Equation [8.4]), they differ only by way of the averaging of f with respect to t. Note that the value $\langle f^h \rangle^\tau$ is not determined on the classes $F^{<0,\alpha_2>}$ for $\alpha_2 < 0$.

At the same time, an important advantage of the two-level method is that the estimate expressing its stability in the energetic norm and the corresponding error estimates do not contain half-sums (i.e., the operators $(\cdot)^{(1/2)}$) and are valid for any nonuniform mesh, $\bar{\omega}^\tau$.

To establish the absolute stability of the two-level finite-element method, introduce the norm $\|w\|_{H_h^{-1}} = \sup_{\varphi \in S_{1,h}} (w, \varphi)_{0,\Omega} / \|\varphi\|_{H^1}$ for $w \in S_{1,h}$. It is obvious that $\|s_0 w\|_{H_h^{-1}} \leq \|w\|_{H^{-1}}$.

Theorem 8.1.

(1) If the function $\mathbf{y} \in S \times S$ satisfies identity (8.1) when we have $(f, \dot{\eta})_{0,Q} = \mathcal{L}_Q(f_1, \dot{\eta})$, $f_1 \in S_1 \otimes L_1(0, T)$, and the identity $(D_t y - \dot{y}, \eta)_{0,Q} = (g, \eta)_{0,Q}$ for all $\eta \in \bar{S}$, then

$$\|\mathbf{y}\|_{C_\tau(H_h^{-1} \times H^0)} \leq \|\mathbf{y_0}\|_{H_h^{-1} \times H^0} + 2\|f_1\|_{L_1(H^1)} + 2\|g\|_{L_1(H^0)}. \tag{8.8}$$

(2) If the function $\mathbf{y} \in S \times S$ satisfies identity (8.1) and the identity $\mathcal{L}_Q(D_t y - \dot{y}, \eta) = \mathcal{L}_Q(g, \eta)$ for all $\eta \in \bar{S}$, then

$$\|\mathbf{y}\|_{C_\tau(H^0 \times H^1)} \leq \|\mathbf{y_0}\|_{H^0 \times H^1} + 2\|f\|_{L_1(H^0)} + 2\|g\|_{L_1(H^1)}. \tag{8.9}$$

This estimate (8.9) remains valid while $\|f\|_{L_1(H^0)}$ is replaced by $\|f\|_{W_1^1(H^{-1})}$.

Here we use norms such that

$$\|\mathbf{y}\|^2_{H_h^{-1} \times H^0} = \|\dot{y}\|^2_{H_h^{-1}} + \|y\|^2_{H^0},$$

$$\|\mathbf{y}\|^2_{H^0 \times H^1} = \|\dot{y}\|^2_{H^0} + \|y\|^2_{H^1}.$$

Proof.

(1) The function $\mathbf{y}$ satisfies the equations

$$B_h \bar{\partial}_t \dot{y}_h + L_h y_h^{(1/2)} = L_h[f_1], \quad B_h \bar{\partial}_t y_h - B_h \dot{y}_h^{(1/2)} = [g^h].$$

Apply the operator L_h^{-1} to the first equation and then multiply it scalarly by $B_h \dot{y}_h^{(1/2)}$; the second equation multiply scalarly by $y_h^{(1/2)}$. Summing the results, we have

$$\left(L_h^{-1} B_h \bar{\partial}_t \dot{y}_h, B_h \dot{y}_h^{(1/2)} \right)_h + \left(B_h \bar{\partial}_t y_h, y_h^{(1/2)} \right)_h$$

$$= \left([f_1], B_h \dot{y}_h^{(1/2)} \right)_h + \left([g^h], y_h^{(1/2)} \right)_h.$$

By virtue of the formulas $(\bar{\partial}_t w_h) w_h^{(1/2)} = 1/2 \bar{\partial}_t (w_h^2)$ and $\|L_h^{-1/2} B_h w_h\|_h = \|w\|_{H_h^{-1}}$, we obtain the inequality

$$\frac{1}{2} \bar{\partial}_t \left(\|\dot{y}\|^2_{H_h^{-1}} + \|y\|^2_{H^0} \right) \leq \| [f_1] \|_{H^1} \|\dot{y}\|_{C_\tau(H_h^{-1})}$$

$$+ \| [g] \|_{H^0} \|y\|_{C_\tau(H^0)}.$$

This yields estimate (8.8).

(2) Let $1 \leq l \leq M$. In the identities for **y** choose $\dot{\eta}_m = \dot{y}_m^{(1/2)}$, $\eta_m = y_m^{(1/2)}$ for $1 \leq m \leq l$ and $\dot{\eta}_m = \eta_m = 0$ for $l < m \leq M$. Summing the results, we obtain the energetic equality and the estimate

$$\frac{1}{2} \left(\|\dot{y}_l\|_{H^0}^2 + \|y_l\|_{H^1}^2 \right) - \frac{1}{2} \left(\|\dot{y}_0\|_{H^0}^2 + \|y_0\|_{H^1}^2 \right)$$

$$= I_\tau^l \{ ([f], \dot{y}^{(1/2)})_{0,\Omega} + \mathcal{L}_\Omega([g], y^{(1/2)}) \}$$

$$\leq \|f\|_{L_1(H^0)} \|\dot{y}\|_{C_\tau(H^0)} + \|g\|_{L_1(H^1)} \|y\|_{C_\tau(H^1)}. \quad (8.10)$$

Hence estimate (8.9) holds true.

It is enough to obtain estimate (8.9) with another norm of f only for $y_0 = 0$, $g = 0$. Then $\dot{y}_m^{(1/2)} = \bar{\partial}_t y_m$ for $1 \leq m \leq M$, and with the help of property (2.4) in equality (8.10) we have

$$I_\tau^l \{ ([f], \dot{y}^{(1/2)})_{0,\Omega} = (f, D_t y)_{0,Q^l} = (f, y)_{0,\Omega|_{t=t_l}} - (D_t f, y)_{0,Q^l}$$

$$\leq (\|f\|_{C(H^{-1})} + \|D_t f\|_{L_1(H^{-1})}) \|y\|_{C(H^1)}.$$

Let us derive a priori error estimates of the order $O(\beta_h^2 + \tau_{\max}^2)$ relying on Theorem 8.1.

Theorem 8.2. For the approximate solution defined by identities (8.1) and (8.2) the a priori error estimates

$$\|u - y\|_{C_\tau(H^0)} + \|s_1 u - y\|_{C_\tau(H^0)} + \|s_0 \dot{u} - \dot{y}\|_{C_\tau(H_h^{-1})}$$

$$\leq c_0 \{ \|s_1 u^{(0)} - y_0\|_{H^0} + \|s_0 u^{(1)} - \dot{y}_0\|_{H_h^{-1}} + \beta_h^2 (\|u^{(0)}\|_{H^2}$$

$$+ \|D_t A u\|_{L_1(H^0)}) + \tau_{\max}^2 (\|D_t^2 u\|_{L_1(H^1)} + \|D_t^3 u\|_{L_1(H^0)}) \}$$

$$\quad (8.11)$$

and

$$\|D_t u - \dot{y}\|_{C_\tau(H^0)} + \|s_1 D_t u - \dot{y}\|_{C_\tau(H^0)} + \|s_1 u - y\|_{C(H^1)}$$

$$\leq c_0 \{ \|s_1 u^{(0)} - y_0\|_{H^1} + \|s_1 u^{(1)} - \dot{y}_0\|_{H^0}$$

$$+ \beta_h^2 (\|u^{(1)}\|_{H^2} + \|D_t A u\|_{L_1(H^0)})$$

$$+ \tau_{\max}^2 (\|D_t^2 u\|_{L_\infty(H^1)} + \|D_t^3 u\|_{L_1(H^1)} + \|D_t^2 A u\|_{L_1(H^0)}) \}$$

$$\quad (8.12)$$

are valid. Here the condition $u \in L_1(H^2)$ is assumed.

Proof. Introduce a vector function $\mathbf{r} = (\dot{r}, r) \in S \times S$ with $r = s_1 s_t u - y$. Let $\dot{\eta}, \eta \in \bar{S}$.

To derive estimate (8.11), set $\dot{r} = s_0 s_t \dot{u} - \dot{y}$. Using properties (2.1) and (2.3) of operators s_0, s_1, s_t and also identities (8.1) and (8.2), we obtain

$$\mathcal{B}^{(1)}(\mathbf{r}, \dot{\eta}) = (D_t s_0 s_t \dot{u}, \eta)_{0,Q} + \mathcal{L}_Q(s_1 s_t u, \eta) - \mathcal{B}^{(1)}(\mathbf{y}, \dot{\eta})$$

$$= (D_t \dot{u}, \dot{\eta})_{0,Q} + \mathcal{L}_Q(s_t u, \dot{\eta}) - \mathcal{B}^{(1)}(\mathbf{u}, \dot{\eta}) = \mathcal{L}_Q(s_t u - u, \dot{\eta})$$

$$(D_t r - \dot{r}, \eta)_{0,Q} = (D_t s_1 s_t u - s_0 s_t \dot{u}, \eta)_{0,Q}$$

$$= (D_t s_1 u - s_t \dot{u}, \eta)_{0,Q} = (D_t s_1 u - D_t u + \dot{u} - s_t \dot{u}, \eta)_{0,Q}$$

where $\mathbf{u} = (\dot{u}, u)$. By virtue of Theorem 8.1, item (1), estimate

$$\|\mathbf{r}\|_{C_\tau(H_h^{-1} \times H^0)} \leq \|\mathbf{r}_0\|_{H_h^{-1} \times H^0} + 2\|s_t u - u\|_{L_1(H^0)}$$

$$+ 2\big(\|D_t u - s_1 D_t u\|_{L_1(H^0)} + \|\dot{u} - s_t \dot{u}\|_{L_1(H^0)}\big)$$

is valid.

To derive estimate (8.11), set $\dot{r} = s_1 s_t \dot{u} - \dot{y}$. In this case we have (see Equations [8.1] and [8.5])

$$\mathcal{B}^{(1)}(\mathbf{r}, \dot{\eta}) = (D_t s_1 \dot{u}, \dot{\eta})_{0,Q} + \mathcal{L}_Q(s_t u, \dot{\eta}) - \mathcal{B}^{(1)}(\mathbf{u}, \dot{\eta})$$

$$= -(D_t \dot{u} - D_t s_1 \dot{u} + A(u - s_t u), \dot{\eta})_{0,Q}$$

$$\mathcal{L}_Q(D_t r - \dot{r}, \eta) = \mathcal{L}_Q(D_t s_1 s_t u - s_1 s_t \dot{u}, \eta) = \mathcal{L}_Q(D_t u - s_t u, \eta).$$

By virtue of Theorem 8.1, item (2), the estimate

$$\|\mathbf{r}\|_{C_\tau(H^0 \times H^1)} \leq \|\mathbf{r}_0\|_{H^0 \times H^1} + 2\big(\|D_t \dot{u} - s_1 D_t \dot{u}\|_{L_1(H^0)}$$

$$+ \|Au - s_t Au\|_{L_1(H^0)}\big) + 2\|D_t u - s_t D_t u\|_{L_1(H^1)}$$

is valid. The derivation of estimates (8.11) and (8.12) is completed in a similar way as in Theorem 3.1.

We indicate error estimates in the data classes for the two-level method.

Theorem 8.3.

(1) Let $\dot{y}_0 = s_0 u^{(1)}$, $y_0 = s_0 u^{(0)}$. The error estimate

$$\|u - y\|_{C_\tau(H^0)} \leq c\Big(T_1\big(\beta_h^2 + \tau_{\max}^2\big)\Big)^{\alpha/3} \|\mathbf{d}\|_\alpha \qquad (8.13)$$

is valid ($0 \leq \alpha \leq 3$). The numbers α_1, α_2 satisfy the same conditions as in Theorem 4.1, but additionally, $\max\{2\alpha_1 - 1, 0\} \leq \alpha_2$.

(2) Let $\dot{y}_0 = s_0 u^{(1)}$, $y_0 = s_1 u^{(0)}$. The error estimate

$$\|D_t u - \dot{y}\|_{C_\tau(H^0)} + \|s_1 u - y\|_{C(H^1)}$$
$$\leq c\big(T_1(\beta_h^2 + \tau_{\max})\big)^{(\alpha-1)/3}\|\mathbf{d}\|_\alpha. \quad (8.14)$$

is valid ($1 \leq \alpha \leq 4$). The numbers α_1, α_2 satisfy the same conditions as in Theorem 4.2, but additionally, $\alpha_2 \geq 1$ for $\alpha_1 > 0$.

(3) Items (1) and (2) remain valid if:

(a) weaken the norm $\|f\|_{F^{(\alpha_1,\alpha_2)}}$ to $\|f\|_{\bar{F}^{(\alpha_1,\alpha_2)}}$ for integer $\alpha_2 \geq 1$;

(b) for $\alpha_1 \leq 0$ replace the norm $\|f\|_{F^{(\alpha_1,\alpha_2)}}$ by $\|f\|_{\bar{F}^{<\alpha_1,\alpha_2>}}$.

Also, for $1 \leq \alpha \leq 5/2$ it is possible to add the summand $\|u - y\|_{C(H^1)}$ to the left-hand part of error estimate (8.14), and if additionally $\beta_h \alpha_h \leq N$, then the latter estimate is valid also for $y_0 = s_0 u^{(0)}$.

Remark 8.1. For $f = 0$ (or for $\alpha = 0, 3$) it is possible to add the summand $\|s_0 D_t u - \dot{y}\|_{C_\tau(H_h^{-1})}$ to the left-hand part of error estimate (8.13).

Proof. Estimates (8.13) (for $f = 0$, or for $\alpha = 0, 3$) and (8.14) are derived by the same method as in Theorems 4.1 and 4.2. Here Propositions 1.1–1.3, Theorems 8.1 and 8.2, and interpolation theory are essentially used.

Let us establish estimate (8.13) also for $f \not\equiv 0$, $0 < \alpha < 3$. It is enough to do this for $u^{(0)} = u^{(1)} = 0$. Let v be the solution to the three-level method with $\sigma = 1/4$ from Section 2. The difference, $v - y$, satisfies equations of (8.6) and (8.7) type, but with the right-hand parts $f^{h,\tau} - \langle f^h \rangle^\tau = \hat{\partial}_t[f^h \chi_\tau]$ and $(\tau_1/2)(f_0^{h,\tau} - [f^h]_1) = [f^h \chi_\tau]_1$. Here, $\chi_\tau(t) = 1/2(t_{m-1} + t_m) - t$ for $t_{m-1} < t < t_m$ ($1 \leq m \leq M$), and $f \in L_1(H^0)$.

According to Theorem 2.1, item (1), and with the help of equalities (2.14) (where for the nonuniform mesh one should replace ∂_t by $\hat{\partial}_t$ and τ by τ_1) and Remark 2.3, we have

$$\|v - y\|_{C(H^0)} \leq \|v - y\|_{C_\tau(H_\tau^0)}$$
$$+ \|I_t(v - y)\|_{C_\tau(H^1)} \leq 2\| [f^h \chi_\tau] \|_{L_1(H^0)}.$$

It is not difficult to verify that

$$\| [f^h \chi_\tau] \|_{L_1(H^0)} \le 1/2 \tau_{\max}^{\lambda+1} \| D_t^\lambda f \|_{L_1(H^0)}, \qquad \lambda = 0, 1.$$

Applying interpolation theory, we derive the estimate

$$\| v - y \|_{C(H^0)} \le c \tau_{\max}^{\alpha_2+1} \| f \|_{F^{(0,\alpha_2)}}, \qquad 0 \le \alpha_2 \le 1.$$

Using Theorem 4.1 to estimate $\| u - v \|_{C_\tau(H^0)}$, we obtain estimate
(8.13) for $\alpha_1 \ge 0$, $\alpha_2 \ge 0$, $2\alpha/3 \le \alpha_2 + 1$. With the help of
interpolation theory, when $-1 < \alpha_1 < 0$, $\alpha_2 = 0$ the last case
reduces to the already considered cases $(\alpha_1, \alpha_2) = (-1, 0), (0, 0)$.

Item 3 can be proved as were the corresponding items of
Theorem 4.3.

9. ABSTRACT SECOND-ORDER HYPERBOLIC EQUATIONS: APPLICATIONS TO DYNAMIC PROBLEMS OF MECHANICS

Let $H^{<0>}, H^{<1>}$ be abstract separable Hilbert spaces, and $H^{<1>}$ be
densely and compactly imbedded into $H^{<0>}$.[3] Let H^k be Hilbert spaces
coinciding with $H^{<k>}$ to within the norm equivalence ($k = 0, 1$). Introduce the Hilbert spaces $H^{<-1>}, H^{-1}$ and the operator A: $H^1 \to H^{-1}$
in accordance with the plan given in Section 1.1 for $\mathcal{L}_\Omega(\cdot, \cdot) \equiv (\cdot, \cdot)_{H^1}$.
Consider the Cauchy problem for the abstract hyperbolic equation

$$D_t^2 u + Au = f, \quad u|_{t=0} = u^{(0)}, \quad D_t u|_{t=0} = u^{(1)}. \tag{9.1}$$

All the basic results in Sections 1–4 and Section 8 are presented in
such a way that they can automatically be transferred into the case
of problem (9.1). This simple observation has important consequences
(see reference 26, Chapter V). The results for problems (1.3) and (1.4)
are generalized for the case of the initial-boundary value problem for a
strongly hyperbolic system of equations of the form

$$\rho D_t^2 \mathbf{u} + L^{(m)} \mathbf{u} = \rho \mathbf{f}$$

$$D_{\mathbf{n}}^l u|_\Gamma = 0, \quad 0 \le l < m, \quad u|_{t=0} = u^{(0)}, \quad D_t u|_{t=0} = u^{(1)}.$$

Here $\mathbf{u} = (u_1, \ldots, u_k)$ and $\mathbf{f} = (f_1, \ldots, f_k)$ are vector functions, $L^{(m)}$
is an operator of a strongly elliptic self-adjoint system of $2m$th order
differential equations ($m \ge 1$), $D_{\mathbf{n}}$ is the derivative with respect to
the outward normal, $\mathbf{n}$, to $\partial\Omega$. The operator $L^{(m)}$ is assumed to be
$H^{<1>} = [\overset{\circ}{W}_2^m (\Omega)]^k$-elliptic.

Thus, we discuss the problems:

(1) for $k = 3$, $m = 1$, $n = 3$, the system of equations of the dynamic (isotropic) elasticity theory,

$$\rho D_t^2 \mathbf{u} - \mu \triangle \mathbf{u} - (\lambda + \mu)D\text{div}\mathbf{u} = \rho \mathbf{f},$$

where $\lambda > 0$, $\mu > 0$, are the Lamé constants, $\triangle = D_i D_i$ is the Laplace operator;

(2) for $k = 1$, $m = 2$, $n = 2$, the equation of the transverse vibrations of nonhomogeneous thin plates (having variable thickness),

$$\rho D_t^2 u + \triangle(a\triangle u) = \rho f,$$

where $N^{-1} \le a(x)$, $\|a\|_{L_\infty(\Omega)} \le N$;

(3) for $k = 1$, $m = 2$, $n = 1$, the generalized equation of the transverse vibrations of nonhomogeneous bars (having variable cross-section),

$$\rho D_t^2 u + \sum_{0 \le l \le m} (-D)^l (a_l D^l u) = \rho f,$$

where $N^{-1} \le a_m(x)$, $\|a_l\|_{L_\infty(\Omega)} \le N$ for $0 \le l \le m$ (for $0 \le l < m$ the last condition can be weakened); it is important that the results of Section 5 are also transferred onto the case (3) for arbitrary $m \ge 2$.

In addition, the abstract approach enables us to consider other boundary conditions. For example, in Equation (1.4) the boundary condition $u|_\Gamma = 0$ can be replaced by

$$a_{ij}(D_j u)\cos \gamma_i + \sigma_0 u|_\Gamma = 0$$

where γ_i is the angle between $\mathbf{n}$ and the axis x_i, and $N^{-1} \le \sigma_0(x)$, $\|\sigma_0\|_{L_\infty(\partial\Omega)} \le N$. In this case, the role of $H^{<1>}$ is played by $W_2^1(\Omega)$.

REFERENCES

1. Amosov, A.A., and Amosova, O.A., Error estimates for FEM schemes constructed for degenerate diffusion equation with discontinuous coefficients, *Sov. Jour. Numer. Anal. Math. Modeling*, v. 1, 163, 1986.

2. Amosov, A.A., and Zlotnik, A.A., Difference schemes of second-order of accuracy for the equations of the one-dimensional motion of a viscous gas, *Comp. Math. and Math. Phys.*, v. 27, No. 4, 46, 1987.

3. Aubin J.–P., *Approximation of Elliptic Boundary–Value Problems*, Wiley-Interscience, New York, 1972.

4. Bachvalov, N.S., Berezin, B. I., and Bessonov, Yu. L., Interpolation theorems for operators subordinate to several derivatives, *Dokl. AN SSSR*, v. 256, 1296, 1981 (in Russian).

5. Baker, G.A., Error estimates for finite element methods for second order hyperbolic equations, *SIAM Jour. Numer. Anal.*, v. 13, 564, 1976.

6. Baker, G.A., and Bramble, J.H., Semidiscrete and single step fully discrete approximations for second order hyperbolic equations, *RAIRO. Anal. Numér.*, v. 13, 75, 1979.

7. Baker, G.A., and Dougalis, V.A., On the L^∞-convergence of Galerkin approximations for second order hyperbolic equations, *Math. Comp.*, v. 34, 401, 1980.

8. Baker, G.A., Dougalis, V.A., and Serbin, S.M., High order accurate two-step approximations for hyperbolic equations, *RAIRO. Anal. Numér.*, v. 13, 201, 1979.

9. Bales, L.A., Higher order single step fully discrete approximations for second order hyperbolic equations with time-dependent coefficients, *SIAM Jour. Numer. Anal.*, v. 23, 27, 1986.

10. Bales, L.A., Dougalis, V.A., and Serbin, S.M., Cosine methods for second order hyperbolic equations with time-dependent coefficients, *Math. Comp.*, v. 45, 67, 1985.

11. Bergh, J., and Löfström, J., *Interpolation Spaces. An Introduction*, Springer–Verlag, Berlin, 1976.

12. Brenner P., Thomeé, V., and Wahlbin, L.B., *Besov Spaces and Applications to Difference Methods for Initial Value Problems*, Springer–Verlag, Berlin, 1975.

13. Burkovskaya, V.L., Djuraev, I.N., and Moskal'kov, M.N., Convergence rate estimates for discretization methods for hyperbolic type equations in a class of generalized solutions, *Dokl AN Ukr.SSR Ser. A*, No. 4, 6, 1986 (in Russian).

14. Dougalis, V.A., Multistep-Galerkin methods for hyperbolic equations, *Math. Comp.*, v. 33, 563, 1980.

15. Dougalis, V.A., and Serbin, S.M., On the superconvergence of Galerkin approximations to second order hyperbolic equations, *SIAM Jour. Numer. Anal.*, v. 17, 431, 1980.

16. Douglas, J., Jr., and Dupont, T., Alternating–direction Galerkin methods on rectangles, *in Proc. of SYNSPADE–II*, Hubbard B., Ed., Academic Press, New York, p. 133, 1971.

17. Douglas, J., Jr., Dupont, T., and Wheeler, M.F., A quasi–projection analysis of Galerkin methods for parabolic and hyperbolic equations, *Math. Comp.*, v. 32, 345, 1978.

18. Dupont, T., L^2-estimates for Galerkin methods for second order hyperbolic equations, *SIAM Jour. Numer. Anal.*, v. 10, 880, 1973.

19. Hille, E., and Phillips, R.S., *Functional Analysis and Semi-Groups*, AMS, Coll. Publ., XXXI, Providence, R.I., 1957.

20. Hoppe, R.H.W., Interpolation of cosine operator functions, *Ann. Mat. Pura ed Appl.*, No. 136, 183, 1984.

21. Iosida K., *Functional Analysis*, Springer–Verlag, Berlin, 1965.

22. Jovanović, B.S., and Jovanović, L.D., Convergence of finite–difference scheme for second–order hyperbolic equations with variable coefficients, *IMA Jour. Numer. Anal.*, v. 7, 39, 1987.

23. Jovanović, B.S., Jovanović, L.D., and Süli E. E., Sur la convergence des schémas aux différences finies pour l'équation des ondes, *Z. angew. Math. Mech.*, v. 66, 308, 1986.

24. Kok, B., and Geveci, T., The convergence of Galerkin approximation schemes for second–order hyperbolic equations with dissipation, *Math. Comp.*, v. 44, 379, 1985.

25. Křižek, M., and Neittaanmäki, P., On superconvergence techniques, *Acta Appl. Math.*, v. 9, 175, 1987.

26. Ladyjenskaya, O.A., *Boundary Value Problems of Mathematical Physics*, Nauka, Moscow, 1973 (in Russian).

27. Ladyjenskaya, O.A., and Uraltseva, N. N., *Linear and Quasilinear Equations of Elliptic Type*, Academic Press, New York, 1968.

28. Lions, J.–L., and Magenes, E., *Problèmes aux Limites non Homogénes et Applications*, Vol. 1, Dunod, Paris, 1968.

29. Marchuk, G.I., *Splitting-up Methods*, Nauka, Moscow, 1988 (in Russian).

30. Nikol'skiĭ, S. M., *Approximation of Functions of Several Variables and Imbedding Theorems*, Springer–Verlag, Berlin, 1975.

31. Oganesjan, L.A., and Rukhovets, L.A., *Variational–Difference Methods for Solving Elliptic Equations*, Acad. of Sci. of Armenian SSR, Erevan, 1979 (in Russian).

32. Piskarev, S.I., Discretization of abstract hyperbolic equations, *Uch. Zapiski Tartusk. Univ.*, v. 500, 3, 1979 (in Russian).

33. Piskarev, S.I., Solving evolution equations of second order in Krein–Fattorini conditions, *Diff. Uravnenija*, v. 21, 1604, 1985 (in Russian); English translation in *Diff. Equations*, v. 21, 1985.

34. Rauch, J., On convergence of the finite element method for the wave equation, *SIAM Jour. Numer. Anal.*, v. 22, 245, 1985.

35. Samarskiĭ, A.A., *Theory of Difference Schemes*, Nauka, Moscow, 1977 (in Russian).

36. Scott, R., Application of Banach space interpolation to finite element theory, in *Functional Analysis Methods in Numerical Analysis*, Lecture Notes in Mathematics, v. 701, 298, Springer–Verlag, Berlin, 1979.

37. Sobolev, S.L., *Selected Questions of Function Spaces and Generalized Functions Theory*, Nauka, Moscow, 1989 (in Russian).

38. Triebel, H., *Interpolation Theory. Function Spaces. Differential Operators*, North–Holland, Amsterdam, 1978.

39. Wheeler, M.F., L_∞ estimates of optimal order for Galerkin methods for one-dimensional second order parabolic and hyperbolic equations, *SIAM Jour. Numer. Anal.*, v. 10, 908, 1973.

40. Zlotnik, A.A., On the convergence rate estimate of one projection-difference scheme for the one-dimensional non-stationary diffusion equation, *Vestnik Moscov Univ. Series XV. Vychisl. Mat. Kibern.*, No. 1, 58, 1979 (in Russian); English translation in *Moscow University Comp. Math. Cybernetics*, v. 3, 1979.

41. Zlotnik, A.A., A projection-difference scheme for the vibrating string equation, *Sov. Math. Dokl.*, v. 20, 290, 1979.

42. Zlotnik, A.A., Principle of choice for functions of bounded variation with values in the Banach space, *Vestnik Moscov Univ. Series 1. Mat. Meh.*, No. 3, 52, 1979 (in Russian); English translation in *Moscow University Math. and Mech.*, 1979.

43. Zlotnik, A.A., On the rate of convergence in $V_2(Q_T)$ of projection-difference schemes for parabolic equations, *Vestnik Moscov Univ. Series XV. Vychisl. Mat. Kibern.*, No. 1, 27, 1980 (in Russian); English translation in *Moscow University Comp. Math. Cybernetics*, v. 4, 1980.

44. Zlotnik, A.A., The rate of convergence of the projection-difference scheme with a splitting operator for parabolic equation, *Comp. Math. and Math. Phys.*, v. 20, No. 2, 155, 1980.

45. Zlotnik, A.A., On the rate of the convergence in $W_{2,h}^1$ of the variational-difference method for elliptic equations, *Sov. Math. Dokl.*, v. 28, 143, 1983.

46. Zlotnik, A.A., Convergence rate in the space $W_{2,h}^1$ of variational-difference schemes for elliptic equations, in *Prikladnaja Matematika*, Frolov, G.D., Ed., MGPI, Moscow, p. 3, 1986 (in Russian).

47. Zlotnik, A.A., Some finite-element and finite-difference methods for solving mathematical physics problems with nonsmooth data in an n-dimensional cube. Part I, *Sov. Jour. Numer. Anal. Math. Modeling*, v. 6, 421, 1991.

48. Zlotnik, A.A., Lower estimates for the error of three-level difference methods for solving the wave equation with data in the Hölder classes, *Mat. Zametki*, v. 51, No. 3, 140, 1992 (in Russian); English translation in *Math. Notes*, v. 51, 1992.

49. Zlotnik, A.A., Properties of the projective grid method with a quasidecoupled operator for second order hyperbolic equations, *Comp. Math. and Math. Phys.*, v. 32, No. 2, 459, 1992.

50. Zlotnik, A.A., The two-level projection-difference method with a splitting operator for the wave equation, *Mat. zametki*, v. 51, No. 4, 23, 1992 (in Russian); English translation in *Math. Notes*, v. 51, 1992.

51. Zlotnik, A.A., "To the Theory of Projection-mesh Methods for Mathematical Physics Problems in the Non-Smooth Data Classes", Doctor of Mathematics dissertation, MPGU, Moscow, 1992 (in Russian).

Fictitious Domain Methods and Computation of Homogenized Properties of Composites with a Periodic Structure of Essentially Different Components

N.S. Bakhvalov and A.V. Knyazev

There are many important physical problems that can be modeled using partial differential equations (PDE) with discontinuous coefficients. Jumps of the coefficients may be very large. We consider one class of these problems concerning composites of essentially different components.

Composites or composite materials are said to be media with a large number of nonhomogeneous inclusions, with small sizes at least in one direction. Stationary states of that media are described by elliptic PDE with highly oscillatory coefficients. Homogenization is said to be a process of finding homogeneous coefficients such that a solution of the original PDE can be approximated by the solution of the same PDE, but with these new homogeneous coefficients instead of old highly oscillatory coefficients. Computation of homogenized coefficients of composites with periodic structure reduces to solving a series of periodic boundary value problems for the original PDE in the domain, called a cell, of periodicity.[1] For composites of essentially different components, the coefficients of the PDE in the domain of periodicity are discontinuous and have large jumps.

0-8493-8947-X/94 /$0.00 + $.50

Another important source of PDE large coefficient jumps is the Fictitious Domain Method (see references 2–8). In this method the original boundary value problem for the PDE is changed into a new boundary value problem in a domain that covers the original one. In the new fictitious part of the domain the coefficients of the PDE are chosen to be near zero for the Neumann original boundary conditions, or very large for the Dirichlet original boundary conditions, such that a solution of the new problem approximates, or even coincides with, the desired solution. Therefore, the Fictitious Domain Method makes it possible to "improve" the shape of the original boundary, but leads to a PDE with very large coefficient jumps.

We consider here a nonstandard variant of the Fictitious Domain Method, namely, with periodic boundary conditions on the fictitious boundary. For this variant the problem produced by the Fictitious Domain Method looks exactly like a problem produced by the homogenization procedure we were talking about, that is, a periodic boundary value problem in a cube for elliptic PDE with discontinuous coefficients that can achieve, in the limit case, zero or infinitely large values.

There are several different difficulties associated with numerical solution of these problems. In the present paper we consider only one, the rate of convergence of preconditioned iterative methods. It is well known that, typically, the larger the coefficient jumps the slower is the convergence. Our main goal is to prove that with a special initial guess the rate of convergence does not depend on the value of the jumps. We do not touch here the mesh approximations of our PDE, but we can expect that for some natural types of approximations all our results still hold and the rate of convergence does not depend on the mesh size parameter.

The central idea of the proof is that with the special initial guess, the errors of all iterative steps belong to a subspace, and in this subspace our problem is well posed independent of the coefficient jumps.[4] We also consider some related questions, for example the dependence of the solution on small coefficients in a subdomain.[9]

Iterative methods with the same properties were developed earlier for the mixed formulation of PDE.[10,11] Our investigations are based on extension theorems and the Korn inequality.[12,13]

For our preliminary work in this area see references 14 and 15. Further analogous results can be found in references 16–19. The present paper is based on its variant in Russian.[20]

Finally, throughout the text we try to simplify the formulations, for example for a factor space we often use an element of a factor class instead of the factor class; without special comments we write equalities that must be viewed in appropriate functional spaces, not in the classical sense; and so forth. However, unique proper rigorous formulations are always possible.

1. HOMOGENIZATION

1.1. PROBLEMS OF HOMOGENIZATION

We consider in the space $\mathbf{R}^s$, typically with $s = 2$ or 3, a problem of homogenization of a composite material with the periodic structure specified by the unit cube $\Box = (0,1)^s \subset \mathbf{R}^s$, the domain of periodicity. The computation of averaged, or effective, characteristics of a composite material with the periodic structure reduces to the solution of s periodic boundary value problems with special forms for the right-hand sides,[2]

$$\frac{\partial}{\partial \xi_i} \left(A_{ij} \frac{\partial (N_k + \xi_k I)}{\partial \xi_j} \right) = 0, \qquad k = 1, \ldots, s, \qquad (1.1)$$

where the ξ with various indices are independent Cartesian variables in $\mathbf{R}^s$, $\boldsymbol{\xi} = (\xi_1, \ldots, \xi_s)^T$, $A_{ij} = A_{ij}(\boldsymbol{\xi})$ are $m \times m$ periodic matrices of composite coefficients subject to averaging, I is the identity matrix, and the $m \times m$ matrix solution, $N_k = N_k(\boldsymbol{\xi})$, must also be periodic.

Here and in subsequent text, summation for the repeated indices i and/or j from 1 to s is understood. The average coefficients are founded using

$$\hat{A}_{i_1 i_2} = \int_\Box A_{i_1 j}(\boldsymbol{\xi}) \frac{\partial (N_{i_2}(\xi) + \xi_{i_2} I)}{\partial \xi_j} \, d\Box, \qquad i_1, i_2 = 1, \ldots, s.$$

In the present paper we consider two basic equations, the diffusion and the elasticity, simultaneously. For the diffusion equation, $A_{ij}(\boldsymbol{\xi})$ and $N_k(\boldsymbol{\xi})$ are scalar functions, and I is the scalar unity. For the elasticity case, $A_{ij}(\boldsymbol{\xi})$ and $N_k(\boldsymbol{\xi})$ are $s \times s$ matrix functions, and I is the $s \times s$ unity matrix. To combine these cases, we define a natural parameter, m, and treat $A_{ij}(\boldsymbol{\xi})$ and $N_k(\boldsymbol{\xi})$ as $m \times m$ matrix functions and I as the $m \times m$ matrix unity.

For the diffusion equation we set $m = 1$. The case $m > 1$ corresponds to an elliptic system. We convert an elliptic system with $m = s$ to the elasticity equation using special requirements, which will be formulated in an explicit form every time we need them.

1.2. ASSUMPTIONS FOR COEFFICIENTS AND RIGHT-HAND SIDES

We denote by A the tensor of the coefficients $A_{ij} = \left(a_{ij}^{kl} \right)$ and assume the tensor, A, to be symmetric, that is,

$$A_{ij} = A_{ij}^T \quad \text{or} \quad a_{ij}^{kl} = a_{ji}^{lk}.$$

For the elasticity case we need the additional conditions

$$a_{ij}^{kl} = a_{kj}^{il} = a_{il}^{kj} = a_{ji}^{lk}.$$

The system of Equations (1.1) breaks up into m independent systems of partial differential equations for the columns of the matrices N_k of the form

$$\frac{\partial}{\partial \xi_i} \left(A_{ij} \frac{\partial \mathbf{u}}{\partial \xi_j} - \mathbf{f}_i \right) = 0, \tag{1.2}$$

where the solution $\mathbf{u} = \mathbf{u}(\boldsymbol{\xi})$ is an m component periodic vector function, and the right-hand sides — periodic vector functions $\mathbf{f}_i$, $i = 1, \ldots, s$, — are described by the rule that every $\mathbf{f}_i$ is the kth column, $k = 1, \ldots, m$ of the matrix A_{ij}. Note, however, that we will not use this restriction and consider Equations (1.2) under general conditions for $\mathbf{f}_i$.

1.3.   THE GENERALIZED FORMULATION

We define the torus, $\mathbf{T}$, the multiplication of s unit circles, by the standard identification of the opposite sides of the periodicity domain $\square$ and, further, consider periodic functions in $\mathbf{R}^s$ as functions defined on the torus $\mathbf{T}$, that is, the vector $\boldsymbol{\xi}$ of independent variables of Cartesian coordinates is now taken on the torus,

$$\boldsymbol{\xi} = (\xi_1, \ldots, \xi_s)^T \in \mathbf{T}. \tag{1.3}$$

It is conventional to consider the generalized formulation of problem (1.2). Thus, let all elements of the matrices $A_{ij}(\boldsymbol{\xi})$ be elements of the space $L_\infty(\mathbf{T})$ and all $\mathbf{f}_i \in \mathbf{L}_2(\mathbf{T}) = \{L_2(\mathbf{T})\}^m$. We define the Hilbert space, $\mathbf{H}$, as the factor space, $\mathbf{H} = \{W_2^1(\mathbf{T})\}^m / \mathbf{R}^m$, with the scalar product

$$\Lambda(\star, \star) = \left\langle \left(\frac{\partial \star}{\partial \xi_i}, \frac{\partial \star}{\partial \xi_i} \right) \right\rangle,$$

where $\langle \star \rangle$ denotes the (Lebesgue) integral of $\star$ over the torus, and $(\star, \star)$ is the natural scalar product of the vectors of $\mathbf{R}^m$.

The generalized solution of problem (1.2) on the torus, $\mathbf{T}$, is said to be a function $\mathbf{v} \in \mathbf{H}_E$ such that

$$\Lambda_A(\mathbf{u}, \mathbf{v}) = \left\langle \left(\mathbf{f}_i, \frac{\partial \mathbf{v}}{\partial \xi_i} \right) \right\rangle,$$

$$\Lambda_A(\mathbf{u}, \mathbf{v}) \stackrel{\text{def}}{=} \left\langle \left(A_{ij}(\boldsymbol{\xi}) \frac{\partial \mathbf{u}}{\partial \xi_j}, \frac{\partial \mathbf{v}}{\partial \xi_i} \right) \right\rangle, \qquad \mathbf{v} \in \mathbf{H}. \tag{1.4}$$

Note that with (piecewise) smooth coefficients, A_{ij}, and right-hand sides $\mathbf{f}_i$ the solution, $\mathbf{u}$, is (piecewise) smooth as well, which fulfills Equation (1.2) in a classical sense in domains of smoothness, and standard conditions

$$[\mathbf{u}] = \left[\left(A_{ij} \frac{\partial \mathbf{u}}{\partial \xi_j} - \mathbf{f}_i \right)_{\mathbf{n}} \right] = 0$$

on surfaces of discontinuity. Note also that an arbitrary generalized solution of Equation (1.2) fulfills Equation (1.2) in the sense of distributions of $\{W_2^{-1}(\mathbf{T})\}^m$.

1.4. AN ISOTROPIC PRECONDITIONER

Usually the matrix functions $A_{ij}(\boldsymbol{\xi}) = \left(a_{ij}^{kl}(\boldsymbol{\xi})\right)$ are subject to the conditions

$$0 < \underline{a} \le \frac{\sum a_{ij}^{kl}(\boldsymbol{\xi})\eta_i^k \eta_j^l}{\sum \eta_i^k \eta_j^l} \le \bar{a} < \infty, \qquad \boldsymbol{\xi} \in \mathbf{T}, \qquad (1.5)$$

for the elasticity case, $\eta_i^k = \eta_k^i$, with constants $\underline{a}$ and $\bar{a}$ independent of $\boldsymbol{\xi}$.

Here and in subsequent text, summation for the repeated indices k and/or l from 1 to m is understood. Also recall the rule that summation for the repeated indices i and/or j from 1 to s is understood. We put the signs $\sum$ when two independent summations are acquired in a single term.

With conditions (1.5), problem (1.4) is well posed, that is, there exists a unique solution $\mathbf{u} \in \mathbf{H}$ and

$$\Lambda(\mathbf{u}, \mathbf{u}) \le \text{const } \langle (\mathbf{f}_i, \mathbf{f}_i) \rangle.$$

Consider the iterative method

$$\Lambda \left(\frac{\mathbf{u}^{n+1} - \mathbf{u}^n}{\tau}, \mathbf{v} \right) + \Lambda_A(\mathbf{u}^n, \mathbf{v}) - \left\langle \left(\mathbf{f}_i, \frac{\partial \mathbf{v}}{\partial \xi_i} \right) \right\rangle = 0,$$

$$\mathbf{v} \in \mathbf{H}, \quad n = 0, 1, \dots, \qquad (1.6)$$

with an arbitrary initial guess of $\mathbf{u}^0 \in \mathbf{H}$.

It is easy to see that with an appropriate $\tau > 0$ iteration approximations $\mathbf{u}^n$ converge to a solution of problem (1.4) in $\mathbf{H}$, with the rate of a geometric progression whose convergence factor can be bounded above by a quantity depending only on $\underline{a}/\bar{a}$.

One has to solve periodic boundary value problems for the Poisson equation to implement iterative method (1.6). Effective numerical methods for such problems are well known (e.g., see reference 3).

1.5. AN ANISOTROPIC PRECONDITIONER

If the ratio $\underline{a}/\bar{a}$ is small due to anisotropy of the material, a generalization is sometimes justified: Assume the $m \times m$ matrices, $E_{ij} = \left(e_{ij}^{kl}\right)$, independent of $\boldsymbol{\xi}$, satisfy the conditions

$$e_{ij}^{kl} = e_{ji}^{lk}, \qquad e_{ij}^{kl}\eta_i^k \eta_j^l > 0 \quad \text{with } \boldsymbol{\eta} \neq 0, \qquad (1.7)$$

and for the elasticity case,

$$e_{ij}^{kl} = e_{kj}^{il} = e_{il}^{kj} = e_{ji}^{lk}, \qquad e_{ij}^{kl}\eta_i^k\eta_j^l > 0, \ \ \boldsymbol{\eta} \neq 0, \ \ \eta_i^k = \eta_k^i.$$

We set

$$\Lambda_E(\star, \star) \overset{def}{=} \left\langle \left(E_{ij}(\boldsymbol{\xi}) \frac{\partial \star}{\partial \xi_j}, \frac{\partial \star}{\partial \xi_i} \right) \right\rangle \quad \text{in } \mathbf{H}.$$

This bilinear form is a new scalar product in the space $\mathbf{H}$ and generates the norm $\sqrt{\Lambda_E(\star, \star)}$ equivalent to the original norm $\sqrt{\Lambda(\star, \star)}$ of the space $\mathbf{H}$. One of the two inequalities of the equivalence is the Korn inequality on torus $\mathbf{T}$. With this new scalar product we denote the function space $\mathbf{H}$ by $\mathbf{H}_E$.

We now redefine constants $\underline{a}$ and $\bar{a}$,

$$0 < \underline{a} \leq \frac{\sum a_{ij}^{kl}(\boldsymbol{\xi})\eta_i^k\eta_j^l}{\sum e_{ij}^{kl}\eta_i^k\eta_j^l} \leq \bar{a} < \infty, \qquad \boldsymbol{\xi} \in \mathbf{T}; \tag{1.8}$$

for the elasticity case, $\eta_i^k = \eta_k^i$.

With conditions (1.8), problem (1.4) is also well posed, that is, there exists a unique solution $\mathbf{u} \in \mathbf{H}_E$ and

$$\Lambda_E(\mathbf{u}, \mathbf{u}) \leq \text{const} \langle (\mathbf{f}_i, \mathbf{f}_i) \rangle.$$

We consider the iterative method with the new bilinear form,

$$\Lambda_E \left(\frac{\mathbf{u}^{n+1} - \mathbf{u}^n}{\tau}, \mathbf{v} \right) + \Lambda_A(\mathbf{u}^n, \mathbf{v}) = \left\langle \left(\mathbf{f}_i, \frac{\partial \mathbf{v}}{\partial \xi_i} \right) \right\rangle,$$

$$\mathbf{v} \in \mathbf{H}, \ n = 0, 1, \ldots, \tag{1.9}$$

with an arbitrary initial guess of $\mathbf{u}^0 \in \mathbf{H}$.

As in Section (1.4), one can prove that with an appropriate $\tau > 0$ iteration approximations $\mathbf{u}^n$ converge to a solution of problem (1.4) in $\mathbf{H}_E$ with the rate of a geometric progression whose convergence factor can be bounded above by a quantity depending only on $\underline{a}/\bar{a}$. In particular,

$$\Lambda_E(\epsilon^n, \epsilon^n) \leq q^{2n} \Lambda_E(\epsilon^0, \epsilon^0),$$

$$q = 1 - \frac{\underline{a}}{\bar{a}}, \quad \epsilon^n = \mathbf{u}^n - \mathbf{u}, \quad \text{if } \tau = \frac{1}{\bar{a}}. \tag{1.10}$$

By choosing E to make the ratio $\underline{a}/\bar{a}$ as close to one as possible, we can improve the convergence estimates as compared with method (1.6). It is also possible to use more complicated and faster methods, for example variation methods, like the conjugate gradient method. We further consider methods of the type in Equation (1.9) only for simplicity.

At every iteration of Equation (1.9) it is necessary to solve periodic boundary value problems for elliptic systems of equations, in particular for the diffusion equation, or for the linear elasticity equations. All the coefficients here are constants, and the Fourier method can be applied even in the case of anisotropy. We also note that Lamé equations on a torus with constant Lamé parameters are equivalent to several Poisson equations.[11,16]

1.6.   JUMPS OF COEFFICIENTS

Composite materials are characterized not so much by the anisotropy of its components as by the inhomogeneity occurring in large changes of the coefficients $a_{ij}^{kl}(\boldsymbol{\xi})$ in corresponding parts of the periodicity cell. In that situation the ratio $\underline{a}/\bar{a}$ is small for every choice of E given in Section 1.5.

We decompose the periodicity cell into two parts of the following types. In the "main part" we assume the coefficients $a_{ij}^{kl}(\boldsymbol{\xi})$ to be of the order of one. In the "inclusions" these coefficients may be very small, for example "like a cavity" inclusion, or very large, for example "almost rigid" inclusion. We also consider limit cases of cavities and rigid inclusions. Such inclusions are typical for composite materials.

Another important source of problems with these properties is the Fictitious Domain Method.[2-8] The cavities correspond to a Neumann-type boundary value problem in the main domain. The rigid inclusions lead to a Dirichlet-type boundary value problem in the main domain. A peculiarity of our approach for the Fictitious Domain Method is periodic boundary conditions on the fictitious boundary.

1.7.   DOMAIN ASSUMPTIONS

Let $\mathcal{D} \subset \mathbf{T}$ be itself a Lipschitz domain, that is, a connected open set with a Lipschitz boundary, or let it consist of a finite number of Lipschitz domains, $\mathcal{D}_p$, with nonintersecting closures.

Set $\mathcal{D}^\perp = \mathbf{T} \setminus \mathcal{D}$ and denote by $\mathcal{D}_q^\perp$ connected components of $\mathcal{D}^\perp$, if it is not connected itself. Take into account that there are a finite number of subdomains $\mathcal{D}_q^\perp$, and every one of them is a Lipschitz domain.

1.8.   MAIN TYPES OF JUMPS OF COEFFICIENTS

Assume these properties of the coefficients $a_{ij}^{kl}(\boldsymbol{\xi})$ in the main part $\mathcal{D}$ of the torus, $\mathbf{T}$,

$$0 < \underline{a}_{\mathcal{D}} \leq \frac{\sum a_{ij}^{kl}(\boldsymbol{\xi})\eta_i^k \eta_j^l}{\sum e_{ij}^{kl}\eta_i^k \eta_j^l} \leq \bar{a}_{\mathcal{D}} < \infty, \qquad \boldsymbol{\xi} \in \mathcal{D}, \qquad (1.11)$$

with $\eta_i^k = \eta_k^i$ for the elasticity case, $\underline{a}_\mathcal{D} \leq 1 \leq \bar{a}_\mathcal{D}$.

Consider the case

$$A_{ij}(\boldsymbol{\xi}) = \omega E_{ij}, \qquad \boldsymbol{\xi} \in \mathcal{D}^\perp, \ 0 \leq \omega \leq 1 \qquad (1.12)$$

with $\omega = 0$ in Section 2, which corresponds to cavities, and with $\omega > 0$ in Section 3. The multiparameteric case

$$A_{ij}(\boldsymbol{\xi}) = \omega_q E_{ij}, \qquad \boldsymbol{\xi} \in \mathcal{D}_q^\perp, \ 0 \leq \omega_q \leq 1 \qquad (1.13)$$

is treated in Section 4. In the present paper we do not consider problem (1.4) with conditions (1.11) and

$$A_{ij}(\boldsymbol{\xi}) = \omega E_{ij}, \qquad \boldsymbol{\xi} \in \mathcal{D}^\perp, \ 1 \leq \omega \leq +\infty, \qquad (1.14)$$

that corresponds to rigid inclusions for $\omega = +\infty$ (see, e.g., references 10, 11, and 14. We do not yet cover the multiparameteric case

$$A_{ij}(\boldsymbol{\xi}) = \omega_q E_{ij}, \qquad \boldsymbol{\xi} \in \mathcal{D}_q^\perp, \ 1 \leq \omega_q \leq +\infty; \qquad (1.15)$$

however, we expect that it is possible.

We also mention the conditions

$$0 < \omega \underline{a}_\mathcal{D} \leq \frac{\sum a_{ij}^{kl}(\boldsymbol{\xi}) \eta_i^k \eta_j^l}{\sum e_{ij}^{kl} \eta_i^k \eta_j^l} \leq \omega \bar{a}_\mathcal{D} < \infty, \qquad \boldsymbol{\xi} \in \mathcal{D}, \qquad (1.16)$$

with $\eta_i^k = \eta_k^i$ for the elasticity case, $\underline{a}_\mathcal{D} \leq 1 \leq \bar{a}_\mathcal{D}$, $A_{ij}(\boldsymbol{\xi}) = E_{ij}$, $\boldsymbol{\xi} \in \mathcal{D}^\perp$, which become conditions (1.11) and (1.12) or (1.11) and (1.14) after dividing by ω.

Note that our further results still hold for matrices E_{ij} dependent on $\boldsymbol{\xi}$ as well, if the condition of Equation (1.5) is fulfilled for E_{ij} instead of the inequality of Equation (1.7). However, a practical implementation of iterative method (1.9) becomes harder; inner iterative procedures should be involved. We do not consider that possibility in the present paper.

2. FICTITIOUS GRADIENTS METHOD: PERFORATED COMPOSITES

2.1. FORMULATION OF THE PROBLEM

We consider problem (1.4), that is,

$$\Lambda_A(\mathbf{u}, \mathbf{v}) = \left\langle \left(\mathbf{f}_i, \frac{\partial \mathbf{v}}{\partial \xi_i} \right) \right\rangle, \qquad \mathbf{v} \in \mathbf{H}, \qquad (2.1)$$

with conditions (1.11) and (1.12), with $\omega = 0$, that is,

$$0 < \underline{a}_\mathcal{D} \leq \frac{\sum a_{ij}^{kl}(\boldsymbol{\xi})\eta_i^k\eta_j^l}{\sum e_{ij}^{kl}\eta_i^k\eta_j^l} \leq \bar{a}_\mathcal{D} < \infty, \qquad \boldsymbol{\xi} \in \mathcal{D}, \qquad (2.2)$$

with $\eta_i^k = \eta_k^i$ for the elasticity case, and

$$A_{ij}(\boldsymbol{\xi}) = 0, \qquad \boldsymbol{\xi} \in \mathcal{D}^\perp.$$

Conditions (2.2) correspond to porous media, with the subdomain $\mathcal{D}^\perp$ constituting a set of cavities. Such media can be treated as perforated composite material. We prove in Section 2.3 that problem (2.1) with Equation (2.2) is the problem of the Fictitious Domain Method applied to the Neumann boundary value problem in $\mathcal{D}$.

Our approach to the solution of Equation (2.1) here has much in common with the familiar method.[2] The main difference is that we suggest periodic boundary conditions on the fictitious boundary, and the traditional choice is Dirichlet or Neumann boundary conditions. This leads to some difficulties with the theory, but leads also to advantages of a practical implementation, because the periodic boundary value problem for the elasticity equations can be solved efficiently.

2.2. MAIN RESULTS

We denote the tensor of the coefficients $A \equiv A^\omega$ by A^0 to underline that $\omega = 0$ in this section. The bilinear form, $\Lambda_{A^0}, (\star, \star)$ has a kernel

$$\mathbf{Ker} = \left\{\mathbf{w} \in \mathbf{H} : \left\langle\left(A_{ij}^0 \frac{\partial \mathbf{w}}{\partial \xi_j}, \frac{\partial \mathbf{v}}{\partial \xi_i}\right)\right\rangle = 0, \qquad \mathbf{v} \in \mathbf{H}\right\}.$$

This consists of vector functions $\mathbf{w} \in \mathbf{H}$, which must obey the equality $\mathbf{w} = \mathbf{c}_p$ with a constant vector, $\mathbf{c}_p$, in each component of connectedness, $\mathcal{D}_p$. For the elasticity case, $\mathbf{w} = \mathbf{c}_p + C_p\boldsymbol{\xi}$ in $\mathcal{D}_p$ with some vector $\mathbf{c}_p$ and an $s \times s$ matrix, $C_p = -C_p^T$, both independent of $\boldsymbol{\xi}$.

Here we comment on the use of the function $C_p\boldsymbol{\xi}$ in $\mathcal{D}_p$. If the $s \times s$ matrix $C = -C^T \neq 0$, then the function $C\boldsymbol{\xi}$ with $\boldsymbol{\xi} \in \mathbf{R}^s$ is not periodic. Therefore, when passing from $\mathbf{R}^s$ to the torus, $\mathbf{T}$, this function changes to a multiplace function with single-valued branches which differ from each other by constant vectors with integer components. Evidently, a restriction, $C\boldsymbol{\xi}\mid_\mathcal{D}$, of the function $C\boldsymbol{\xi}$ to an open subset, $\mathcal{D} \subset \mathbf{T}$, is multivalued also. We consider the restriction $C\boldsymbol{\xi}\mid_\mathcal{D}$ as a restriction of a function in the space $\mathbf{H}$, and only the continuous single-valued branch of the function $C\boldsymbol{\xi}\mid_\mathcal{D}$ can play that role. This motivates Definition 2.1.

Definition 2.1. A "continuous restriction," $C\boldsymbol{\xi}\mid_\mathcal{D}$, is called a continuous single-valued branch of the multiplace function, $C\boldsymbol{\xi}\mid_\mathcal{D}$

in $\mathcal{D}$, if such a branch exists for given matrix C and given open subset $\mathcal{D} \subset \mathbf{T}$.

Using this definition, we note that subspace **Ker** for the elasticity case consists of vector functions $\mathbf{w} \in \mathbf{H}$, which must obey the equality $\mathbf{w} = \mathbf{c}_p + C_p \boldsymbol{\xi}$ in $\mathcal{D}_p$ with some vector $\mathbf{c}_p$ and an $s \times s$ matrix $C_p = -C_p^T$, both independent of $\boldsymbol{\xi}$. The restriction $C\boldsymbol{\xi}\mid_{\mathcal{D}}$ must be continuous in each component of connectedness, $\mathcal{D}_p$. The condition of continuity of the restriction $C\boldsymbol{\xi}\mid_{\mathcal{D}}$ is fulfilled for all $s \times s$ matrices $C_p = -C_p^T$ independent of $\boldsymbol{\xi}$, if the domain $\mathcal{D}_p$ does not surround the torus, $\mathbf{T}$.

Definition 2.2. An open subset, $\mathcal{D} \subset \mathbf{T}$, is said to not "surround" the torus, $\mathbf{T}$, if all components of connectedness of the image of $\mathcal{D}$, when the torus, $\mathbf{T}$, is periodically extended on $\mathbf{R}^s$, are bounded in $\mathbf{R}^s$.

If $\mathcal{D}_p$ surrounds the torus, $\mathbf{T}$, then the condition of continuity of the restriction $C\boldsymbol{\xi}\mid_{\mathcal{D}}$ leads to additional constraints on the matrix $C_p = -C_p^T$ — it must have a kernel. For example, if $\mathcal{D}_p = \mathbf{T}$ or $\mathcal{D}_p = \{\boldsymbol{\xi} \in \mathbf{T} : 0 < \xi_1 < 1/2\}$, then $C_p = 0$. In general, the case of surrounding the constraints seems to be that some linear combinations of columns of C_p have integer coefficients equal to zero, and the values of the integer coefficients are defined by the directions of surrounding.

Let $\mathbf{v} \in \mathbf{Ker}$ in Equation (2.1); then

$$\left\langle \left(\mathbf{f}_i, \frac{\partial \mathbf{v}}{\partial \xi_i} \right) \right\rangle = 0, \qquad \mathbf{v} \in \mathbf{Ker} \tag{2.3}$$

is a necessary condition on the right-hand sides, $\mathbf{f}_i$, for the existence of a solution of problem (2.1) in the space $\mathbf{H}$. We will prove in Theorem 2.1 that it is a sufficient condition as well.

If problem (2.1) with Equation (2.2) has a solution $\mathbf{u} \in \mathbf{H}$ for given $\mathbf{f}_i$, then any function of the type $\mathbf{u} + \mathbf{w}$, $\mathbf{w} \in \mathbf{Ker}$ is also a solution. We will choose the normal solution from these.

Definition 2.3. The "normal" solution is a solution with minimal norm in $\mathbf{H}_E$ among all solutions for given $\mathbf{f}_i$.

Lemma 2.1. Let problem (2.1), with Equation (2.2), have a solution in $\mathbf{H}$ for given $\mathbf{f}_i$. Then the normal solution exists, is unique, and lies in the subspace

$$\mathbf{Im} = \{\mathbf{w} \in \mathbf{H} : \Lambda_E(\mathbf{w}, \mathbf{v}) = 0, \ \mathbf{v} \in \mathbf{Ker}\}.$$

The proof of Lemma 2.1 is based on the decomposition $\mathbf{H}_E = \mathbf{Ker} \oplus \mathbf{Im}$ and is a standard one.

Theorem 2.1. Assume conditions (2.2) and (2.3) are met. Let matrices E_{ij} satisfy the requirements of Section 1.5, and a set $\mathcal{D}$ complies with the requirements of Section 1.7. Then problem (2.1) has a unique normal solution, $\mathbf{u} \in \mathbf{H}$, and

$$\Lambda_E(\mathbf{u}, \mathbf{u}) \leq \text{const} \langle (\mathbf{f}_i, \mathbf{f}_i) \rangle. \tag{2.4}$$

Theorem 2.2 states that iterative method (1.9) can be applied for effective solution of problem (2.1) with Equation (2.2), as well as of problem (1.4) with Equation (1.8); but a special initial guess must be used.

Theorem 2.2. Let the conditions of Theorem 2.1 be satisfied. We consider iterative method (1.9),

$$\Lambda_E\left(\frac{\mathbf{u}^{n+1} - \mathbf{u}^n}{\tau}, \mathbf{v}\right) + \Lambda_{A^0}(\mathbf{u}^n, \mathbf{v}) = \left\langle\left(\mathbf{f}_i, \frac{\partial \mathbf{v}}{\partial \xi_i}\right)\right\rangle,$$

$$\mathbf{v} \in \mathbf{H}, n = 0, 1, \ldots \tag{2.5}$$

with the initial guess, $\mathbf{u}^0$, a solution of

$$\Lambda_E(\mathbf{u}^0, \mathbf{v}) = \left\langle\left(\mathbf{g}_i, \frac{\partial \mathbf{v}}{\partial \xi_i}\right)\right\rangle, \qquad \mathbf{v} \in \mathbf{H}, \tag{2.6}$$

where $\mathbf{g}_i \in \mathbf{L_2(T)}$ are arbitrary functions such that

$$\left\langle\left(\mathbf{g}_i, \frac{\partial \mathbf{v}}{\partial \xi_i}\right)\right\rangle = 0, \qquad \mathbf{v} \in \mathbf{Ker}.$$

Compare this with Equation (2.3), for example with $\mathbf{g}_i \equiv \mathbf{0}$.

For an appropriate $\tau > 0$, iteration approximations, $\mathbf{u}^n$, converge to the normal solution of problem (2.1) in $\mathbf{H}_E$, with the rate of a geometric progression whose convergence factor can be bounded above by a quantity depending only on $\kappa \underline{a}_{\mathcal{D}}/\bar{a}_{\mathcal{D}}$. In particular,

$$\Lambda_E(\epsilon^n, \epsilon^n) \leq q^{2n}\Lambda_E(\epsilon^0, \epsilon^0), \qquad q = 1 - \frac{\kappa \underline{a}_{\mathcal{D}}}{\bar{a}_{\mathcal{D}}}, \quad \text{if } \tau = \frac{1}{\bar{a}_{\mathcal{D}}}. \tag{2.7}$$

Here $\kappa > 0$ is the constant of the following proposition of extension in $\mathbf{H}_E$ from $\mathcal{D}$ to $\mathbf{T}$.

Proposition 2.1. For any function $\mathbf{v} \in \mathbf{H}$ there exists a function $\mathbf{w} \in \mathbf{H}$ such that

$$\int_{\mathcal{D}}\left(E_{ij}\frac{\partial \mathbf{v}}{\partial \xi_j}, \frac{\partial \mathbf{v}}{\partial \xi_i}\right)d\mathcal{D} \geq \kappa \int_{\mathbf{T}}\left(E_{ij}\frac{\partial \mathbf{w}}{\partial \xi_j}, \frac{\partial \mathbf{w}}{\partial \xi_i}\right)d\mathbf{T}, \qquad \mathbf{w} - \mathbf{v} \in \mathbf{Ker}.$$

This proposition plays an important role. The same statement is not true, in general, if the boundary of $\mathcal{D}$ is not Lipschitz. We will prove the proposition in the last section of this paper.

We now consider a special case of constant coefficients, $A_{ij}(\boldsymbol{\xi})$, in $\mathcal{D}$.

Theorem 2.3.   Let the conditions of Theorem 2.2 be satisfied. In addition, we assume that

$$A_{ij}^0(\boldsymbol{\xi}) = E_{ij}, \qquad \boldsymbol{\xi} \in \mathcal{D} \tag{2.8}$$

and take $\mathbf{g}_i = \mathbf{f}_i$ in Equation (2.6) to find an initial guess.

$$\text{Then} \quad \text{supp } \mathbf{r}^n = \text{supp} \; \frac{\partial}{\partial \xi_i} \left(E_{ij} \frac{\partial (\mathbf{u}^{n+1} - \mathbf{u}^n)}{\partial \xi_j} \right) \subseteq \partial \mathcal{D}, \quad (2.9)$$

$$\text{where} \qquad \mathbf{r}^n = \frac{\partial}{\partial \xi_i} \left(A_{ij} \frac{\partial \mathbf{u}^n}{\partial \xi_j} - \mathbf{f}_i \right) \in \{W_2^{-1}(\mathbf{T})\}^m$$

are residuals.

Proofs of all these theorems are based on the following key statement.

Lemma 2.2.   Let the conditions of Theorem 2.2 be satisfied. Then

(1) the initial guess of Equation (2.6) $\mathbf{u}^0 \in \mathbf{Im}$;

(2) in the subspace $\mathbf{Im}$

$$0 < \kappa \underline{a}_{\mathcal{D}} \le \frac{\Lambda_A(\mathbf{v}, \mathbf{v})}{\Lambda_E(\mathbf{v}, \mathbf{v})} \le \bar{a}_{\mathcal{D}} < \infty, \qquad \mathbf{v} \in \mathbf{Im}; \quad (2.10)$$

(3) the subspace $\mathbf{Im}$ is an invariant subspace of the operator $L : \mathbf{H} \to \mathbf{H}$ defined by

$$\Lambda_E(L\mathbf{w}, \mathbf{v}) = \Lambda_A(\mathbf{w}, \mathbf{v}), \qquad \mathbf{w}, \mathbf{v} \in \mathbf{H}. \tag{2.11}$$

2.3.   CONNECTION WITH THE FICTITIOUS DOMAIN METHOD

We now check that problem (2.1), with Equation (2.2), can be viewed as a problem of the Fictitious Domain Method applied to the Neumann boundary value problem in $\mathcal{D}$. To make our consideration simpler we assume in this section that for all f_i,

$$\mathbf{f}_i = \mathbf{f}_i(\boldsymbol{\xi}) = 0, \qquad \boldsymbol{\xi} \in \mathcal{D}^\perp. \tag{2.12}$$

Then problem (2.1) takes the form

$$\int_{\mathcal{D}} \left(A_{ij}^0(\boldsymbol{\xi}) \frac{\partial \mathbf{u}}{\partial \xi_j}, \frac{\partial \mathbf{v}}{\partial \xi_i} \right) d\mathcal{D} = \int_{\mathcal{D}} \left(\mathbf{f}_i, \frac{\partial \mathbf{v}}{\partial \xi_i} \right) d\mathcal{D}, \qquad \mathbf{v} \in \mathbf{H}. \qquad (2.13)$$

The subset $\mathcal{D} \subset \mathbf{T}$ is either connected by itself or consists of a finite number of connected components, $\mathcal{D}_p$, whose closures do not intersect, as was assumed in Section 1.7. For the first case,

$$\{W_2^1(\mathbf{T})\}^m \mid_{\mathcal{D}} = \{W_2^1(\mathcal{D})\}^m;$$

for the second case, the space $\{W_2^1(\mathbf{T})\}^m \mid_{\mathcal{D}}$ is a direct multiplication of spaces $\{W_2^1(\mathcal{D}_p)\}^m$ for all different p. Therefore, problem (2.13) is a variation formulation of the next Neumann boundary value problem,

$$\frac{\partial}{\partial \xi_i} \left(A_{ij}^0 \frac{\partial \mathbf{u}}{\partial \xi_j} - \mathbf{f}_i \right) = 0, \qquad \boldsymbol{\xi} \in \mathcal{D}, \left. \left(A_{ij}^0 \frac{\partial \mathbf{u}}{\partial \xi_j} - \mathbf{f}_i \right) \right|_{\mathbf{n} \mid \partial \mathcal{D}} = 0. \qquad (2.14)$$

For a nonconnected $\mathcal{D}$ this problem falls into several independent analogous problems in every domain $\mathcal{D}_p$. In other words, the solution of problem (2.1) with Equation (2.2) is, at the same time, a generalized solution in $\mathcal{D}$ of the Neumann boundary value problems (2.14). As we have already noted, an arbitrary function $\mathbf{w} \in \mathbf{Ker}$ can be added to a solution of problem (2.1) with Equation (2.2), and the sum also will be a solution. For the elasticity case, all kinds of functions $\mathbf{w} \in \mathbf{Ker}$ in $\mathcal{D}_p$ describe all possible shifts and rotations of $\mathcal{D}_p$ in $\mathbf{T}$. This corresponds to the well-known fact of linear elasticity theory,[12] that shifts and rotations are trivial solutions of the Neumann boundary value problem.

2.4.   CONDITIONS FOR THE RIGHT-HAND SIDES

We now consider conditions for the right-hand sides $\mathbf{f}_i$ more carefully.

The Neumann problem (2.14) has a solution for arbitrary right-hand sides $\mathbf{f}_i$, except for the elasticity case. For that case the same statement is true, if $\mathcal{D} = \mathbf{T}$. It is also well known in the elasticity case[12] that the necessary and sufficient condition of solvability of Equation (2.14) is the vanishing of the average moment of applied forces, that is,

$$F_p - F_p^T = 0, \qquad (2.15)$$

where the $s \times s$ matrix, F_p, is an average in $\mathcal{D}_p$ of an $s \times s$ matrix composed of the right-hand sides $\mathbf{f}_i$,

$$F_p = \int_{\mathcal{D}_p} (\mathbf{f}_1 \dots \mathbf{f}_m) \, d\mathcal{D},$$

if the $\mathcal{D}_p$ do not surround the torus, $\mathbf{T}$. For our problem (2.1) with Equation (2.2) we stated already in Theorem 2.1 that Equation (2.3) is the desired condition.

Assume Equation (2.12) for simplicity during the rest of this section, and prove that condition (2.3) always holds, except in the elasticity case. For that case we discover a specific analog of Equation (2.3) for general domains $\mathcal{D}_p$ and recognize that it is the same as Equation (2.15), if $\mathcal{D}$ is connected and does not surround the torus, $\mathbf{T}$.

In view of Equation (2.12), condition (2.3) is equivalent to a set of independent conditions,

$$\int_{\mathcal{D}_p} \left(\mathbf{f}_i, \frac{\partial \mathbf{v}}{\partial \xi_i} \right) d\mathcal{D}_p = 0, \qquad \mathbf{v} \in \mathbf{Ker}. \tag{2.16}$$

If $\mathbf{v} \in \mathbf{Ker}$, then $\mathbf{v} = \mathbf{c}_p$ in $\mathcal{D}_p$ with a constant vector $\mathbf{c}_p$, except in the elasticity case. Substituting $\mathbf{c}_p$ for $\mathbf{v}$ in Equation (2.16), we conclude that condition (2.16) is fulfilled.

The elasticity case is more complicated. In Equation (2.3) we substitute $\mathbf{v} = \mathbf{c}_p + C_p \boldsymbol{\xi}$ in $\mathcal{D}_p$, with a vector $\mathbf{c}_p$ and an $s \times s$ matrix, $C_p = -C_p^T$, both independent of $\boldsymbol{\xi}$. We differentiate (formally) with respect to ξ_i

$$\frac{\partial \mathbf{w}}{\partial \xi_i} = \frac{\partial (C_p \boldsymbol{\xi})}{\partial \xi_i}$$

equal to the ith column of matrix C_p. Therefore, condition (2.16) can be written in equivalent algebraic form,

$$\mathrm{tr} \left(F_p^T C_p \right) = 0.$$

Taking into account the equality $C_p = -C_p^T$ and the properties of the trace operation, we rewrite this as

$$\mathrm{tr} \left(\left\{ F_p - F_p^T \right\} C_p \right) = 0. \tag{2.17}$$

Here, F_p is an average in $\mathcal{D}_p$ of the $s \times s$ matrix composed of the right-hand sides $\mathbf{f}_i$, as in Equation (2.15), and C_p is an arbitrary $s \times s$ matrix, $C_p = -C_p^T$, independent of $\boldsymbol{\xi}$ such that the restriction $C_p \boldsymbol{\xi} \mid_{\mathcal{D}_p}$ is continuous (see Definition 2.1).

A collection of conditions (2.17) with Equation (2.12) for all different $\mathcal{D}_p$ is the desired concrete variant of the solvability condition (2.3) for the elasticity case. It is thus a necessary and sufficient condition of $\mathbf{f}_i$ for (generalized) solvability of the Neumann boundary value problem (2.14) for the elasticity case.

We illustrate the condition by several examples:

(1) Let $\mathcal{D}_p$ not surround the torus, $\mathbf{T}$ (see Definition 2.2). For example, let $\mathcal{D}_p = \{\boldsymbol{\xi} : \xi_i\xi_i < 1/4\}$. Then $C_p = -C_p^T$ can be arbitrary. We take $C_p = F_p - F_p^T$ in Equation (2.17) and are led to Equation (2.15).

(2) An opposite extreme case is when $\mathcal{D}_p$ surrounds the torus $\mathbf{T}$ in every Cartesian direction except perhaps one. For example, let $\mathcal{D}_p = \{\boldsymbol{\xi} : 0 < \xi_1 < 1/2\}$. Then $C_p = 0$, that is, the domain $\mathcal{D}_p$ cannot be rotated in $\mathbf{T}$. Condition (2.17) becomes trivially fulfilled for arbitrary $\mathbf{f}_i$.

(3) Let $\mathcal{D}_p = \{\boldsymbol{\xi} : 0 < \xi_1, \xi_2 < 1/2\}$, $s > 2$. Then in the matrix C_p there are only two nonvanishing elements, which have indexes (1,2) and (2,1). Using Equation (2.17), we conclude that

$$\int_{\mathcal{D}_p} (f_{12} - f_{21})\,d\mathcal{D}_p = 0.$$

Note that for the special right-hand sides, $\mathbf{f}_i$, of Equation (1.3), which come from the homogenization process, the equalities $f_{ik} = f_{ki}$ are true for the elasticity case. Therefore, condition (2.17) is fulfilled independently of the shape of $\mathcal{D}_p$.

2.5. PROOFS

Proof of Lemma 2.2.

(1) By definition of the subspace **Im** it is sufficient to check the equality

$$\Lambda_E(\mathbf{u}^0, \mathbf{v}) = 0, \qquad \mathbf{v} \in \mathbf{Ker}.$$

But it readily apparent from Equations (2.6) for $\mathbf{u}^0$.

(2a) Except the elasticity case: We first prove that the right inequality of Equation (2.10) holds even for an arbitrary $\mathbf{v} \in \mathbf{H}$. Taking into account Equation (2.2), we have

$$\Lambda_{A^0}(\mathbf{v}, \mathbf{v}) = \int_{\mathcal{D}} \left(A_{ij}^0(\boldsymbol{\xi}) \frac{\partial \mathbf{v}}{\partial \xi_j}, \frac{\partial \mathbf{v}}{\partial \xi_i} \right) d\mathcal{D}$$

$$\leq \bar{a}_{\mathcal{D}} \int_{\mathcal{D}} \left(E_{ij} \frac{\partial \mathbf{v}}{\partial \xi_j}, \frac{\partial \mathbf{v}}{\partial \xi_i} \right) d\mathcal{D}.$$

By requirements imposed on matrices E_{ij} in Section 1.5, the last integral cannot decrease when the domain $\mathcal{D}$ of integration is changed to $\mathbf{T}$, which concludes the proof of the right inequality.

The first step in checking the left inequality is the same,

$$\Lambda_{A^0}(\mathbf{v}, \mathbf{v}) \geq \underline{a}_{\mathcal{D}} \int_{\mathcal{D}} \left(E_{ij} \frac{\partial \mathbf{v}}{\partial \xi_j}, \frac{\partial \mathbf{v}}{\partial \xi_i} \right) d\mathcal{D}, \qquad \mathbf{v} \in \mathbf{H}.$$

The second (and last) step is to establish that

$$\int_{\mathcal{D}} \left(E_{ij} \frac{\partial \mathbf{v}}{\partial \xi_j}, \frac{\partial \mathbf{v}}{\partial \xi_i} \right) d\mathcal{D} \geq \kappa \int_{\mathbf{T}} \left(E_{ij} \frac{\partial \mathbf{v}}{\partial \xi_j}, \frac{\partial \mathbf{v}}{\partial \xi_i} \right) d\mathbf{T},$$

$$\mathbf{v} \in \mathbf{Im}. \qquad (2.18)$$

For this we use the proposition of extension for the function $\mathbf{v}$, that is, there is a function $\mathbf{w} \in \mathbf{H}$ such that

$$\int_{\mathcal{D}} \left(E_{ij} \frac{\partial \mathbf{v}}{\partial \xi_j}, \frac{\partial \mathbf{v}}{\partial \xi_i} \right) d\mathcal{D} \geq \kappa \int_{\mathbf{T}} \left(E_{ij} \frac{\partial \mathbf{w}}{\partial \xi_j}, \frac{\partial \mathbf{w}}{\partial \xi_i} \right) d\mathbf{T},$$

$$\mathbf{v} \in \mathbf{Im}.$$

and $\mathbf{w} - \mathbf{v} \in \mathbf{Ker}$. But our function $\mathbf{v}$ belongs to $\mathbf{Im}$, which is the orthogonal complement of $\mathbf{Ker}$ in $\mathbf{H}_E$; thus,

$$\int_{\mathbf{T}} \left(E_{ij} \frac{\partial \mathbf{w}}{\partial \xi_j}, \frac{\partial \mathbf{w}}{\partial \xi_i} \right) d\mathbf{T} = \Lambda_E(\mathbf{w}, \mathbf{w}) \geq \Lambda_E(\mathbf{v}, \mathbf{v}),$$

$$\mathbf{v} \in \mathbf{Im}.$$

Inequality (2.18), and thus Equation (2.10), are proved.

(b) For the elasticity case: Recall that for the elasticity case indexes i, j, k, and l vary from 1 to $s = m$; consider components v_l of the displacement vector $\mathbf{v} \in \mathbf{H}$, $\mathbf{v} = (v_1, \ldots, v_s)^T$ and define components

$$\epsilon_{lj} = \frac{1}{2} \left(\frac{\partial v_l}{\partial \xi_j} + \frac{\partial v_j}{\partial \xi_l} \right)$$

of the symmetric tensor of deformations ϵ. Using the symmetry of the tensor, A, of the elastic modulus, we obtain well-known equalities for components σ_{ki} of the stress tensor,

$$\sigma_{ki} = a_{ij}^{kl} \epsilon_{lj} = a_{ij}^{kl} \frac{\partial v_l}{\partial \xi_j}. \qquad (2.19)$$

From Equation (2.19) it follows that

$$\Lambda_{A^0}(\mathbf{v}, \mathbf{v}) = \int_{\mathcal{D}} \left(A_{ij}^0(\boldsymbol{\xi}) \frac{\partial \mathbf{v}}{\partial \xi_j}, \frac{\partial \mathbf{v}}{\partial \xi_i} \right) d\mathcal{D}$$

$$= \int_{\mathcal{D}} a_{ij}^{kl} \epsilon_{lj} \epsilon_{ki} d\mathcal{D}. \qquad (2.20)$$

Taking into account conditions (2.2), we find that

$$\underline{a}_\mathcal{D} \le \frac{\int_\mathcal{D} a_{ij}^{kl} \epsilon_{lj} \epsilon_{ki} d\mathcal{D}}{\int_\mathcal{D} e_{ij}^{kl} \epsilon_{lj} \epsilon_{ki} d\mathcal{D}} \le \bar{a}_\mathcal{D}. \tag{2.21}$$

Tensor $\mathbf{E}$ is symmetric, as is tensor $\mathbf{A}$; therefore, by analogy with Equations (2.19) and (2.20), the equalities

$$e_{ij}^{kl} \epsilon_{lj} = e_{ij}^{kl} \frac{\partial v_l}{\partial \xi_j},$$

$$\int_\mathcal{D} e_{ij}^{kl} \epsilon_{lj} \epsilon_{ki} d\mathcal{D} = \int_\mathcal{D} \left(E_{ij} \frac{\partial \mathbf{v}}{\partial \xi_j}, \frac{\partial \mathbf{v}}{\partial \xi_i} \right) d\mathcal{D}$$

hold. The remaining proof is the same as the previous one for the "except elasticity" case.

(3) We will see that $L\mathbf{w} \in \mathbf{Im}$ even for an arbitrary function, $\mathbf{w} \in \mathbf{H}$. We must check the equality $\Lambda_E(L\mathbf{w}, \mathbf{v}) = 0$ for $\mathbf{v} \in \mathbf{Ker}$. But this equality follows directly from the definition in Equation (2.11) of the operator L, because the subspace $\mathbf{Ker}$ is a kernel of the symmetric bilinear form, $\Lambda_{A^0}(\star, \star)$.

Note that operator $L : \mathbf{H} \to \mathbf{H}$ is selfadjoint in $\mathbf{H}_E$ and inequalities (2.11) are equivalent to

$$\kappa \underline{a}_\mathcal{D} I \le L \le \bar{a}_\mathcal{D} I \quad \text{in } \mathbf{Im} \subset \mathbf{H}_E. \tag{2.22}$$

We conclude that the image, $\mathbf{Im}\, L$, of the operator L is closed in $\mathbf{H}$ and $\mathbf{Im}\, L = \mathbf{Im}$, and the subspace $\mathbf{Ker}$ is a kernel of operator L.

Proof of Theorem 2.1. We represent vector $\mathbf{v} \in \mathbf{H}$ of Equation (2.1) as an orthogonal in $\mathbf{H}_E$ sum,

$$\mathbf{v} = \mathbf{v}_K + \mathbf{v}_I, \qquad \mathbf{v}_K \in \mathbf{Ker}, \ \ \mathbf{v}_I \in \mathbf{Im},$$

and use the additive property. Terms with $\mathbf{v}_K$ all vanish, and problem (2.1) takes the form

$$\Lambda_A(\mathbf{u}, \mathbf{v}) = \left\langle \left(\mathbf{f}_i, \frac{\partial \mathbf{v}}{\partial \xi_i} \right) \right\rangle, \qquad \mathbf{v} \in \mathbf{Im}. \tag{2.23}$$

A solution, $\mathbf{u}$, can be found in the subspace $\mathbf{Im}$, as was stated in Lemma 2.1. By Lemma 2.2 the bilinear form, $\Lambda_{A^0}(\star, \star)$ in the subspace $\mathbf{Im}$ is $\mathbf{H}_E$ bounded and $\mathbf{H}_E$ coercive. The linear functional, $\langle (\mathbf{f}_i, \partial \mathbf{v}/\partial \xi_i) \rangle$, $\mathbf{v} \in \mathbf{H}$, is $\mathbf{H}_E$ bounded because

$$\left| \left\langle \left(\mathbf{f}_i, \frac{\partial \mathbf{v}}{\partial \xi_i} \right) \right\rangle \right|^2 \leq \langle (\mathbf{f}_i, \mathbf{f}_i) \rangle \, \Lambda \, (\mathbf{u}, \mathbf{v})$$

$$\leq c \langle (\mathbf{f}_i, \mathbf{f}_i) \rangle \, \Lambda_E \, (\mathbf{u}, \mathbf{v}), \qquad \mathbf{v} \in \mathbf{H},$$

where for the elasticity case c is a constant of Korn's inequality for functions in a torus, $\mathbf{T}$.[12] Using standard arguments,[21] we conclude that problem (2.23) is correct in $\mathbf{H}_E$ and the constant in Equation (2.4) is equal to the ratio $c/\kappa \underline{a}_{\mathcal{D}}$.

We now prove the statement once more, taking advantage of an operator language. Equation (2.1) can be written in equivalent operator form as

$$L\mathbf{u} = \mathbf{w}, \qquad \mathbf{u} \in \mathbf{Im}, \ \mathbf{w} \in \mathbf{Im}, \tag{2.24}$$

where $\mathbf{w}$ is a canonical representation in $\mathbf{H}_E$ of the linear bounded functional $\langle (\mathbf{f}_i, \partial \mathbf{v}/\partial \xi_i) \rangle$, $\mathbf{v} \in \mathbf{H}$. By Lemma 2.2, the operator $L : \mathbf{Im} \to \mathbf{Im}$ has bounded in $\mathbf{H}_E$ an inverse with a norm less than or equal to the ratio $1/\kappa \underline{a}_{\mathcal{D}}$.

Proof of Theorem 2.2. First we check $\mathbf{u}^n \in \mathbf{Im}$, $n = 0, 1, \ldots$, using induction. By Lemma 2.2 we have $\mathbf{u}^0 \in \mathbf{Im}$. Let $\mathbf{u}^n \in \mathbf{Im}$. Setting $\mathbf{v} \in \mathbf{Ker}$ in Equation (2.5) yields

$$\Lambda_E(\mathbf{u}^{n+1} - \mathbf{u}^n, \mathbf{v}) = 0, \qquad \mathbf{v} \in \mathbf{Ker},$$

when using the definition of $\mathbf{Ker}$ and condition (2.3). This means that $\mathbf{u}^{n+1} - \mathbf{u}^n \in \mathbf{Im}$ by the definition of $\mathbf{Im}$.

Now define the error function, $\epsilon^n = \mathbf{u}^n - \mathbf{u}$ of the nth iteration for iterative process (2.5), where $\mathbf{u}$ is the normal solution of Equation (2.1). Lemma 2.1 states that the normal solution is $\mathbf{u} \in \mathbf{Im}$, therefore $\epsilon^n \in \mathbf{Im}$ as well. It is easy to rewrite iterative process (2.5) for errors $\epsilon^n \in \mathbf{Im}$ in the subspace $\mathbf{Im}$, that is,

$$\Lambda_E \left(\frac{\epsilon^{n+1} - \epsilon^n}{\tau}, \mathbf{v} \right) + \Lambda_{A^0} (\epsilon^n, \mathbf{v}) = 0,$$

$$\mathbf{v} \in \mathbf{H}, \ n = 0, 1, \ldots, \ \epsilon^0 \in \mathbf{Im}. \tag{2.25}$$

We can also convert iterations (2.25) to the operator form,

$$\epsilon^{n+1} = (I - \tau L)\epsilon^n, \qquad n = 0, 1, \ldots, \ \epsilon^0 \in \mathbf{Im}, \tag{2.26}$$

where operator $I - \tau L$ of the iteration step from ϵ^n to ϵ^{n+1} is a compression in $\mathbf{Im} \subset \mathbf{H}_E$, that is,

$$\| I - \tau L \|_E \leq q < 1 \quad \text{in } \mathbf{Im}$$

with some q if τ is an appropriate one. For example, $q = 1 - \kappa \underline{a}_D / \bar{a}_D$ if $\tau = 1/\bar{a}_D$. It follows from Equation (2.22) directly and completes the proof.

Proof of Theorem 2.3. We first note that

$$\tau \mathbf{r}^n = \frac{\partial}{\partial \xi_i} \left(E_{ij} \frac{\partial(\mathbf{u}^{n+1} - \mathbf{u}^n)}{\partial \xi_j} \right).$$

We now verify that conditions (2.2), without Equation (2.8), imply the inclusion

$$\operatorname{supp} \frac{\partial}{\partial \xi_i} \left(E_{ij} \frac{\partial(\mathbf{u}^{n+1} - \mathbf{u}^n)}{\partial \xi_j} \right) \subseteq \bar{\mathcal{D}}, \qquad (2.27)$$

and moreover, an arbitrary function $\mathbf{w} \in \mathbf{Im}$ can be used here instead of the difference $\mathbf{u}^{n+1} - \mathbf{u}^n$. That inclusion means

$$\Lambda_E(\mathbf{w}, \mathbf{v}) = \int_{\mathbf{T}} \left(E_{ij} \frac{\partial \mathbf{w}}{\partial \xi_j}, \frac{\partial \mathbf{v}}{\partial \xi_i} \right) d\mathbf{T} = 0 \qquad (2.28)$$

with a smooth function

$$\mathbf{v} \in \mathbf{H}, \qquad \operatorname{supp} \mathbf{v} \subseteq \mathbf{T} \setminus \mathcal{D}.$$

But any of the functions $\mathbf{v} \in \mathbf{Ker}$ and desired Equation (2.28) hold because of $\mathbf{w} \in \mathbf{Im}$.

It remains to find that condition (2.8) leads to

$$\operatorname{supp} \frac{\partial}{\partial \xi_i} \left(E_{ij} \frac{\partial(\mathbf{u}^{n+1} - \mathbf{u}^n)}{\partial \xi_j} \right) \subseteq \mathbf{T} \setminus \mathcal{D}. \qquad (2.29)$$

If so, then Equations (2.27) and (2.29) constitute Equation (2.9).

We will prove Equation (2.29) for a more general case in the next section (see the proof of Theorem 3.4).

3. COMPOSITES WITH INCLUSIONS OF A SOFT MATERIAL

3.1 FORMULATION OF THE PROBLEM

We consider Equation (1.4):

$$\Lambda_A(\mathbf{u}, \mathbf{v}) = \left\langle \left(\mathbf{f}_i, \frac{\partial \mathbf{v}}{\partial \xi_i} \right) \right\rangle, \qquad \mathbf{v} \in \mathbf{H}, \qquad (3.1)$$

with conditions (1.11) and (1.12), with $\omega > 0$, that is,

$$0 < \underline{a}_{\mathcal{D}} \le \frac{\sum a_{ij}^{kl}(\boldsymbol{\xi}) \eta_i^k \eta_j^l}{\sum e_{ij}^{kl} \eta_i^k \eta_j^l} \le \bar{a}_{\mathcal{D}} < \infty, \qquad \boldsymbol{\xi} \in \mathcal{D}, \ \bar{a}_{\mathcal{D}} \ge 1, \qquad (3.2)$$

with $\eta_i^k = \eta_k^i$ for the elasticity case, and

$$A_{ij}(\boldsymbol{\xi}) = \omega E_{ij}, \qquad \boldsymbol{\xi} \in \mathcal{D}^\perp, \ 0 < \omega \le 1.$$

For the elasticity case, conditions (3.2), with small ω, correspond to inclusions $\mathcal{D}^\perp$ of a soft material. If the material is anisotropic, then anisotropic axes must have the same directions in all inclusions $\mathcal{D}_q^\perp$.

Note that for $\omega > 0$ conditions (3.2) lead to Equation (1.8), and all the results of Section 1.5 hold. But $\underline{a}$ tends to zero with ω, therefore there is no convergence of iterative method (1.9) uniformly of $\omega \to 0$. The uniformity of $\omega \to 0$ correctness estimate for problem (3.1) with arbitrary $\mathbf{f}_i$ also is absent. Evidently, a norm of the solution of Equation (3.1) can tend to infinity as $\omega \to 0$.

The main goal of this section is to prove, however, a uniformity of $\omega \to 0$ convergence estimate of iterative method (1.9), with a special initial guess and a correctness estimate for the problem, with some restrictions on $\mathbf{f}_i$.

3.2   RESULTS

An analog of Theorem 2.1 for $\omega > 0$ is

Theorem 3.1.   Let

$$\mathbf{g}_i = \begin{cases} \text{an arbitrary function of } \mathbf{L}_2(\mathcal{D}) \text{ in } \mathcal{D}, \\[2mm] \dfrac{\mathbf{f}_i}{\omega} \text{ in } \mathcal{D}^\perp. \end{cases} \qquad (3.3)$$

For the elasticity case, the functions $\mathbf{g}_i$ must satisfy in every domain, $\mathcal{D}_p$, the conditions (compare with Equation [2.15])

$$\mathrm{tr} \left\{ (F_p - \omega\, G_p)\, C_p \right\} = 0, \qquad (3.4)$$

where F_p and G_p are averages in $\mathcal{D}_p$ of the matrices, with $\mathbf{f}_i$ and $\mathbf{g}_i$ as columns, and C_p is an arbitrary $s \times s$ matrix, $C_p = -C_p^T$, such that the restriction $C_p \boldsymbol{\epsilon} \, |_{\mathcal{D}_p}$ is continuous.

We consider problem (3.1) with conditions (3.2). Let matrices E_{ij} satisfy requirements of Section 1.5 and a set, $\mathcal{D}$, comply with the requirements of Section 1.7. Then problem (3.1) has a unique solution, $\mathbf{u} \in \mathbf{H}$, and

$$\Lambda_E(\mathbf{u}, \mathbf{u}) \le \text{const} \ \{\langle (\mathbf{f}_i, \mathbf{f}_i) \rangle + \langle (\mathbf{g}_i, \mathbf{g}_i) \rangle \}, \tag{3.5}$$

where the constant does not depend on ω.

As an example, we now explain condition (3.4) for a domain, $\mathcal{D}_p$, not surrounding the torus, $\mathbf{T}$. For this domain condition (3.4) is equivalent to

$$F_p - \omega\, G_p = F_p^T - \omega\, G_p^T. \tag{3.6}$$

(see Section 2.4). If $F_p = F_p^T$, then one can take $\mathbf{g}_i \equiv 0$ in $\mathcal{D}$ to satisfy Equation (3.6). If $F_p \ne F_p^T$, then the choice $\mathbf{g}_i = \mathbf{f}_i/\omega$ in $\mathcal{D}$ is possible to satisfy Equation (3.6). But for this choice, $\langle (\mathbf{g}_i, \mathbf{g}_i) \rangle$ of Equation (3.5) tends to infinity as ω tends to zero, if the $\mathbf{f}_i$ do not depend on ω. This associates with the statement that for the limit case, $\omega = 0$, condition (2.15), that is, $F_p = F_p^T$, is necessary if the domain $\mathcal{D}_p$ does not surround the torus, $\mathbf{T}$. Of interest is a solution $\mathbf{u}$ dependence of $\omega \to 0$.

Theorem 3.2. Consider problem (3.1) with conditions (3.2). Let matrices E_{ij} satisfy requirements of Section 1.5; a set, $\mathcal{D}$, complies with the requirements of Section 1.7. Let also the functions $\mathbf{f}_i$ be independent on ω, and, for the elasticity case, the functions $\mathbf{f}_i$ satisfy in every domain $\mathcal{D}_p$ the conditions

$$\text{tr}\{F_p\, C_p\} = 0, \tag{3.7}$$

where F_p is an average in $\mathcal{D}_p$ of the matrix, with $\mathbf{f}_i$ as columns. C_p is an arbitrary $s \times s$ matrix, $C_p = -C_p^T$, such that the restriction $C_p \boldsymbol{\xi} \,|_{\mathcal{D}_p}$ is continuous. Then

$$\mathbf{u} = \omega^{-1}\, \mathbf{u}_K + O(1) \quad \text{in } \mathbf{H}_E, \ \ \omega \to 0, \tag{3.8}$$

where $\mathbf{u}_K \in \mathbf{Ker}$ is a (unique) solution of the problem

$$\int_{\mathcal{D}^\perp} \left(E_{ij} \frac{\partial \mathbf{u}_K}{\partial \xi_j}, \frac{\partial \mathbf{v}}{\partial \xi_i} \right) d\mathcal{D}^\perp = \int_{\mathcal{D}^\perp} \left(\mathbf{f}_i, \frac{\partial \mathbf{v}}{\partial \xi_i} \right) d\mathcal{D}^\perp,$$

$$\mathbf{v} \in \mathbf{Ker}. \tag{3.9}$$

The subspace $\mathbf{Ker} \subset \mathbf{H}$ was defined in Section 2.2.

In the particular case $\mathbf{f}_i \equiv 0$ in $\mathcal{D}^\perp$, we have $\mathbf{u}_K \equiv 0$ and then

$$\mathbf{u} = \overset{0}{\mathbf{u}}_I + O(\omega) \quad \text{in } \mathbf{H}_E, \ \ \omega \to 0, \tag{3.10}$$

where $\overset{0}{\mathbf{u}}_I \in \mathbf{Im}$ is the normal in $\mathbf{H}_E$ solution of the problem

$$\int_{\mathcal{D}} \left(A_{ij} \frac{\partial \overset{0}{\mathbf{u}}_I}{\partial \xi_j}, \frac{\partial \mathbf{v}}{\partial \xi_i} \right) d\mathcal{D} = \int_{\mathcal{D}} \left(\mathbf{f}_i, \frac{\partial \mathbf{v}}{\partial \xi_i} \right) d\mathcal{D},$$

$$\mathbf{v} \in \mathbf{H}, \qquad (3.11)$$

that is, problem (3.1) with $\omega = 0$ in Equation (3.2) (see Section 3.2).

We note that the same representation of the solution for small ω was found for the Dirichlet boundary value problem of the diffusion equation.[9]

Remark 3.1. Using $\{\overset{0}{W_2^1} (\mathcal{D})\}^m \subseteq \mathbf{Ker}$, we find that the solution, $\mathbf{u}_K$, of problem (3.9) is, in $\mathcal{D}^{\perp}$, also a (generalized) solution of the boundary value problem

$$\frac{\partial}{\partial \xi_i} \left(E_{ij} \frac{\partial \mathbf{u}_K}{\partial \xi_j} - \mathbf{f}_i \right) = 0, \qquad \xi \in \mathcal{D}^{\perp}, \qquad (3.12)$$

with the next boundary conditions on every connected component, Γ, of the boundary $\partial \mathcal{D}^{\perp}$,

$$\mathbf{u}_K \mid_{\Gamma} = \mathbf{c}_{\Gamma} \quad \text{or for the elasticity case,} \ = \mathbf{c}_{\Gamma} + C_{\Gamma} \xi \mid_{\Gamma}, \quad (3.13)$$

where the vector $\mathbf{c}_{\Gamma}$ and the $s \times s$ matrix $C_{\Gamma} = -C_{\Gamma}^T$ are both independent of ξ. Vectors $\mathbf{c}_{\Gamma}$ and matrices C_{Γ} cannot, in general, be chosen arbitrarily. They are defined by the condition $\mathbf{u}_K \in \mathbf{Ker}$, that is, $\mathbf{u}_K$ in $\mathcal{D}_p$ is equal to $\mathbf{c}_p$, or, for the elasticity case, to $\mathbf{c}_p + C_p \xi$, with a constant vector, $\mathbf{c}_p$, and a constant $s \times s$ matrix, $C_p = -C_p^T$, such that the restriction $C_p \xi \mid_{\mathcal{D}_p}$ is continuous.

If the set $\mathcal{D}$ is connected, then there is only one p, $p = 1$, and vectors $\mathbf{c}_{\Gamma}$ and matrices C_{Γ} must be the same for all different components Γ of the boundary $\partial \mathcal{D} = \partial \mathcal{D}^{\perp}$. In that case, the difference between the solution, $\mathbf{u}_K$, of problem (3.9) and a solution of problem (3.12) with homogeneous Dirichlet boundary conditions on $\partial \mathcal{D}^{\perp}$ is simply a constant, or a constant plus a rotation for the elasticity case.

In other words, we can consider problem (3.1) with Equation (3.2) for small ω as a problem of the Fictitious Domain Method applied to the homogeneous Dirichlet boundary value problem for Equation (3.12) in a simple connected domain, $\mathcal{D}^{\perp}$, or in a finite collection of a simple connected domains, $\mathcal{D}_q^{\perp}$. We can find a solution for problem (3.1) with Equation (3.2) on torus $\mathbf{T}$, subtract an appropriate constant vector, or a shift plus a rotation for the elasticity case, and get an $O(\omega)$ in $\mathbf{H}_E$ approximation to a solution of problem (3.12) with homogeneous Dirichlet boundary conditions. We can even take the limit case, $\omega = 0$, to obtain an exact solution of problem (3.12) with homogeneous Dirichlet boundary conditions.

In the next theorem we state that iterative method (1.9) can be applied for an effective solution of problem (3.1) with Equation (3.2), as well as of problem (1.4) with Equation (1.8). But a special initial guess must be chosen by analogy with Section 3.2.

Theorem 3.3. Let the conditions of Theorem 3.1 be satisfied. We consider iterative method (1.9):

$$\Lambda_E \left(\frac{\mathbf{u}^{n+1} - \mathbf{u}^n}{\tau}, \mathbf{v} \right) + \Lambda_A(\mathbf{u}^n, \mathbf{v}) = \left\langle \left(\mathbf{f}_i, \frac{\partial \mathbf{v}}{\partial \xi_i} \right) \right\rangle,$$

$$\mathbf{v} \in \mathbf{H}, \quad n = 0, 1, \dots \quad (3.14)$$

with the initial guess $\mathbf{u}^0$, a solution of

$$\Lambda_E(\mathbf{u}^0, \mathbf{v}) = \left\langle \left(\mathbf{g}_i, \frac{\partial \mathbf{v}}{\partial \xi_i} \right) \right\rangle, \qquad \mathbf{v} \in \mathbf{H}. \quad (3.15)$$

For an appropriate $\tau > 0$, iteration approximations $\mathbf{u}^n$ converge to a solution of problem (3.1) in $\mathbf{H}_E$ with the rate of a geometric progression whose convergence factor can be bounded above by a quantity depending only on $\kappa \underline{a}_{\mathcal{D}}/\bar{a}_{\mathcal{D}}$, where $\kappa > 0$ is the constant of Proposition 2.1 of extension in $\mathbf{H}_E$ from $\mathcal{D}$ to $\mathbf{T}$. In particular,

$$\Lambda_E(\mathbf{u}^n - \mathbf{u}, \mathbf{u}^n - \mathbf{u},) \leq q^{2n} \Lambda_E \left(\mathbf{u}^0 - \mathbf{u}, \mathbf{u}^0 - \mathbf{u} \right),$$

$$q = 1 - \frac{\kappa \underline{a}_{\mathcal{D}}}{\bar{a}_{\mathcal{D}}}, \quad \text{if } \tau = \frac{1}{\bar{a}_{\mathcal{D}}}. \quad (3.16)$$

For the initial error, $\mathbf{u}^0 - \mathbf{u}$, we have

$$\Lambda_E(\mathbf{u}^0 - \mathbf{u}, \mathbf{u}^0 - \mathbf{u}) \leq \text{const} \ \{\langle (\mathbf{f}_i, \mathbf{f}_i) \rangle + \langle (\mathbf{g}_i, \mathbf{g}_i) \rangle\}, \quad (3.17)$$

where the constant does not depend on ω.

Remark 3.2. Let $\mathcal{D}$ be connected, and for the elasticity case also let none of the connected components, $\mathcal{D}_q^\perp$, of the complement $\mathcal{D}^\perp$ surround the torus $\mathbf{T}$. Then $\mathbf{Ker} = \{\overset{0}{W_2^1}(\mathcal{D})\}^m$, and statements of Theorem 3.1 and Theorem 3.3 still hold, if even arbitrary constant vectors, $\mathbf{c}_q$, were added to the functions, $\mathbf{g}_i$, of Equation (3.3) in $\mathcal{D}_q^\perp$, for example vectors

$$\mathbf{c}_q = -\frac{\int_{\mathcal{D}_q^\perp} \mathbf{f}_i d\mathcal{D}_q^\perp}{\omega \text{ mes } \mathcal{D}_q^\perp}.$$

If the right-hand side $\mathbf{f}_i$ do not depend on $\boldsymbol{\xi}$ in $\mathcal{D}^\perp$, then by using Equation (3.3) and adding these constant vectors to $\mathbf{g}_i$ of Equation (3.3) we find that $\mathbf{g}_i \equiv 0$ are possible as well. Hence, the estimate, Equation (3.5), becomes an ordinary inequality of correctness, and Equation (3.15) leads to the trivial initial guess, $\mathbf{u}^0 \equiv 0$.

We now consider the special case of constant coefficients $A_{ij}(\boldsymbol{\xi})$ in $\mathcal{D}$, as in the previous section.

Theorem 3.4. Let the conditions of Theorem 3.3 be satisfied. We consider the particular case of condition (3.2),

$$A_{ij}(\boldsymbol{\xi}) = E_{ij}, \qquad \boldsymbol{\xi} \in \mathcal{D},$$
$$A_{ij}(\boldsymbol{\xi}) = \omega E_{ij}, \qquad \boldsymbol{\xi} \in \mathcal{D}^\perp, \ 0 < \omega \leq 1, \qquad (3.18)$$

and the particular choice in Equation (3.3),

$$\mathbf{g}_i = \begin{cases} \mathbf{f}_i & \text{in } \mathcal{D}, \\ \dfrac{\mathbf{f}_i}{\omega} & \text{in } \mathcal{D}^\perp \end{cases} \qquad (3.19)$$

to find an initial guess, $\mathbf{u}^0$. Then

$$\text{supp } \mathbf{r}^n = \text{supp } \frac{\partial}{\partial \xi_i}\left(E_{ij}\frac{\partial(\mathbf{u}^{n+1} - \mathbf{u}^n)}{\partial \xi_j}\right) \subseteq \partial\mathcal{D}, \qquad (3.20)$$

where $\qquad \mathbf{r}^n = \dfrac{\partial}{\partial \xi_i}\left(A_{ij}\dfrac{\partial \mathbf{u}^n}{\partial \xi_j} - \mathbf{f}_i\right) \in \{W_2^{-1}(\mathbf{T})\}^m$

are residuals.

Proofs of these theorems are based on the next important statements.

Lemma 3.1. Consider problem (3.1) with conditions (3.2). Let matrices E_{ij} satisfy the requirements of Section 1.5, and let a set $\mathcal{D}$ comply with the requirements of Section 1.7. Then the decomposed representation

$$\Lambda_A(\mathbf{w}, \mathbf{v}) = \Lambda_A(\mathbf{w}_I, \mathbf{v}_I) + \omega \Lambda_E(\mathbf{w}_K, \mathbf{v}_K),$$
$$\mathbf{w} = \mathbf{w}_I + \mathbf{w}_K \in \mathbf{H}, \qquad \mathbf{v} = \mathbf{v}_I + \mathbf{v}_K \in \mathbf{H},$$
$$\mathbf{w}_I, \mathbf{v}_I \in \mathbf{Im}, \ \mathbf{w}_K, \mathbf{v}_K \in \mathbf{Ker}. \qquad (3.21)$$

does exist.

Lemma 3.2. Let the conditions of Theorem 3.3 be satisfied. Then

(1) the difference $\mathbf{u}^0 - \mathbf{u} \in \mathbf{Im}$ of the initial guess of Equation (3.15) and the solution of problem (3.1);

(2) in the subspace $\mathbf{Im}$ we have

$$0 < \kappa \underline{a}_{\mathcal{D}} \leq \frac{\Lambda_A(\mathbf{v}, \mathbf{v})}{\Lambda_E(\mathbf{v}, \mathbf{v})} \leq \bar{a}_{\mathcal{D}} < \infty, \qquad \mathbf{v} \in \mathbf{Im}; \qquad (3.22)$$

(3) subspace $\mathbf{Im}$ is an invariant subspace of the operator $L : \mathbf{H} \to \mathbf{H}$ defined by the rule (see Equation [2.10]),

$$\Lambda_E(L\mathbf{w}, \mathbf{v}) = \Lambda_A(\mathbf{w}, \mathbf{v}), \qquad \mathbf{w}, \mathbf{v} \in \mathbf{H}. \qquad (3.23)$$

3.3 PROOFS

Proof of Lemma 3.1. Set

$$\Lambda_{A^0}(\mathbf{w}, \mathbf{v}) = \int_{\mathcal{D}} \left(A_{ij} \frac{\partial \mathbf{w}}{\partial \xi_j}, \frac{\partial \mathbf{v}}{\partial \xi_i} \right) d\mathcal{D},$$

and note that the bilinear form $\Lambda_{A^0}(\star, \star)$ is equal to $\Lambda_A(\star, \star)$ with $\omega = 0$, and has a kernel $\mathbf{Ker}$, as described in the previous section. Thus,

$$\Lambda_A(\mathbf{w}, \mathbf{v}) = \Lambda_{A^0}(\mathbf{w}, \mathbf{v}) + \omega \int_{\mathcal{D}^\perp} \left(E_{ij} \frac{\partial \mathbf{w}}{\partial \xi_j}, \frac{\partial \mathbf{v}}{\partial \xi_i} \right) d\mathcal{D}^\perp. \qquad (3.24)$$

We substitute the orthogonal in $\mathbf{H}_E$ representations,

$$\mathbf{w} = \mathbf{w}_I + \mathbf{w}_K \in \mathbf{H}, \quad \mathbf{v} = \mathbf{v}_I + \mathbf{v}_K \in \mathbf{H},$$

$$\mathbf{w}_I, \mathbf{v}_I \in \mathbf{Im}, \quad \mathbf{w}_K, \mathbf{v}_K \in \mathbf{Ker}$$

into the right-hand side of Equation (3.22). The first term

$$\Lambda_{A^0}(\mathbf{w}, \mathbf{v}) = \Lambda_{A^0}(\mathbf{w}_I, \mathbf{v}_I),$$

because $\mathbf{Ker}$ is a kernel of this bilinear form. For the second term we must prove that

$$\int_{\mathcal{D}^\perp} \left(E_{ij} \frac{\partial \mathbf{w}}{\partial \xi_j}, \frac{\partial \mathbf{v}}{\partial \xi_i} \right) d\mathcal{D}^\perp = \int_{\mathcal{D}^\perp} \left(E_{ij} \frac{\partial \mathbf{w}_I}{\partial \xi_j}, \frac{\partial \mathbf{v}_I}{\partial \xi_i} \right) d\mathcal{D}^\perp$$

$$+ \int_{\mathcal{D}^\perp} \left(E_{ij} \frac{\partial \mathbf{w}_K}{\partial \xi_j}, \frac{\partial \mathbf{v}_K}{\partial \xi_i} \right) d\mathcal{D}^\perp.$$

This equality follows from

$$\int_{\mathcal{D}^\perp} \left(E_{ij} \frac{\partial \mathbf{w}_I}{\partial \xi_j}, \frac{\partial \mathbf{v}_K}{\partial \xi_i} \right) d\mathcal{D}^\perp = 0, \qquad \mathbf{w}_I \in \mathbf{Im}, \ \mathbf{v}_K \in \mathbf{Ker}.$$

(3.25)

Equality (3.25) in its turn is checked directly,

$$\int_{\mathcal{D}^\perp} \left(E_{ij} \frac{\partial \mathbf{w}_I}{\partial \xi_j}, \frac{\partial \mathbf{v}_K}{\partial \xi_i} \right) d\mathcal{D}^\perp = \Lambda_E(\mathbf{w}_I, \mathbf{v}_K)$$

$$- \int_{\mathcal{D}} \left(E_{ij} \frac{\partial \mathbf{w}_I}{\partial \xi_j}, \frac{\partial \mathbf{v}_K}{\partial \xi_i} \right) d\mathcal{D} = 0.$$

Here both terms are equal to zero, due to $\mathbf{H}_E$ orthogonality of $\mathbf{w}_I$ and $\mathbf{v}_K$, and $\mathbf{v}_K \in \mathbf{Ker}$.

Proof of Lemma 3.2.

(1) By definition of the subspace **Im**, we verify the equality

$$\Lambda_E(\mathbf{u}^0 - \mathbf{u}, \mathbf{v}) = 0, \qquad \mathbf{v} \in \mathbf{Ker}.$$

Use the immediate consequence of Lemma 3.1,

$$\Lambda_A(\mathbf{w}, \mathbf{v}) = \omega \Lambda_E(\mathbf{w}, \mathbf{v}), \qquad \mathbf{w} \in \mathbf{H}, \mathbf{v} \in \mathbf{Ker}. \quad (3.26)$$

Substitute into Equation (3.26), $\mathbf{w} = \mathbf{u}$, the solution of Equation (3.1), and obtain

$$\Lambda_E(\mathbf{u}, \mathbf{v}) = \frac{1}{\omega} \Lambda_A(\mathbf{u}, \mathbf{v}) = \left\langle \left(\frac{\mathbf{f}_i}{\omega}, \frac{\partial \mathbf{v}}{\partial \xi_i} \right) \right\rangle, \qquad \mathbf{v} \in \mathbf{Ker}.$$

Subtracting by parts this equality from Equation (3.15) yields

$$\Lambda_E(\mathbf{u}^0 - \mathbf{u}, \mathbf{v}) = \left\langle \left(\mathbf{g}_i - \frac{\mathbf{f}_i}{\omega}, \frac{\partial \mathbf{v}}{\partial \xi_i} \right) \right\rangle, \qquad \mathbf{v} \in \mathbf{Ker}.$$

From Equation (3.3) the right-hand side, $\mathbf{g}_i - \mathbf{f}_i/\omega \equiv 0$, in $\mathcal{D}^\perp$ and integrals

$$\int_{\mathcal{D}_p} \left(\mathbf{g}_i - \frac{\mathbf{f}_i}{\omega}, \frac{\partial \mathbf{v}}{\partial \xi_i} \right) d\mathcal{D}_p, \qquad \mathbf{v} = \mathbf{c}_p, \ \text{ or } \ \mathbf{c}_p + C_p \boldsymbol{\xi}$$

for the elasticity case, vanish because of condition (3.4) (see Section 2.4).

(2) Consider Equation (3.25), with $\mathbf{w} = \mathbf{v}$. Lower and upper estimates for the first term on the right-hand side of Equation (3.24) were obtained through Lemma 2.2. By $0 < \omega \leq 1 \leq \bar{a}_{\mathcal{D}}$, the last term is estimated trivially,

$$0 \leq \omega \int_{\mathcal{D}^{\perp}} \left(E_{ij} \frac{\partial \mathbf{v}}{\partial \xi_j}, \frac{\partial \mathbf{v}}{\partial \xi_i} \right) d\mathcal{D}^{\perp}$$

$$\leq \bar{a}_{\mathcal{D}} \int_{\mathcal{D}^{\perp}} \left(E_{ij} \frac{\partial \mathbf{v}}{\partial \xi_j}, \frac{\partial \mathbf{v}}{\partial \xi_i} \right) d\mathcal{D}^{\perp}.$$

(3) By definition in Equation (3.20) of the operator, L, we must check the last equality of

$$\Lambda_E(L\mathbf{w}, \mathbf{v}) = \Lambda_A(\mathbf{w}, \mathbf{v}) = 0, \qquad \mathbf{w} \in \mathbf{Im}, \mathbf{v} \in \mathbf{Ker}. \tag{3.27}$$

But this is just a particular case of equality Equation (3.26) for $\mathbf{w} \in \mathbf{Im}$.

We note that, as in Section 3.2, the operator $L : \mathbf{H} \to \mathbf{H}$ is selfadjoint in $\mathbf{H}_E$ and inequalities (3.22) are equivalent to the operator inequalities in $\mathbf{H}_E$,

$$\kappa \underline{a}_{\mathcal{D}} I \leq L \leq \bar{a}_{\mathcal{D}} I \quad \text{in } \mathbf{Im} \subset \mathbf{H}_E. \tag{3.28}$$

Proof of Theorem 3.1. We introduce function $\mathbf{u}^0$ as a solution of problem (3.15), with $\mathbf{g}_i$ of Equation (3.3), and consider the difference, $\mathbf{u}^0 - \mathbf{u}$, of this function and the solution $\mathbf{u}$ of the original problem, Equation (3.1). We write

$$\Lambda_A(\mathbf{u}^0 - \mathbf{u}, \mathbf{v}) = \Lambda_A(\mathbf{u}^0, \mathbf{v}) - \left\langle \left(\mathbf{f}_i, \frac{\partial \mathbf{v}}{\partial \xi_i} \right) \right\rangle,$$

$$\mathbf{v} \in \mathbf{H}, \qquad (3.29)$$

and we present vector $\mathbf{v} \in \mathbf{H}$ here as an orthogonal in $\mathbf{H}_E$ sum,

$$\mathbf{v} = \mathbf{v}_K + \mathbf{v}_I, \qquad \mathbf{v}_K \in \mathbf{Ker}, \quad \mathbf{v}_I \in \mathbf{Im}.$$

We will now show that all terms with $\mathbf{v}_K$ of Equation (3.29) vanish. For the left-hand side, $\Lambda_A(\mathbf{u}^0 - \mathbf{u}, \mathbf{v}) = 0$ because of Equation (3.27), as $\mathbf{u}^0 - \mathbf{u} \in \mathbf{Im}$ by Lemma 3.2 and $\mathbf{v}_K \in \mathbf{Ker}$. For the right-hand side our checking becomes more complicated. Using Equation (3.26), we rewrite the first term with $\mathbf{v} = \mathbf{v}_K$,

$$\Lambda_A(\mathbf{u}^0, \mathbf{v}_K) = \omega \Lambda_E(\mathbf{u}^0, \mathbf{v}_K) = \omega \left\langle \left(\mathbf{g}_i, \frac{\partial \mathbf{v}}{\partial \xi_i} \right) \right\rangle,$$

$$\mathbf{v}_K \in \mathbf{Ker},$$

using the definition in Equation (3.15) of $\mathbf{u}^0$ for the last equality. Making that substitution, we obtain the right-hand side of Equation (3.29) in the form

$$\left\langle \left(\omega \mathbf{g}_i - \mathbf{f}_i, \frac{\partial \mathbf{v}_K}{\partial \xi_i} \right) \right\rangle.$$

This value is zero owing to condition (3.4).

Thus, we just proved that it is sufficient to put $\mathbf{v} \in \mathbf{Im}$ in Equation (3.29). Lemma 3.2 states that $\mathbf{u}^0 - \mathbf{u} \in \mathbf{Im}$, and the symmetric bilinear form $\Lambda_A(\star,\star)$ is $\mathbf{H}_E$ bounded and $\mathbf{H}_E$ coercive in the subspace $\mathbf{Im}$ uniformity of $\omega > 0$.

We now estimate uniformity of ω, the $\mathbf{H}_E$ norm of the linear functional of the right-hand side of Equation (3.29). We will not use $\mathbf{v} \in \mathbf{Im}$ here.

For the first term,

$$|\Lambda_A(\mathbf{u}^0, \mathbf{v})|^2 \leq \Lambda_A(\mathbf{u}^0, \mathbf{u}^0)\Lambda_A(\mathbf{v}, \mathbf{v})$$

$$\leq \bar{a}_D^2 \Lambda_E(\mathbf{u}^0, \mathbf{u}^0)\Lambda_E(\mathbf{v}, \mathbf{v}),$$

where $$\Lambda_E(\mathbf{u}^0, \mathbf{u}^0) \leq \text{const } \langle(\mathbf{g}_i, \mathbf{g}_i)\rangle \qquad (3.30)$$

with a constant independent of ω.

For the second term,

$$\left| \left\langle \left(\mathbf{f}_i, \frac{\partial \mathbf{v}}{\partial \xi_i} \right) \right\rangle \right|^2 \leq \text{const } \langle(\mathbf{f}_i, \mathbf{f}_i)\rangle \Lambda_E(\mathbf{v}, \mathbf{v}), \qquad \mathbf{v} \in \mathbf{H},$$

with a constant independent of ω, as well.

Now combine these estimates,

$$\left| \Lambda_A(\mathbf{u}^0, \mathbf{v}) - \left\langle \left(\mathbf{f}_i, \frac{\partial \mathbf{v}}{\partial \xi_i} \right) \right\rangle \right|^2 \leq \text{const } \{\langle(\mathbf{f}_i, \mathbf{f}_i)\rangle$$

$$+ \langle(\mathbf{g}_i, \mathbf{g}_i)\rangle\}\Lambda_E(\mathbf{v}, \mathbf{v}).$$

Therefore, problem (3.29) is well posed in the subspace $\mathbf{Im}$ of $\mathbf{H}_E$ uniformly of ω,

$$\Lambda_E(\mathbf{u}^0 - \mathbf{u}, \mathbf{u}^0 - \mathbf{u}) \leq \text{const } \{\langle(\mathbf{f}_i, \mathbf{f}_i)\rangle + \langle(\mathbf{g}_i, \mathbf{g}_i)\rangle\}, \qquad (3.31)$$

with a constant independent of ω.

By triangle inequality, we conclude from Equation (3.31) and using Equation (3.30) that Equation (3.5), the desired estimate, holds.

Proof of Theorem 3.2. There is a special representation, Equation (3.21), of the bilinear form $\Lambda_A(\star, \star)$. We write the same representation for the linear functional of the right-hand side,

$$\left\langle \left(\mathbf{f}_i, \frac{\partial \mathbf{v}}{\partial \xi_i} \right) \right\rangle = \left\langle \left(\mathbf{f}_i, \frac{\partial \mathbf{v}_I}{\partial \xi_i} \right) \right\rangle + \left\langle \left(\mathbf{f}_i, \frac{\partial \mathbf{v}_K}{\partial \xi_i} \right) \right\rangle,$$

$$\mathbf{v} = \mathbf{v}_k + \mathbf{v}_I, \quad \mathbf{v}_I \in \mathbf{Im}, \quad \mathbf{v}_K \in \mathbf{Ker}. \tag{3.32}$$

Then the original problem falls into two independent problems,

$$\Lambda_A(\mathbf{u}_I, \mathbf{v}_I) = \left\langle \left(\mathbf{f}_i, \frac{\partial \mathbf{v}_I}{\partial \xi_i} \right) \right\rangle, \quad \mathbf{u}_I, \mathbf{v}_I \in \mathbf{Im}, \tag{3.33}$$

$$\Lambda_A(\mathbf{u}_K, \mathbf{v}_K) = \left\langle \left(\mathbf{f}_i, \frac{\partial \mathbf{v}_K}{\partial \xi_i} \right) \right\rangle, \quad \mathbf{u}_K, \mathbf{v}_K \in \mathbf{Ker}, \tag{3.34}$$

and the solution $\mathbf{u}$ of Equation (3.1) is an $\mathbf{H}_E$ orthogonal sum,

$$\mathbf{u} = \mathbf{u}_I + \frac{\mathbf{u}_K}{\omega}, \tag{3.35}$$

where $\mathbf{u}_I$ and $\mathbf{u}_K$ are solutions of Equations (3.33) and (3.34).

We first consider Equation (3.8). We will prove that problem (3.34) is equivalent to Equation (3.9) with conditions (3.7), has a unique solution, and that the term $\mathbf{u}_I$ of Equation (3.35) is bounded in $\mathbf{H}_E$ with uniformity of ω. These statements lead to Equation (3.8).

The equivalence of Equations (3.34) and (3.9) is verified directly. Condition (3.7) is involved for the elasticity case in the last of the equalities,

$$\int_{\mathcal{D}} \left(\mathbf{f}_i, \frac{\partial \mathbf{v}}{\partial \xi_i} \right) d\mathcal{D} = \sum_p \int_{\mathcal{D}_p} \left(\mathbf{f}_i, \frac{\partial \mathbf{v}}{\partial \xi_i} \right) d\mathcal{D}_p = 0,$$

$$\mathbf{v} = \mathbf{c}_p + C_p \xi,$$

where a vector, $\mathbf{c}_p$, and an $s \times s$ matrix, $C_p = -C_p^T$, both independent of $\boldsymbol{\xi}$, and restriction $C\boldsymbol{\xi} \mid_{\mathcal{D}}$ is continuous in each component of connectedness, $\mathcal{D}_p$ (see Section 2.4).

Equation (3.34) is evidently well posed; a unique solution exists and is equal to the $\mathbf{H}_E$ orthoprojection to the subspace $\mathbf{Ker}$, of the canonical representation in $\mathbf{H}_E$ of the linear bounded functional $\langle (\mathbf{f}_i, \partial \mathbf{v}_K / \partial \xi_i) \rangle$, $\mathbf{v}_K \in \mathbf{H}_E$.

To obtain, uniformly of ω, an estimate for the $\mathbf{H}_E$ norm of the function $\mathbf{u}_I$, consider Equation (3.33). As noted in the proof of Theorem 3.1, by Lemma 3.2 the symmetric bilinear form, $\Lambda_A(\star, \star)$, is $\mathbf{H}_E$ bounded and $\mathbf{H}_E$ coercive in the subspace $\mathbf{Im}$, with constants $\bar{a}_{\mathcal{D}}$ and $\kappa \underline{a}_{\mathcal{D}}$, that is, a uniformity of $\omega > 0$. Therefore,

$$\Lambda_E(\mathbf{u}_I, \mathbf{u}_I) \leq \text{const } \langle (\mathbf{f}_i, \mathbf{f}_i) \rangle, \tag{3.36}$$

where const and $\mathbf{f}_i$ are independent of ω, and Equation (3.8) is completely proved.

We now consider the particular case $\mathbf{f}_i \equiv 0$ in $\mathcal{D}^\perp$, when $\mathbf{u}_K \equiv 0$, and verify that the term in the series of powers of ω for $\mathbf{u}_I$, which corresponds to the zero power, is the function $\overset{0}{\mathbf{u}}_I$ of Equation (3.11). This statement constitutes the second part of the theorem.

It was discovered in Theorem 2.1 that under conditions (3.7) problem (3.11) is well posed in the subspace $\mathbf{Im} \subset \mathbf{H}_E$, has a unique solution $\overset{0}{\mathbf{u}}_I \in \mathbf{Im}$ by Lemma 2.1, and it is possible to choose $\mathbf{v} \in \mathbf{Im}$ in Equation (3.11) instead of $\mathbf{v} \in \mathbf{H}$. With $\mathbf{f}_i \equiv 0$ in $\mathcal{D}^\perp$ we obtain from Equation (3.11) and Equation (3.33) the equation for the difference $\overset{0}{\mathbf{u}}_I - \mathbf{u}_I \in \mathbf{Im}$,

$$\Lambda_A(\overset{0}{\mathbf{u}}_I - \mathbf{u}_I, \mathbf{v}) = \omega \int_{\mathcal{D}^\perp} \left(E_{ij} \frac{\partial \overset{0}{\mathbf{u}}_I}{\partial \xi_j}, \frac{\partial \mathbf{v}}{\partial \xi_i} \right) d\mathcal{D}^\perp,$$

$$\mathbf{v} \in \mathbf{Im}. \tag{3.37}$$

The right-hand side of Equation (3.37) contains a linear functional bounded in $\mathbf{H}_E$ by $O(\omega)$. However, as mentioned several times already, an equation with the bilinear form $\Lambda_A(\star, \star)$ is well posed in the subspace $\mathbf{Im} \subset \mathbf{H}_E$ uniformly of ω. Hence, $\overset{0}{\mathbf{u}}_I - \mathbf{u}_I = O(\omega)$ (see Equation [3.10]).

Proof of Theorem 3.3. This proof coincides very closely with that for Theorem 2.2. The estimate, Equation (3.17), was already derived as Equation (3.31) in the proof of Theorem 3.1.

Proof of Theorem 3.4. First note that

$$\tau \mathbf{r}^n = \frac{\partial}{\partial \xi_i} \left(E_{ij} \frac{\partial (\mathbf{u}^{n+1} - \mathbf{u}^n)}{\partial \xi_j} \right),$$

exactly as in the proof of Theorem 2.3. Also, as found there, using the inclusion $\{\overset{0}{W_2^1}(\mathcal{D}^\perp)\}^m \subseteq \mathbf{Ker}$,

$$\text{supp}\ \frac{\partial}{\partial \xi_i}\left(E_{ij}\frac{\partial \mathbf{v}}{\partial \xi_j}\right) \subseteq \overline{\mathcal{D}}, \qquad \mathbf{v} \in \mathbf{Im}.$$

As in Lemma 3.2, $\mathbf{u}^0 - \mathbf{u} \in \mathbf{Im}$, and from $\mathbf{u}^n - \mathbf{u} \in \mathbf{Im}$ it follows that $\mathbf{u}^{n+1} - \mathbf{u} \in \mathbf{Im}$. Therefore, $\mathbf{u}^{n+1} - \mathbf{u}^n \in \mathbf{Im}$, and the difference, $\mathbf{u}^{n+1} - \mathbf{u}^n$, can be substituted in the previous inclusion, $\{\overset{0}{W_2^1}(\mathcal{D}^\perp)\}^m \subseteq \mathbf{Ker}$, instead of $\mathbf{w} \in \mathbf{Im}$, that is,

$$\text{supp}\ \frac{\partial}{\partial \xi_i}\left(E_{ij}\frac{\partial(\mathbf{u}^{n+1}-\mathbf{u}^n)}{\partial \xi_j}\right) \subseteq \overline{\mathcal{D}}. \qquad (3.38)$$

We now verify that

$$\text{supp}\ \frac{\partial}{\partial \xi_i}\left(E_{ij}\frac{\partial(\mathbf{u}^{n+1}-\mathbf{u}^n)}{\partial \xi_j}\right) \subseteq \overline{\mathcal{D}^\perp} = \mathbf{T}\setminus\mathcal{D}. \qquad (3.39)$$

Let $\mathbf{Ker}^\perp \subset \mathbf{H}$ be a subspace that consists of vector functions equal to $\mathbf{c}_q$, or, for the elasticity case, to $\mathbf{c}_q + C_q\boldsymbol{\xi}$ in every $\mathcal{D}_q^\perp$, where a vector $\mathbf{c}_q$ and an $s \times s$ matrix, $C_q = -C_q^T$, are both independent of $\boldsymbol{\xi} \in \mathcal{D}_q^\perp$. Also, restriction $C_q\boldsymbol{\xi}\ |_{\mathcal{D}}$ is continuous in each component of connectedness, $\mathcal{D}_q^\perp$, of $\mathcal{D}^\perp$. This definition differs from the definition of $\mathbf{Ker}$ only in that the domains $\mathcal{D}_q^\perp$ play the role of the domains $\mathcal{D}_p$.

Let $\mathbf{Im}^\perp$ be an orthogonal complement of $\mathbf{Ker}^\perp$ in $\mathbf{H}_E$. Following the proof of Theorem 2.3, and using $\{\overset{0}{W_2^1}(\mathcal{D})\}^m \subseteq \mathbf{Ker}^\perp$, we find that

$$\text{supp}\ \frac{\partial}{\partial \xi_i}\left(E_{ij}\frac{\partial \mathbf{v}}{\partial \xi_j}\right) \subseteq \overline{\mathcal{D}^\perp} = \mathbf{T}\setminus\mathcal{D}, \qquad \mathbf{v} \in \mathbf{Im}^\perp.$$

By the conditions of the theorem, $A_{ij}(\boldsymbol{\xi}) = E_{ij}$ and $\mathbf{g}_i = \mathbf{f}_i$ in every domain, $\mathcal{D}_p$. By analogy with the arguments of Lemma 3.2 we can state that $\mathbf{u}^0 - \mathbf{u} \in \mathbf{Im}^\perp$, and from $\mathbf{u}^n - \mathbf{u} \in \mathbf{Im}^\perp$ it follows $\mathbf{u}^{n+1} - \mathbf{u} \in \mathbf{Im}^\perp$. Therefore, $\mathbf{u}^{n+1} - \mathbf{u}^n \in \mathbf{Im}^\perp$, and Equation (3.39) holds.

Note in this proof of Equation (3.39) that condition $A_{ij}(\boldsymbol{\xi}) = \omega E_{ij}$ in $\mathcal{D}^\perp$ with $\omega > 0$ was not involved. We conclude that Equation (3.39) is also true for $\omega = 0$ (see Equation [2.29]).

4. COMPOSITES WITH INCLUSIONS OF SOFT MATERIALS AND WITH CAVITIES

4.1   FORMULATION OF THE PROBLEM

We consider problem (1.4):

$$\Lambda_A(\mathbf{u}, \mathbf{v}) = \left\langle \left(\mathbf{f}_i, \frac{\partial \mathbf{v}}{\partial \xi_i} \right) \right\rangle, \qquad \mathbf{v} \in \mathbf{H}, \tag{4.1}$$

with conditions (1.11) and (1.13), that is,

$$0 < \underline{a}_\mathcal{D} \le \frac{\sum a_{ij}^{kl}(\boldsymbol{\xi}) \eta_i^k \eta_j^l}{\sum e_{ij}^{kl} \eta_i^k \eta_j^l} \le \bar{a}_\mathcal{D} < \infty, \qquad \boldsymbol{\xi} \in \mathcal{D}, \ \bar{a}_\mathcal{D} \ge 1, \tag{4.2}$$

with $\eta_i^k = \eta_k^i$ for the elasticity case,

$$A_{ij}(\boldsymbol{\xi}) = \omega_q E_{ij}, \qquad \boldsymbol{\xi} \in \mathcal{D}_q^\perp, \ 0 \le \omega_q \le 1.$$

For the elasticity case, conditions (4.2) with small ω_q correspond to inclusions $\mathcal{D}^\perp$ of soft materials. The materials must be of the same type, that is, their tensors of elastic modulus may differ by constant multiplicands, ω_q. If the inclusion materials are anisotropic, then anisotropic axes must have the same directions in all inclusions, $\mathcal{D}_q^\perp$. $\omega_q = 0$ corresponds to a cavity $\mathcal{D}_q^\perp$, as in Section 2.

We studied the case where all $\omega_q = \omega$ in Section 3. Here we consider peculiarities of multiparameter problem (4.2). There will be specific situations when for simplicity we assume that

$$\omega_{q_1} \ne \omega_{q_2}, \ q_1 \ne q_2. \tag{4.3}$$

4.2   A NEW CONDITION

Here we define "Condition of representation of constants, or, for the elasticity case, shifts and rotations in $\mathcal{D} \subset \mathbf{T}$", which play an important role in proving the statements of this section. We constructed an example showing, that the condition is necessary for Theorem 4.1 and for the estimate (Equation [4.11]) of Theorem 4.2. We expect, however, that Theorem 4.2, except for Equation (4.11), and Theorem 4.3 can be proved without that condition, as well as can Lemma 4.1. But this is beyond the scope of the present paper.

Assume Equation (4.3) holds. For every $\mathcal{D}_q^\perp \subset \mathbf{T}$ we denote $\mathbf{Ker}_q \subset \mathbf{H}$, a subspace of vector functions independent of $\boldsymbol{\xi}$, or, for the elasticity case, equal to a shift plus a rotation in every connected component of the set $\mathbf{T} \setminus \overline{\mathcal{D}_q^{\perp}}$. If $\mathcal{D}^{\perp}$ is connected, then $\mathcal{D}^{\perp} = \mathcal{D}_q^{\perp}$ and $\mathbf{Ker} = \mathbf{Ker}_q$ (see Section 2 for the definition of $\mathbf{Ker}$). Here we consider the case when $\mathcal{D}^{\perp}$ is not connected. Therefore, there are different subspaces, $\mathbf{Ker}_q$, for different connected components, $\mathcal{D}_q^{\perp}$ of $\mathcal{D}^{\perp}$.

Definition 4.1.   The condition of representation of constants or, for the elasticity case, shifts and rotations in $\mathcal{D} \subset \mathbf{T}$, is

$$\mathbf{Ker} = \sum_q \mathbf{Ker}_q. \tag{4.4}$$

If Equation (4.3) does not hold, that is, several of ω_q are the same, then the corresponding domains $\mathcal{D}_q^{\perp}$ must be united into a single set $\mathcal{D}_q^{\perp}$. Further, the definition of $\mathbf{Ker}_q$ and condition (4.4) are still the same. For example, if all ω_q are the same, we unite all $\mathcal{D}_q^{\perp}$ into one $\mathcal{D}_q^{\perp}$. However, then $\mathcal{D}_q^{\perp} = \mathcal{D}^{\perp}$, and Equation (4.4) holds trivially.

Another simple possibility to satisfy condition (4.4) is a connected domain, $\mathcal{D}$, except for the elasticity case. For the elasticity case, both the domain $\mathcal{D}$ and its image of the canonical expansion of $\mathbf{T}$ into $\mathbf{R}^s$ must be connected to prove Equation (4.4) simply. In a general (except the elasticity) case, it is easy to prove

$$\mathbf{Ker} \supseteq \sum_q \mathbf{Ker}_q, \tag{4.5}$$

but for the elasticity case we did not check Equation (4.5).

A simple example that condition (4.4) is not always true has been kindly presented by S.P. Novikov.[22] Let $s = 2$, $m = 1$, and

$$\mathcal{D} = \left\{ \boldsymbol{\xi} \in \mathbf{T} : 0 < \xi_1 < \frac{1}{4} \cup \frac{1}{2} < \xi_1 < \frac{3}{4} \right\},$$

that is, $\mathcal{D}$ consists of two connected domains, both of which surround the torus, $\mathbf{T}$, along the direction of the ξ_2-axis, and $\mathcal{D}^{\perp}$ has the same structure. Functions of $\mathbf{Ker}$ may achieve two different constant values in $\mathcal{D}_1$ and $\mathcal{D}_2$; thus, functions of $\mathbf{Ker}_1$ and $\mathbf{Ker}_2$ and their sum can only achieve the same constant values in $\mathcal{D}_1$ and $\mathcal{D}_2$.

4.3   RESULTS

A multiparameter analog of Theorem 3.1 is Theorem 4.1.

Theorem 4.1.   Let

$$\mathbf{g}_i = \begin{cases} \text{an arbitrary function of } \mathbf{L}_2(\mathcal{D}) \text{ in } \mathcal{D}, \\ \mathbf{0} \text{ in } \mathcal{D}_q^{\perp}, \qquad \text{where } \omega_q = 0, \\ \dfrac{\mathbf{f}_i}{\omega_q} \text{ in } \mathcal{D}_q^{\perp}, \qquad \text{where } \omega_q > 0. \end{cases} \tag{4.6}$$

For the elasticity case, the functions $\mathbf{f}_i$ and $\mathbf{g}_i$ must satisfy in every domain $\mathcal{D}_p$ and $\mathcal{D}_q^\perp$ the conditions (see Equation [2.15]),

$$\operatorname{tr} FC = 0, \quad \operatorname{tr} GC = 0, \tag{4.7}$$

where $s \times s$ matrices F and G are averages in a given domain $\mathcal{D}_p$ or $\mathcal{D}_q^\perp$ of the matrices, with $\mathbf{f}_i$ and $\mathbf{g}_i$ as columns, and C is an arbitrary $s \times s$ matrix, $C_p = -C_p^T$, such that the restriction $C\xi$ on a given domain, $\mathcal{D}_p$ or $\mathcal{D}_q^\perp$, is continuous.

We consider problem (4.1) with conditions (4.2). Assume that Equation (4.4) is true. Let the matrices E_{ij} satisfy the requirements of Section 1.5 and a set, $\mathcal{D}$, comply with the requirements of Section 1.7. Let also all $\mathbf{f}_i \equiv 0$ in the domains $\mathcal{D}_q^\perp$ with $\omega_q = 0$. Then Equation (4.1) has a unique normal in $\mathbf{H}_E$ solution, $\mathbf{u} \in \mathbf{H}_E$, and

$$\Lambda_E(\mathbf{u}, \mathbf{u}) \le \text{const} \left\{ \langle (\mathbf{f}_i, \mathbf{f}_i) \rangle + \langle (\mathbf{g}_i, \mathbf{g}_i) \rangle \right\}, \tag{4.8}$$

with the constant independent of the collection $\{\omega_q\}$.

We did not try to generalize Theorem 3.2 on a solution dependence of $\{\omega_q\}$, $\omega_g \to 0$, but we are sure that is possible. All other statements are extended with few modifications.

Theorem 4.2.   Let the conditions of Theorem 4.1 be satisfied. Consider iterative method (1.9):

$$\Lambda_E \left(\frac{\mathbf{u}^{n+1} - \mathbf{u}^n}{\tau}, \mathbf{v} \right) + \Lambda_A(\mathbf{u}^n, \mathbf{v}) = \left\langle \left(\mathbf{f}_i, \frac{\partial \mathbf{v}}{\partial \xi_i} \right) \right\rangle,$$

$$\mathbf{v} \in \mathbf{H}, \quad n = 0, 1, \dots, \tag{4.9}$$

with the initial guess $\mathbf{u}^0$, a solution of

$$\Lambda_E(\mathbf{u}^0, \mathbf{v}) = \left\langle \left(\mathbf{g}_i, \frac{\partial \mathbf{v}}{\partial \xi_i} \right) \right\rangle, \qquad \mathbf{v} \in \mathbf{H}. \tag{4.10}$$

For an appropriate $\tau > 0$, iteration approximations $\mathbf{u}^n$ converge to a solution of problem (4.1) in $\mathbf{H}_E$, with the rate of a geometric progression whose convergence factor can be bounded above by a quantity depending only on $\kappa \underline{a}_\mathcal{D} / \bar{a}_\mathcal{D}$, where $\kappa > 0$ is the constant of Proposition 2.1 of extension in $\mathbf{H}_E$ from $\mathcal{D}$ to $\mathbf{T}$. In particular,

$$\Lambda_E(\epsilon^n, \epsilon^n) \le q^{2n} \Lambda_E(\epsilon^0, \epsilon^0), \qquad q = 1 - \kappa \frac{\underline{a}_\mathcal{D}}{\bar{a}_\mathcal{D}} \ \text{ if } \tau = \frac{1}{\bar{a}_\mathcal{D}}.$$

$$\tag{4.11}$$

For the initial error $\mathbf{u}^0 - \mathbf{u}$ we have the estimate

$$\Lambda_E(\mathbf{u}^0 - \mathbf{u}, \mathbf{u}^0 - \mathbf{u}) \leq \text{const } \{\langle(\mathbf{f}_i, \mathbf{f}_i)\rangle + \langle(\mathbf{g}_i, \mathbf{g}_i)\rangle\}, \qquad (4.12)$$

where the constant does not depend on $\{\omega_q\}$.

Remark 4.1. Let $\mathcal{D}$ be connected, and for the elasticity case, also let none of the connected components $\mathcal{D}_q^\perp$ of the complement, $\mathcal{D}^\perp$, surround the torus, **T**. Then $\mathbf{Ker} = \{\overset{0}{W_2^1}(\mathcal{D})\}^m$, and the statements of Theorem 4.1 and Theorem 4.2 still hold, if even arbitrary constant vectors, $\mathbf{c}_q$, were added to the functions $\mathbf{g}_i$ of Equation (4.6) in $\mathcal{D}_q^\perp$, for example vectors

$$\mathbf{c}_q = -\frac{\int_{\mathcal{D}_q^\perp} \mathbf{f}_i \, d\mathcal{D}_q^\perp}{\omega_q \text{ mes } \mathcal{D}_q^\perp}.$$

If the right-hand sides, $\mathbf{f}_i$, are not dependent on $\boldsymbol{\xi}$ in $\mathcal{D}^\perp$, then using Equation (4.6) and adding these constant vectors to $\mathbf{g}_i$ of Equation (4.6) we have that $\mathbf{g}_i \equiv 0$ are possible as well. Hence, the estimate (Equation [4.5]) becomes an ordinary, well-posed inequality, and Equation (4.15) leads to the trivial initial guess, $\mathbf{u}^0 \equiv 0$.

Theorem 4.3. Let the conditions of Theorem 4.2 be satisfied. Consider a particular case of condition (4.2),

$$A_{ij}(\boldsymbol{\xi}) = \omega_p E_{ij}, \qquad \boldsymbol{\xi} \in \mathcal{D}_p, \; \underline{a}_{\mathcal{D}} \leq \omega_p \leq \bar{a}_{\mathcal{D}} \geq 1,$$
$$A_{ij}(\boldsymbol{\xi}) = \omega_q E_{ij}, \qquad \boldsymbol{\xi} \in \mathcal{D}_q^\perp, \; 0 \leq \omega_q \leq 1, \qquad (4.13)$$

with sets of constants $\{\omega_p\}$, $\{\omega_p\}$, and a particular choice in Equation (4.3),

$$\mathbf{g}_i = \begin{cases} \dfrac{\mathbf{f}_i}{\omega_p} & \text{in } \mathcal{D}_p, \\[2mm] \mathbf{0} & \text{in } \mathcal{D}_q^\perp \text{ with } \omega_q = 0, \\[2mm] \dfrac{\mathbf{f}_i}{\omega_q} & \text{in } \mathcal{D}^\perp \text{ with } \omega_q > 0, \end{cases} \qquad (4.14)$$

to find an initial guess, $\mathbf{u}^0$, of Equation (4.10). Then

$$\text{supp } \mathbf{r}^n = \text{supp } \frac{\partial}{\partial \xi_i} \left(E_{ij} \frac{\partial(\mathbf{u}^{n+1} - \mathbf{u}^n)}{\partial \xi_j} \right) \subseteq \partial \mathcal{D}, \qquad (4.15)$$

where $\qquad \mathbf{r}^n = \dfrac{\partial}{\partial \xi_i} \left(A_{ij} \dfrac{\partial \mathbf{u}^n}{\partial \xi_j} - \mathbf{f}_i \right) \in \{W_2^{-1}(\mathbf{T})\}^m$

are residuals.

Proofs are based on the following important statements.

Lemma 4.1. Let the conditions of Theorem 4.2 be satisfied and the subspace **Im** be defined as in Section 2. Then

(1) the difference $\mathbf{u}^0 - \mathbf{u} \in \mathbf{Im}$ of the initial guess (Equation [4.10]) and the solution of problem (4.1);

(2) in the subspace **Im** we have

$$0 < \kappa \underline{a}_{\mathcal{D}} \leq \frac{\Lambda_A(\mathbf{v}, \mathbf{v})}{\Lambda_E(\mathbf{v}, \mathbf{v})} \leq \bar{a}_{\mathcal{D}} < \infty, \qquad \mathbf{v} \in \mathbf{Im}; \quad (4.16)$$

(3) subspace **Im** is an invariant subspace of the operator $L : \mathbf{H} \to \mathbf{H}$ defined by the rule (see Equation [2.10])

$$\Lambda_E(L\mathbf{w}, \mathbf{v}) = \Lambda_A(\mathbf{w}, \mathbf{v}), \qquad \mathbf{w}, \mathbf{v} \in \mathbf{H}. \quad (4.17)$$

4.4   PROOFS

As usual, this last section contains proofs.

Proof of Lemma 4.1. For every subspace $\mathbf{Ker}_q$ defined in Section 4.2, let $\mathbf{Im}_q$ be the $\mathbf{H}_E$ orthogonal complement to $\mathbf{Ker}_q$. By Equation (4.4) $\mathbf{Ker} = \sum_q \mathbf{Ker}_q$; therefore,

$$\mathbf{Im} = \bigcap_q \mathbf{Im}_q. \quad (4.18)$$

By analogy with Equation (3.26) we have

$$\Lambda_A(\mathbf{w}, \mathbf{v}) = \omega_q \Lambda_E(\mathbf{w}, \mathbf{v}), \qquad \mathbf{w} \in \mathbf{H}, \ \mathbf{v} \in \mathbf{Ker}_q, \quad (4.19)$$

which can be checked immediately, for example,

$$\Lambda_A(\mathbf{w}, \mathbf{v}) - \omega_q \Lambda_E(\mathbf{w}, \mathbf{v})$$

$$= \int_{\mathbf{T} \backslash \overline{\mathcal{D}_q^\perp}} \left([A_{ij} - \omega_q E_{ij}] \frac{\partial \mathbf{w}}{\partial \xi_j}, \frac{\partial \mathbf{v}}{\partial \xi_i} \right) d\boldsymbol{\xi} = 0.$$

(1) By definition of the subspace $\mathbf{Im}_q$ we verify the equality

$$\Lambda_E(\mathbf{u}^0 - \mathbf{u}, \mathbf{v}) = 0, \qquad \mathbf{v} \in \mathbf{Ker}_q$$

to prove that $\mathbf{u}^0 - \mathbf{u} \in \mathbf{Im}_q$. The collection of equalities for all $\mathbf{Ker}_q$ in accordance with Equation (4.18) leads to the desired statement.

We first consider a domain $\mathcal{D}_q^\perp$ with $\omega_q > 0$. We substitute in Equation (4.19) $\mathbf{w} = \mathbf{u}$, the solution to Equation (4.1), and obtain

$$\Lambda_E(\mathbf{u}, \mathbf{v}) = \frac{1}{\omega_q} \Lambda_A(\mathbf{u}, \mathbf{v}) = \left\langle \left(\frac{\mathbf{f}_i}{\omega_q}, \frac{\partial \mathbf{v}}{\partial \xi_i} \right) \right\rangle,$$

$$\mathbf{v} \in \mathbf{Ker}_q.$$

Subtracting by parts this equality from Equation (4.10) for the initial guess, $\mathbf{u}^0$, yields

$$\Lambda_E(\mathbf{u}^0 - \mathbf{u}, \mathbf{v}) = \left\langle \left(\mathbf{g}_i - \frac{\mathbf{f}_i}{\omega_q}, \frac{\partial \mathbf{v}}{\partial \xi_i} \right) \right\rangle, \qquad \mathbf{v} \in \mathbf{Ker}_q.$$

By the definition of $\mathbf{g}_i$ in Equation (4.6), in the right-hand side $\mathbf{g}_i - \mathbf{f}_i/\omega_q \equiv 0$ in $\mathcal{D}_q^\perp$, and we should check that

$$\int_{\hat{\mathcal{D}}} \left(\mathbf{g}_i - \frac{\mathbf{f}_i}{\omega}, \frac{\partial \mathbf{v}}{\partial \xi_i} \right) d\hat{\mathcal{D}}, \qquad \mathbf{v} = \mathbf{c}_p,$$

or $\mathbf{c}_p + C_p \boldsymbol{\xi}$ for the elasticity case, where $\hat{\mathcal{D}}$ is every connected component of the set $\mathbf{T} \setminus \overline{\mathcal{D}_q^\perp}$, and the $s \times s$ matrix C is independent of $\boldsymbol{\xi}$ such that the restriction $C\boldsymbol{\epsilon} \mid_{\hat{\mathcal{D}}}$ is continuous. Except for the elasticity case, this is trivial. For the elasticity case it follows from condition (4.7), that is, $\hat{\mathcal{D}}$ consists of one or more domains, $\mathcal{D}_p$ and/or $\mathcal{D}_q^\perp$, of the original decomposition of the torus, $\mathbf{T}$. The restriction $C\boldsymbol{\xi} \mid_{\hat{\mathcal{D}}}$ is continuous, thus the restrictions on subdomains of $\hat{\mathcal{D}}$ are continuous as well. In all these subdomains, condition (4.7) is valid with this matrix C. That completes the proof with $\omega_q > 0$.

Now let $\omega_q = 0$. Then $\mathbf{u} \in \mathbf{Im}_q$ as an $\mathbf{H}_E$ normal solution, see Lemma 2.1. We verify that $\mathbf{u}^0 \in \mathbf{Im}_q$,

$$\Lambda_E(\mathbf{u}^0, \mathbf{v}) = \left\langle \left(\mathbf{g}_i, \frac{\partial \mathbf{v}}{\partial \xi_i} \right) \right\rangle = \int_{\mathbf{T} \setminus \overline{\mathcal{D}_q^\perp}} \left(\mathbf{g}_i, \frac{\partial \mathbf{v}}{\partial \xi_i} \right) d\boldsymbol{\xi} = 0,$$

$$\mathbf{v} \in \mathbf{Ker}_q.$$

We invoke here the definition of $\mathbf{u}^0$ in Equation (4.10) and the condition $\mathbf{g}_i \equiv 0$ in $\mathcal{D}_q^\perp$ of Equation (4.6) and Equation (4.7).

(2) We denote by $\Lambda_{A^0}(\star, \star)$ the bilinear form $\Lambda_A(\star, \star)$ with all $\omega_q = 0$ and consider the representation (see Equation [3.24])

$$\Lambda_A(\mathbf{v}, \mathbf{v}) = \Lambda_{A^0}(\mathbf{v}, \mathbf{v}) + \sum_q \omega_q \int_{\mathcal{D}_q^\perp} \left(E_{ij} \frac{\partial \mathbf{v}}{\partial \xi_j}, \frac{\partial \mathbf{v}}{\partial \xi_i} \right) d\mathcal{D}_q^\perp.$$

Lower and upper bounds for the first term in the right-hand side were obtained from Lemma 2.2 for $\mathbf{v} \in \mathbf{Im}$. By $0 < \omega_q \leq 1 \leq \bar{a}_\mathcal{D}$, the last term is estimated trivially, as in Lemma 3.2.

(3) By the definition of the operator L in Equation (4.17), the subspace $\mathbf{Im}_q$ is its invariant subspace iff

$$\Lambda_E(L\mathbf{w}, \mathbf{v}) = \Lambda_A(\mathbf{w}, \mathbf{v}) = 0, \qquad \mathbf{w} \in \mathbf{Im}_q, \mathbf{v} \in \mathbf{Ker}_q.$$
$$(4.20)$$

But this is a particular case of Equation (4.19) for $\mathbf{w} \in \mathbf{Im}_q$. Therefore, all subspaces $\mathbf{Im}_q$ are invariant with respect to the operator L, hence their intersection $\mathbf{Im} = \cap \, \mathbf{Im}_q$ is an invariant subspace of L as well.

Proof of Theorem 4.1. We invoke function $\mathbf{u}^0$, a solution of problem (4.10) with $\mathbf{g}_i$ of Equation (4.6), and consider the difference $\mathbf{u}^0 - \mathbf{u}$ of this function and the solution, $\mathbf{u}$, of the original problem (4.1). As in the proof of Theorem 3.1, we have

$$\Lambda_A(\mathbf{u}^0 - \mathbf{u}, \mathbf{v}) = \Lambda_A(\mathbf{u}^0, \mathbf{v}) - \left\langle \left(\mathbf{f}_i, \frac{\partial \mathbf{v}}{\partial \xi_i} \right) \right\rangle, \qquad \mathbf{v} \in \mathbf{H}$$
$$(4.21)$$

and present vector $\mathbf{v} \in \mathbf{H}$ here as an orthogonal in $\mathbf{H}_E$ sum,

$$\mathbf{v} = \mathbf{v}_K + \mathbf{v}_I, \qquad \mathbf{v}_K \in \mathbf{Ker}, \ \mathbf{v}_I \in \mathbf{Im}.$$

We now see that all terms with $\mathbf{v}_K$ in Equation (4.21) vanish. Due to Equation (4.4), every function $\mathbf{v}_K \in \mathbf{Ker}$ can be written as a sum of functions $\mathbf{v}_q \in \mathbf{Ker}_q$. We use the additive property and find that all terms with $\mathbf{v}_q$ vanish.

For the left-hand side, $\Lambda_A(\mathbf{u}^0 - \mathbf{u}, \mathbf{v}_q) = 0$ because of Equation (4.20), as $\mathbf{u}^0 - \mathbf{u} \in \mathbf{Im} \subseteq \mathbf{Im}_q$ by Lemma 4.1, Equation (4.18), and $\mathbf{v}_q \in \mathbf{Ker}_q$. In the right-hand side both terms vanish independently by Equation (4.7). We rewrite the first term according to Equation (4.19),

$$\Lambda_A(\mathbf{u}^0, \mathbf{v}_q) = \omega_q \Lambda_E(\mathbf{u}^0, \mathbf{v}_q) = \omega_q \left\langle \left(\mathbf{g}_i, \frac{\partial \mathbf{v}_q}{\partial \xi_i} \right) \right\rangle = 0,$$

$$\mathbf{v}_q \in \mathbf{Ker}_q.$$

Therefore, we have proved that it is sufficient to take $\mathbf{v} \in \mathbf{Im}$ in Equation (4.21). Much of what follows is the same as the proof of Theorem 3.1.

Proofs of Theorems 4.2 and 4.3 are very similar to the proofs of Theorems 3.3 and 3.4 and are not shown here.

5. ON A FUNCTION EXTENSION ON A TORUS

We consider here the possibility of function extension from the set $\mathcal{D} \subset \mathbf{T}$ on the whole torus, $\mathbf{T}$. We assume that the conditions of Section 1.7 for $\mathcal{D}$ are fulfilled, that is, $\mathcal{D}$ is itself a Lipschitz domain, or consists of a finite number of Lipschitz domains, $\mathcal{D}_p$, with nonintersecting closures. We start from the classical theorem of extensions.[13]

> **Theorem 5.1.** Let $\Omega \subset \mathbf{R}^s$ be a bounded Lipschitz domain. Then there is a constant, $c_1(\Omega) > 0$, such that for a function $v \in W_2^1(\Omega)$ there exists a function $w \in W_2^1(\mathbf{R}^s)$ with finite support such that $w - v \equiv 0$ in Ω and
>
> $$\int_{\mathbf{R}^s} \left(w^2 + \frac{\partial w}{\partial \xi_i} \frac{\partial w}{\partial \xi_i} \right) d\xi \le c_1(\Omega) \int_{\Omega} \left(v^2 + \frac{\partial v}{\partial \xi_i} \frac{\partial v}{\partial \xi_i} \right) d\xi. \quad (5.1)$$

This theorem leads to the equality $W_2^1(\Omega) = W_2^1(\mathbf{R}^s) \mid_\Omega$. Therefore, we can exchange $v \in W_2^1(\Omega)$ for $v \in W_2^1(\mathbf{R}^s)$ in the formulation of the theorem.

> **Corollary 5.1.** Let $\Omega \subset \mathbf{R}^s$ be a bounded Lipschitz domain. Then there is a constant $c_1(\Omega) > 0$ such that for a function $v \in W_2^1(\mathbf{R}^s)$ with finite support there exists a function $w \in W_2^1(\mathbf{R}^s)$ with finite support such that $w - v \equiv 0$ in Ω and
>
> $$\int_{\mathbf{R}^s} \left(w^2 + \frac{\partial w}{\partial \xi_i} \frac{\partial w}{\partial \xi_i} \right) d\xi \le c_1(\Omega) \int_{\Omega} \left(v^2 + \frac{\partial v}{\partial \xi_i} \frac{\partial v}{\partial \xi_i} \right) d\xi.$$

Note that in the corollary there is a possibility to choose $w \equiv v$, because both functions now are of the same functional space. But it is not evident at all that such a choice affects a constant, c_1, which is independent of w.

The next example shows that the statement of the corollary can break down if the boundary of Ω is not Lipschitz.

> **Example 5.1.** Let $\Omega = \Omega_0 \cup \Omega_1 \subset \mathbf{R}^2$, where Ω_0 and Ω_1 are unit squares and have only one common vertex, $\bar{\Omega}_0 \cap \bar{\Omega}_1 = \{\text{a vertex}\}$. We consider the functional space $W_2^1(\Omega_0) \times W_2^1(\Omega_1)$, with $W_2^1(\Omega_p) = W_2^1(\mathbf{R}^2) \mid_{\Omega_p}, p = 0, 1$. This space is complete, because its multiplicands are complete, as is well known. We have

$W_2^1(\mathbf{R}^2)\,|_{\mathbf{\Omega}} \subset W_2^1(\mathbf{\Omega}_0) \times W_2^1(\mathbf{\Omega}_1)$. Let

$$
u \equiv \begin{cases} 0 & \text{in } \mathbf{\Omega}_0, \\ 1 & \text{in } \mathbf{\Omega}_1. \end{cases}
$$

Then $u \in W_2^1(\mathbf{\Omega}_0) \times W_2^1(\mathbf{\Omega}_1), \quad \text{but } u \notin W_2^1(\mathbf{R}^2)\,|_{\mathbf{\Omega}}$.

Now consider any sequence $\{u^n\} \subset W_2^1(\mathbf{R}^2)$ of functions with finite support such that $u^n\,|_{\mathbf{\Omega}} \to u$ in $W_2^1(\mathbf{\Omega}_0) \times W_2^1(\mathbf{\Omega}_1)$. Assume that the sequence is fundamental in the space $W_2^1(\mathbf{R}^2)$. Then it must converge in $W_2^1(\mathbf{R}^2)$ to a function of $W_2^1(\mathbf{R}^2)$, whose restrictions on $\mathbf{\Omega}$ have to coincide with u; but that is impossible. Therefore, the sequence is not fundamental in the space $W_2^1(\mathbf{R}^2)$, that is, there exists a sequence $\{v^n\} : v^n = u^{i_n} - u^{j_n}$ with $i_n, j_n \to \infty$ as $n \to \infty$, that does not tend to zero in the space $W_2^1(\mathbf{R}^2)$.

On the other hand, the sequence $\{u^n\,|_{\mathbf{\Omega}}\}$ is fundamental in the space $W_2^1(\mathbf{\Omega}_0) \times W_2^1(\mathbf{\Omega}_1)$ as a converged sequence and, hence, the sequence $\{v^n\,|_{\mathbf{\Omega}}\}$ must tend to zero in the space $W_2^1(\mathbf{\Omega}_0) \times W_2^1(\mathbf{\Omega}_1)$. This contradicts the statement of Corollary 5.1 for the given $\mathbf{\Omega}$ with an irregular boundary. It also means that the set $W_2^1(\mathbf{R}^2)\,|_{\mathbf{\Omega}}$ is not closed in the subspace $W_2^1(\mathbf{\Omega}_0) \times W_2^1(\mathbf{\Omega}_1)$.

By analogy with Theorem 5.1 and Corollary 5.1 it is possible to prove the following theorem on extension of functions on a torus.

Theorem 5.2. Let $\mathbf{\Omega}$ and $\mathbf{\Omega}'$ be bounded Lipschitz domains on the torus, $\overline{\mathbf{\Omega}} \subset \mathbf{\Omega}' \subseteq \mathbf{T}$.

Then there is a constant, $c_2(\mathbf{\Omega}, \mathbf{\Omega}') > 0$, such that for a function $v \in W_2^1(\mathbf{T})$ there exists a function $w \in W_2^1(\mathbf{T})$ with support in $\mathbf{\Omega}'$ such that $w - v \equiv 0$ in $\mathbf{\Omega}$ and

$$
\int_{\mathbf{T}} \left(w^2 + \frac{\partial w}{\partial \xi_i}\, \frac{\partial w}{\partial \xi_i} \right) d\mathbf{T} \leq c_2(\mathbf{\Omega}, \mathbf{\Omega}') \int_{\mathbf{\Omega}} \left(v^2 + \frac{\partial v}{\partial \xi_i}\, \frac{\partial v}{\partial \xi_i} \right) d\mathbf{\Omega}.
$$

$$(5.2)$$

Now change the W_2^1 norm to the W_2^1 seminorm and consider the case of a nonconnected $\mathcal{D}$ to obtain the next simplest variant of the extension theorem that we use for the diffusion equation.

Theorem 5.3.   Let $\mathcal{D} \subset \mathbf{T}$ be itself a Lipschitz domain, or consist of a finite number of Lipschitz domains, $\mathcal{D}_p \subset \mathbf{T}$, with nonintersecting closures.

Then there is a constant, $c_3(\mathcal{D}) > 0$, such that for a function $v \in W_2^1(\mathbf{T})$ there exists a function $w \in W_2^1(\mathbf{T})$ such that grad $w-$ grad $v \equiv 0$ in $\mathcal{D}$ and

$$
\int_{\mathbf{T}} \frac{\partial w}{\partial \xi_i}\, \frac{\partial w}{\partial \xi_i}\, d\mathbf{T} \leq c_3(\mathcal{D}) \int_{\mathcal{D}} \frac{\partial v}{\partial \xi_i}\, \frac{\partial v}{\partial \xi_i}\, d\mathcal{D}.
$$

$$(5.3)$$

Proof of Theorem 5.3. As $\overline{\mathcal{D}_p} \cap \overline{\mathcal{D}_q} = \emptyset, p \neq q$, there exists the same number of Lipschitz domains $\mathcal{D}_p'$ such that

$$\overline{\mathcal{D}_p} \subset \mathcal{D}_p', \ \overline{\mathcal{D}_p'} \cap \overline{\mathcal{D}_q'} = \emptyset, p \neq q.$$

For the function $v \in W_2^1(\mathbf{T})$, let

$$v_p = v - \int_{\mathcal{D}_p} \frac{v \, d\mathcal{D}_p}{mes\mathcal{D}_p} \qquad \text{in } \mathcal{D}_p.$$

Evidently, $\int_{\mathcal{D}_p} v_p d\mathcal{D}_p = 0$; then, by the Poincaré inequality,

$$\int_{\mathcal{D}_p} v_p^2 \, d\mathcal{D}_p \leq c_p \int_{\mathcal{D}_p} \frac{\partial v_p}{\partial \xi_i} \frac{\partial v_p}{\partial \xi_i} \, d\mathcal{D}_p.$$

Therefore,

$$\int_{\mathcal{D}_p} \left(v_p^2 + \frac{\partial v_p}{\partial \xi_i} \frac{\partial v_p}{\partial \xi_i} \right) d\mathcal{D}_p \leq (1 + c_p) \int_{\mathcal{D}_p} \frac{\partial v_p}{\partial \xi_i} \frac{\partial v_p}{\partial \xi_i} \, d\mathcal{D}_p.$$

Through Theorem 5.2 there exists a constant, $c_2(\mathcal{D}_p, \mathcal{D}_p') > 0$, and there is a function, $w_p \in W_2^1(\mathbf{T})$, with support in $\mathcal{D}_p'$ such that $w_p - v_p \equiv 0$ in $\mathcal{D}_p$ and

$$\int_{\mathcal{D}_p'} \left(w_p^2 + \frac{\partial w_p}{\partial \xi_i} \frac{\partial w_p}{\partial \xi_i} \right) d\mathcal{D}_p'$$

$$\leq c_2(\mathcal{D}_p, \mathcal{D}_p') \int_{\mathcal{D}_p} \left(v_p^2 + \frac{\partial v_p}{\partial \xi_i} \frac{\partial v_p}{\partial \xi_i} \right) d\mathcal{D}_p$$

$$\leq (1 + c_p) c_2(\mathcal{D}_p, \mathcal{D}_p') \int_{\mathcal{D}_p} \frac{\partial v_p}{\partial \xi_i} \frac{\partial v_p}{\partial \xi_i} \, d\mathcal{D}_p. \quad (5.4)$$

We set $w = \sum_p w_p$. In every domain $\mathcal{D}_p$ we have

$$\text{grad } w = \text{ grad } w_p = \text{ grad } v_p = \text{ grad } v.$$

We conclude that Equation (5.4) leads to Equation (5.3), with

$$c_3(\mathcal{D}) = \max_p |(1 + c_p) c_2(\mathcal{D}_p, \mathcal{D}_p')|.$$

We first note that it is possible to add arbitrary constants to the functions v and w in $\mathbf{T}$, that is, Theorem 5.3 holds for the factor space, $W_2^1(\mathbf{T})/\mathbf{R}$, instead of the ordinary Sobolev space, $W_2^1(\mathbf{T})$, as well.

Second, Theorem 5.3 can easily be generalized for a vector function by using the component by component extension.

Taking into account an equivalence of the $\mathbf{H}$ and $\mathbf{H}_E$ norms we conclude, except the elasticity case, that Corollary 8.1 on extension holds in the form of Theorem 5.4.

Theorem 5.4.   Let $\mathcal{D} \subset \mathbf{T}$ be itself a Lipschitz domain, or consist of a finite number of Lipschitz domains, $\mathcal{D}_p \subset \mathbf{T}$, with nonintersecting closures. Let matrices E_{ij} fulfill the conditions of Section 1.5 except for the elasticity case.

There is a constant, $\kappa = \kappa(\mathcal{D}, E) > 0$, such that for a function, $\mathbf{v} \in \{W_2^1(\mathbf{T})\}^m$, there exists a function, $\mathbf{w} \in \{W_2^1(\mathbf{T})\}^m$, such that

$$\kappa \int_{\mathbf{T}} \left(E_{ij} \frac{\partial \mathbf{w}}{\partial \xi_i}, \frac{\partial \mathbf{w}}{\partial \xi_i} \right) d\mathbf{T} \leq \int_{\mathcal{D}} \left(E_{ij} \frac{\partial \mathbf{v}}{\partial \xi_i}, \frac{\partial \mathbf{v}}{\partial \xi_i} \right) d\mathcal{D} \qquad (5.5)$$

and

$$\frac{\partial v_k}{\partial \xi_i} - \frac{\partial w_k}{\partial \xi_i} \equiv 0 \quad \text{in } \mathcal{D}, \qquad i = 1, \ldots, s, \ k = 1, \ldots, m. \quad (5.6)$$

As in the previous scalar functions case, we again note that it is possible to add arbitrary vectors independent of $\boldsymbol{\xi}$ to the functions $\mathbf{v}$ and $\mathbf{v}$ in $\mathbf{T}$, that is, Theorem 5.4 also holds for the factor space, $\mathbf{H} = \{W_2^1(\mathbf{T})\}^m/\mathbf{R}^m$, as well as the vector Sobolev space, $\{W_2^1(\mathbf{T})\}^m$. This comment is still valid for the elasticity case, as seen in Theorem 5.5.

Theorem 5.5.   Let $\mathcal{D} \subset \mathbf{T}$ be itself a Lipschitz domain, or consist of a finite number of Lipschitz domains, $\mathcal{D}_p \subset \mathbf{T}$, with nonintersecting closures. Let matrices E_{ij} fulfill the conditions of Section 1.5 for the elasticity case.

There is a constant, $\kappa = \kappa(\mathcal{D}, E) > 0$, such that for a function, $\mathbf{v} \in \{W_2^1(\mathbf{T})\}^m$, there exists a function, $\mathbf{w} \in \{W_2^1(\mathbf{T})\}^m$, such that

$$\kappa \int_{\mathbf{T}} \left(E_{ij} \frac{\partial \mathbf{w}}{\partial \xi_i}, \frac{\partial \mathbf{w}}{\partial \xi_i} \right) d\mathbf{T} \leq \int_{\mathcal{D}} \left(E_{ij} \frac{\partial \mathbf{v}}{\partial \xi_i}, \frac{\partial \mathbf{v}}{\partial \xi_i} \right) d\mathcal{D} \quad (5.7)$$

and

$$\frac{\partial v_i}{\partial \xi_k} + \frac{\partial v_k}{\partial \xi_i} = \frac{\partial w_i}{\partial \xi_k} + \frac{\partial w_k}{\partial \xi_i} \quad \text{in } \mathcal{D}, \qquad i, k = 1, \ldots, s. \quad (5.8)$$

We note that Equations (5.5) and (5.7) are exactly the same. Our proof of this theorem is based on the next particular case of the known more general statement.[12]

Theorem 5.6. Let $\Omega \subset \mathbf{R}^s$ be a bounded Lipschitz domain, and $\mathbf{V}_\Omega \subset \{W_2^1(\Omega)\}^s$ be such a subspace that conditions

$$\mathbf{v} \in \mathbf{V}_\Omega, \epsilon_{ik}(\mathbf{v}) \stackrel{def}{=} \frac{1}{2} \left(\frac{\partial v_i}{\partial \xi_k} + \frac{\partial v_k}{\partial \xi_i} \right) \equiv 0 \quad \text{in } \Omega,$$

$$i, k = 1, \ldots, s.$$

lead to $\mathbf{v} \equiv 0$ in Ω.

Then there is a constant, $c_4(\Omega) > 0$, such that

$$\int_\Omega \left[(\mathbf{v}, \mathbf{v}) + \left(\frac{\partial \mathbf{v}}{\partial \xi_i}, \frac{\partial \mathbf{v}}{\partial \xi_i} \right) \right] d\Omega$$

$$\leq c_4(\Omega) \int_\Omega \epsilon_{ik}(\mathbf{v}) \epsilon_{ik}(\mathbf{v}) \, d\Omega, \qquad \mathbf{v} \in \mathbf{V}_\Omega. \qquad (5.9)$$

Using the same arguments as in the proof of Theorem 5.6,[12] it is easy to prove the analogous statement in Corollary 5.2 for the torus, $\mathbf{T}$, instead of $\mathbf{R}^s$.

Corollary 5.2. Let $\Omega \subset \mathbf{T}$ be a Lipschitz domain, and $\mathbf{V}_\Omega \subset \{W_2^1(\Omega)\}^s$ is such a subspace that conditions

$$\mathbf{v} \in \mathbf{V}_\Omega, \epsilon_{ik}(\mathbf{v}) \stackrel{def}{=} \frac{1}{2} \left(\frac{\partial v_i}{\partial \xi_k} + \frac{\partial v_k}{\partial \xi_i} \right) \equiv 0 \quad \text{in } \Omega, \ i, k = 1, \ldots, s.$$

lead to $\mathbf{v} \equiv 0$ in Ω.

Then there is a constant, $c_4(\Omega) > 0$, such that

$$\int_\Omega \left[(\mathbf{v}, \mathbf{v}) + \left(\frac{\partial \mathbf{v}}{\partial \xi_i}, \frac{\partial \mathbf{v}}{\partial \xi_i} \right) \right] d\Omega$$

$$\leq c_4(\Omega) \int_\Omega \epsilon_{ik}(\mathbf{v}) \epsilon_{ik}(\mathbf{v}) \, d\Omega, \qquad \mathbf{v} \in \mathbf{V}_\Omega. \qquad (5.10)$$

We now are ready to prove Theorem 5.5.

Proof of Theorem 5.5. Note that the condition

$$\epsilon_{ik}(\mathbf{v}) \stackrel{def}{=} \frac{1}{2} \left(\frac{\partial v_i}{\partial \xi_k} + \frac{\partial v_k}{\partial \xi_i} \right) \equiv 0 \quad \text{in } \mathcal{D}, \qquad i, k = 1, \ldots, s,$$

$$(5.11)$$

with $\mathbf{v} \in \mathbf{H}$, coincides with the definition of the subspace **Ker** of Section 2.2, that is, this condition defines in **H** vector functions $\mathbf{w} \in \mathbf{H}$, which must obey the equality $\mathbf{w} = \mathbf{c}_p + C_p \boldsymbol{\xi}$ in $\mathcal{D}_p$ with a vector $\mathbf{c}_p$ and an $s \times s$ matrix, $C_p = -C_p^T$, both independent of

$\boldsymbol{\xi}$. Restriction $C_p \boldsymbol{\xi} \mid_{\mathcal{D}_p}$ must be continuous in each component of connectedness, $\mathcal{D}_p$ of $\mathcal{D}$. The original condition (5.11) for functions $\mathbf{v} \in \{W_2^1(\mathbf{T})\}^s$ involves also functions $\mathbf{v} \equiv \mathbf{c} \in \mathbf{R}^s$ in $\mathbf{T}$.

Let $\mathbf{V} \subset \{W_2^1(\mathbf{T})\}^s$ be, for example, $\{W_2^1(\mathbf{T})\}^s$ orthogonal complement to the subspace of all functions $\mathbf{v} \in \{W_2^1(\mathbf{T})\}^s$ described by condition (5.11) mentioned previously. Then every function $\mathbf{v} \in \{W_2^1(\mathbf{T})\}^s$ has a unique representation

$$\mathbf{v} = \mathbf{v_V} + \mathbf{v_K}, \qquad \mathbf{v_V} \in \mathbf{V}, \ \mathbf{v_K} = \mathbf{c}_p + C_p \boldsymbol{\xi} \text{ in } \mathcal{D}_p,$$

with vectors $\mathbf{c}_p$ and $s \times s$ matrices $C_p = -C_p^T$ for every $\mathcal{D}_p$, all independent of $\boldsymbol{\xi}$. Restrictions $C_p \boldsymbol{\xi} \mid_{\mathcal{D}_p}$ must be continuous in each component of connectedness, $\mathcal{D}_p$ of $\mathcal{D}$. Then

$$\epsilon_{ik}(\mathbf{v}) = \epsilon_{ik}(\mathbf{v_V}) \quad \text{in } \mathcal{D}.$$

We apply Theorem 5.6, with $\boldsymbol{\Omega} = \mathcal{D}_p$ and $\mathbf{V_\Omega} = \mathbf{V} \mid_{\mathcal{D}_p}$ for the function $\mathbf{v}_p = \mathbf{v_V} \mid_{\mathcal{D}_p} \in \{W_2^1(\mathcal{D}_p)\}^s$, and find that

$$\int_{\mathcal{D}_p} \left[(\mathbf{v}_p, \mathbf{v}_p) + \left(\frac{\partial \mathbf{v}_p}{\partial \xi_i}, \frac{\partial \mathbf{v}_p}{\partial \xi_i} \right) \right] d\mathcal{D}_p$$

$$\leq c_4(\mathcal{D}_p) \int_{\mathcal{D}_p} \epsilon_{ik}(\mathbf{v}_p)\epsilon_{ik}(\mathbf{v}_p) \, d\mathcal{D}_p. \quad (5.12)$$

Further, we use the vector variant of Theorem 5.2 to extend the function $\mathbf{v}_p$ from $\mathcal{D}_p$ to $\mathbf{T}$ by a function, $\mathbf{w}_p \in \{W_2^1(\mathbf{T})\}^s$, such that

$$\mathbf{w}_p - \mathbf{v}_p \equiv 0 \text{ in } \mathcal{D}_p, \qquad \text{supp } \mathbf{w}_p \subset \mathcal{D}_p',$$

and

$$\int_{\mathcal{D}_p'} \left[(\mathbf{w}_p, \mathbf{w}_p) + \left(\frac{\partial \mathbf{w}_p}{\partial \xi_i}, \frac{\partial \mathbf{w}_p}{\partial \xi_i} \right) \right] d\mathcal{D}_p'$$

$$\leq c_2(\mathcal{D}_p, \mathcal{D}_p') \int_{\mathcal{D}_p} \left[(\mathbf{v}_p, \mathbf{v}_p) + \left(\frac{\partial \mathbf{v}_p}{\partial \xi_i}, \frac{\partial \mathbf{v}_p}{\partial \xi_i} \right) \right] d\mathcal{D}_p'.$$

$$(5.13)$$

Finally, we set

$$\mathbf{w} = \sum_p \mathbf{w}_p \in \{W_2^1(\mathbf{T})\}^s.$$

Then $\qquad \epsilon_{ik}(\mathbf{w}) = \epsilon_{ik}(\mathbf{v}_p) = \epsilon_{ik}(\mathbf{v}_p) = \epsilon_{ik}(\mathbf{v}),$

as in Equation (5.8). The inequalities (Equations [5.12] and [5.13]) lead to Equation (5.5) for the particular case of isotropic media. The general anisotropic case follows directly from the isotropic one.

The last step of this proof leads to a strong dependence on the coefficient κ of matrices E_{ij}. And κ tends to zero when tensor E becomes degenerate or has large coefficients. One especially interesting example of large coefficients is the case of almost incompressible media. Using a new approach, we proved that it is possible to generalize Theorem 5.5 and other results to such a case.[16] We treated the limit case of the Stokes' equations for incompressible media as well.

REFERENCES

1. Bakhvalov, N.S., and Panasenko, G.P., *Homogenization: Averaging of Processes in Periodic Media*, Kluwer, Dordrecht, Boston, 1989.

2. Astrakhantsev, G.P., Fictitious domain methods for elliptic equations of second order with natural boundary conditions, *Comp. Math. and Math. Phys.*, v. 18, 118, 1978.

3. D'yakonov, E.G., *Minimization of Computational Work. Asymptotically Optimal Algorithms for Elliptic Problems*, Nauka, Moscow, 1989 (in Russian).

4. Kuznetsov, Yu.A., Numerical methods in subspaces, in *Comp. Processes and Systems No. 3*, Marchuk, G.I., Ed., Nauka, Moscow, 265, 1985 (in Russian).

5. Marchuk, G.I., *Methods of Numerical Mathematics*, Springer-Verlag, New York, 1975.

6. Marchuk, G.I., Kuznetsov, Yu.A., and Matsokin, A.M., Fictitious domain and domain decomposition methods, *Sov. Jour. Numer. Anal. Math. Modeling*, v. 1, No. 1, 1, 1986.

7. Rivkind, V.Ya., Approximative method of solving the Dirichlet problem and on convergence estimates of solutions of finite-difference equations to solutions of elliptic equations with discontinuous coefficients, *Vestnik LGU*, No. 13, 37, 1982 (in Russian).

8. Saul'ev, V.K., On solving boundary value problems with high performance computers by a fictitious domain method, *Siberian Math. Jour.*, v. 4, No. 4, 912, 1963 (in Russian).

9. Lions, J.L., *Perturbations singulièresdans les problèmes aus limites et en contrôle optimal*, Lecture Notes in Mathematics, No. 323, Springer-Verlag, Berlin, 1973.

10. Bakhvalov, N.S., Kobelkov, G.M., and Chizhonkov, E.V., An iterative method for solving elliptic problems with a rate of convergence that does not depend on the range of the coefficients, Preprint No. 190, Dept. Numer. Math. USSR Ac. Sci., Moscow, 1988 (in Russian).

11. Kobelkov, G.M., Fictitious domain method and the solution of elliptic equations with highly varying coefficients, *Sov. Jour. Numer. Anal. Math. Modeling*, v. 2, No. 6, 407, 1987.

12. Litvinov, V.G., *Optimization for Elliptic Boundary Value Problems with Applications in Mechanics*, Nauka, Moscow, 1987 (in Russian).

13. Besov, O.V., Il'in, V.P., and Nikol'skii, S.M., *Integral Representations of Functions and Imbedding Theorems*, vols. 1, 2, Wiley, New York, 1979.

14. Bakhvalov, N.S., and Knyazev, A.V., A new iterative algorithm for solving problems of the fictitious flow method for elliptic equations, *Sov. Math. Dokl.*, v. 41, No. 3, 481, 1990.

15. Bakhvalov, N.S., and Knyazev, A.V., Efficient computation of averaged characteristics of composites of a periodic structure of essentially different materials, *Sov. Math. Dokl.*, v. 42, No. 1, 57, 1991.

16. Bakhvalov, N.S., and Knyazev, A.V., An efficient iterative method for solving the Lamé equations for almost incompressible media and Stokes' equations, *Sov. Math. Dokl.*, v. 44, No. 1, 4, 1992.

17. Bakhvalov, N.S., Knyazev, A.V., and Kobel'kov, G.M., Iterative methods for solving equations with highly varying coefficients, in *Proc. IV International Symposium on Domain Decomposition Methods for Partial Differential Equations, 1990*, SIAM, Philadelphia, 197, 1991.

18. Knyazev, A.V., Iterative solution of PDE with strongly varying coefficients: algebraic version, in *Iterative Methods in Linear Algebra. Proc. IMACS Symp. Iterative Methods in Linear Algebra*, Brussels, 1991, Beauwens, R., and de Groen, P., Eds., Elsevier, Amsterdam, 85, 1992.

19. Bogachev, K.Yu., Iterative methods of solving quasilinear elliptic problems in domains of complicated shape, *Sov. Math. Dokl.*, v. 45, No. 1, 152, 1992.

20. Bakhvalov, N.S., and Knyazev, A.V., Methods of effective computation of homogenized properties for the composites with a periodic structure which consist of essentially different components, in *Comp. Processes and Systems No. 8*, Marchuk, G.I., Ed., Nauka, Moscow, 52, 1991 (in Russian).

21. Kato, T., *Perturbation Theory for Linear Operators*, Springer-Verlag, New York, 1976.

22. Novikov, S.P., Private communication.

INDEX

Adaptation procedure, 15
anisotropic preconditioners,
    225–227
approximate solutions
    to Cauchy problems, 48, 49
    for implicit difference stability
        equations, 53–56
    for stability equations, 51–52
a priori error estimates. *See* error
    estimates
A-stability of difference schemes,
    123–125
asymptotically nearly optimal
    preconditioner, 1
asymptotically optimal
    preconditioners, 1, 4
asymptotic theory, 129, 138–139
average relative rate of variation,
    49

Babushka's model, 78, 79
Babushka-Brezzi inequality, 85, 99
Banach space, 158, 180
block elimination of unknowns,
    20–21
block-triangular factorization, 21
block-triangular factorized
    operators, 22
bordering method, 21
boundary conditions, periodic, in
    Stokes problem, 94
boundary layers, passing, 73–75
boundary value problems. *See also*
    Navier-Stokes equations,
        boundary-value problems
    as first-order equations, 210
    initial-boundary, 156–164, 216

bounded variation, 158–159
Bubnov-Galerkin-Petrov method,
    211

Cauchy problems, 46–47, 50,
    117–118
    for abstract hyperbolic
        equations, 216
    approximate solution, 48, 49
CGM method, 93
C_h error estimates, 193–195
composites
    defined, 222–223
    fictitious domain method,
        221–259
    function extension on torus,
        259–265
    homogenization, 223–228
    with inclusions of soft
        material, 239–251
    with inclusions of soft
        materials and cavities,
        251–256
    perforated: fictitious gradient
        method, 228–239
composite triangulations, 24–25
computational cost, 71
conjugate gradient method,
    226–227
continuous restriction, 229–230
convergence rate estimates
    applications to dynamic problems
        of mechanics, 216–218
    finite-element method with
        splitting operator, 204–210
    fractional error estimates for
        nonsmooth data classes,
        179–189

convergence rate estimates (Cont.):
  initial boundary value for second-
    order multidimensional hyper-
    bolic equations, 156–164
  second-order accuracy a priori
    error estimates, 174–179
  second-order hyperbolic equa-
    tions of general form, 196–204
  three-level finite-element method
    with weight, 165–174
  two-level finite element method,
    210–216
  $W_{2,h}^1$ and C_h error estimates for
    one-dimensional case, 189–195
cooperative operators, 22
cou estimation, 73–75
Courant quantity, 47
Crank-Nicholson-Galerkin method,
  211
cutting (domain-decompensation)
  method, 23

DeMirchyan-Rakitsky scheme,
  139–144
difference equations, explicit.
  See explicit difference methods
diffusion equation, 223, 227
Dirac delta function, 181
Dirichlet problems, 93, 227
domain-decompensation (cutting)
  method, 23
DUMKA code, 73–75

Eigenfunctions, in Navier-Stokes
  equations, 96
eigenvalues, in Navier-Stokes
  equations, 96
elasticity equations, 97–99, 217,
  223, 227
elliptic self-adjoint systems, 216
energy conservation law, 170
ϵ regularizations, 97–100
error estimates
  fractional for nonsmooth data
    classes, 179–189
  a priori, 208, 213
  a priori second-order, 174–179,
    208

$W_{2,h}^1$ for one-dimensional case,
  189–195
Euler method, 47, 146–153
explicit difference methods
  for approximation and stability,
    51–53
  vs. implicit difference schemes,
    53–56
  determination of parameters,
    65–72
  example of stable algorithms,
    72–73
  vs. implicit schemes for linear
    equations, 50–51
  in Navier-Stokes problems,
    102–104
  on passing boundary layers,
    73–75
  optimum vs. implicit, 71
  for polynomials, 56–65
  problem statement and assump-
    tions, 46–47
  simple, 47
  stiff system computations, 75–79
  with time-variable steps, 47–50

Fictitious domain method
  composites with soft inclusions
    formulation of problem,
      239–240
    results, 240–251
  composites with soft inclusions
    and cavities
    condition of representation
      constants, 252–256
    formulation of problem, 252
    proofs, 256–259
  fictitious gradients: perforated
    composites
    conditions for right-hand side,
      233–239
    connection with fictitious
      domain method, 232–233
    formulation of problem,
      228–229
    main results, 229–232
  function extension on torus,
    259–265

fictitious domain method (Cont.):
  homogenization problems
    anisotropic preconditioner,
      225–227
    assumptions for coefficients
      and right-hand sides,
      223–224
    domain assumptions, 227
    generalized formulation,
      224–225
    isotropic preconditioner, 225
    jumps of coefficients, 227–228
fictitious grid region methods, 24
finite-difference methods
  counterpart to Gronwall's lemma,
    197
  explicit, 102–104
  in Navier-Stokes equations, 88–90
  for nonstationary Navier-Stokes
    problems, 113
finite-dimensional analogue of
  Navier-Stokes problem,
  108–112
finite-element methods
  convergence rate estimates,
    155–220. *See also* convergence
    rate estimates
  counterpart of Laplace operator,
    207
  in Navier-Stokes equations, 90
  three-level with weight, 165–174
finite-element subspaces
  angle estimation, 26–34
  recurrent splitting, 24
Fourier coefficients, for initial-
  boundary value problems,
  160–161
Fourier method, 163–164
fourth-order equations, elliptic,
  38–40

Galerkin-Petrov (Bubnov) method,
  211
Gaussian symmetrization, 5
Gear's algorithm, 131
Gram matrices, 4–5, 32–34
grid problems, Stokes, 92–94
grid systems, nonlinear elliptic,

1–40. *See also* nonlinear
  elliptic grid systems
Gronwall's lemma, 197

Hyperbolic equations, initial-
  boundary value problem for
  multidimensional, 156–164

Implicit difference equations
  in Navier-Stokes problems, 104
  optimum vs. explicit, 71
  for stability and approximation,
    53–56
incompressibility equation, 97
  slight, 97
initial-boundary value problems,
  generalized solution, 159–160,
    162
  dominating mixed smoothness,
    163
  integral identity, 162
inner layer problems, 130
integral identity, 162, 166
interpolation inequality, 179–181
isotropic preconditioners, 225
iterative methods
  convergence of two-stage, 13–16
  for explicit difference equations,
    50
  fictitious domain, 225
  fully implicit one-step, 83
  inner, 11–13, 112
  lower layer of, 112
  modified simple, 15–18
  in Navier-Stokes equations, 89–90
  Newton-Raphson, 15
  two-stage for nonlinear systems,
    11–13
  Yanenko, 91

Jacobian vector, 46
jumps of coefficients, 227–228

Kinematic viscosity, 106

Laplace operators, 207
Lebedev, V.I., 125
level energetic norm, 168

linearization
  in explicit difference equations,
    46
  of given operator, 2
  Newton's, 15–16
linear stiff systems, 2–11, 118–123

Matrices
  block-triangular factorization of,
    21
  Gram, 32–34
  Shur, 21–22
  stiff matrix A, 118–123
mean step, 75
  maximum, 75
Medovikov, A.A., 63
mesh operators, 165–166
method of quick motions filtering,
    139–144
modified Richardson method, 3–5,
    6–11, 19
modified simple iterations, 15–18
multigrid acceleration procedure,
    1–2
multigrid predictor-corrector
    method, 1–2
multiparametric case, 228

Navier-Stokes equations
  analogue of explicit-difference
    scheme, 102–104
  Babushka-Brezzi inequality,
    85–88, 99
  canonical form in, 92
  coefficient of kinematic viscosity,
    106
  Dirichlet problem, 93
  eigenfunctions in, 95
  eigenvalue problem in, 95
  ϵ regularizations, 100–101
  finite-difference schemes, 88–90,
    113–114
  finite-dimensional analogue, 108
  finite-element method, 89–91
  fully implicit one-step iterative
    method, 83–85
  implicit difference scheme,
    104–106

incompressibility equation, 95
inner iterative method, 113
iterative method, 89
lower layer of iterative method,
    112–113
  with periodic boundary
    conditions, 94–95
  Poisson equation and, 93
  Richardson method, 93
  seminorm in, 83
  solenoidal functions, 107
  theory of elasticity equations,
    95–96
  in velocity-pressure variables
    nonlinear equations, 106–114
    nonstationary Stokes problem,
      97–105
    stationary Stokes problem,
      82–95
  Wandermond determinant in, 102
  weak solutions, 95–100
  Yanenko method, 91
Navier-Stokes system, 37–38,
    106–114
Newton-Kantorovich (Newton-
    Raphson) method, 2
Newton-Raphson (Newton-
    Kantorovich) method, 2
Newton's linearization, 15–16
Nitsche-(Aubin-Oganisjian-
    Rukhovets) technique, 174–179
nonhomogeneous thin plates, 217
nonlinear elliptic grid systems
  examples
    fourth-order elliptic equations,
      38–40
    quasilinear second-order elliptic
      equations and systems,
      34–38
  modified simple iterations, 16–18
  optimal and nearly optimal
    preconditioning, 24
    angle estimation, 28–34
    basic preconditioner
      construction, 19
    composite grids with local
      refinements, 24–28

nonlinear elliptic grid systems
(Cont.):
  preconditioning and
    symmetrization, 2–11
    modified Richardson methods
    in general setting, 6–11
    notation, 2–3
    spectrally equivalent operators,
    3
    symmetrization, 5–6
  two-stage iterative methods,
    11–18
    convergence of, 13–14
    inner iterations, 11–13
    Newton's linearization, 15–16
nonlinear equations, Navier-Stokes,
  106–114
nonstationary problems, straight,
  74
nuclear reactors, slow processes,
  144–146

One-dimensional problems, with
  discontinuous effects, 156
operators
  block-triangular factorized, 22
  cooperative, 22
  mesh, 165–166
  model, 2
  monotone nonlinear, 35–37
  nonnegative, 87
  for optimal and nearly optimal
    preconditioning, 19–24
  splitting with finite-element
    method, 204–210
  transfer, 92
  transition, 51, 52
ordinary differential equations
  applications
    DeMirchyan-Rakitsky scheme,
    139–144
    Euler's implicit scheme
    analyzed, 146–153
    nuclear reactor slow processes,
    144–146
    A-stability difference schemes,
    123–125
    Cauchy problem, 117–118

linear stiff systems, 118–123
regular stiff systems, 131–138
singularly perturbated systems,
  126–131
stiff systems defined, 117–118

Perforated composites, 228–239
piecewise differential functions, 181
Poisson equation, 93
polynomials
  Chebyshev, 57, 58
  explicit difference equations,
    56–65
  Medovikov parameters, 63
preconditioners
  anisotropic, 225–227
  asymptotically nearly optimal, 1
  asymptotically optimal, 1, 4
  basic approaches to constructing,
    19–24
  isotropic, 225
  for linear systems, 2–11
  nearly optimal, 4
  two-stage, 13

Quantity estimation
  cou, 73–75
  r, 73–75
Quasilinear second-order elliptic
  equations and systems, 34–38
quick motions filtering method,
  139–144

Refinement of grids or triangulations, 25
regular stiff systems, 131–138
  reduced to singularly pertur-
    bated, 135–137
  trajectory, 137
Richardson method, 93
  modified, 2–5, 6–11, 19

Second-order equations
  hyperbolic of general form,
    196–204
  initial-boundary value problems
    for multidimensional
    hyperbolic, 156–164

second-order equations (Cont.):
   quasilinear elliptic, 34–38
Shur matrix, 21–22
singularly perturbated systems,
      126–128
   and asymptotic theory, 129
   compared with regular, 132
   DeMirchyan-Rakitsky scheme,
      139–144
   inner layer, 130
   nuclear reactor slow processes,
      144–146
   numerical integration of, 128–131
   regular system reduced to,
      135–137
   slow-motion stage, 129–130
   unstable branch, 130
solenoidal functions, 107
spectral equivalence, near, 4
spectrally equivalent operators, 3–5
splitting operators, 204–210
stability
   absolute, 171, 205, 212
   A-stability of difference schemes,
      123–125
   defined, 53
   explicit difference equations,
      51–53
   implicit difference equations,
      53–56
   T-sequence method, 72–73
stiff systems defined, 47, 117–118
Stokes problem, 82. *See also*
      Navier-Stokes equations
   grid, 92–94
   with periodic boundary
      conditions, 94

straight nonstationary problems, 74
stream function-vorticity variables,
      81
superconvergence, 177, 210
symmetrization
   Gaussian, 5
   for nonlinear elliptic grid
      systems, 5–6

Time-step size restrictions, 47
torus calculations. *See* composites,
      fictitious domain method
trajectories, 137–138
transition operator
   for explicit difference equations,
      51–52
   for implicit difference equations,
      53
transverse vibrations, 217
triangulations
   composite, 24–25
   refinement of, 25
   of the region omega, 24
   topologically equivalent, 25–26
T-sequence method, 72–73

Velocity-pressure variables, 82. *See
      also* Navier-Stokes equations
vibrations, transverse, 217
viscosity, kinematic, 106

Wandermond determinant, 102
$W_{2,h}^1$ error estimates, 189–193

Yanenko, N.N., 91